Civil Engineering Hydraulics

J. R. D. Francis

Late Professor of Hydraulics, Imperial College of Science and Technology, University of London

and P. Minton

Formerly Assistant Director of Civil Engineering Department, Imperial College of Science and Technology, University of London

Edward Arnold

© Mrs S Francis and P. Minton 1984
Based on material first published in Great Britain 1958 by
Edward Arnold (Publishers) Ltd, 41 Bedford Square, London WC1B 3DQ
as Fluid Mechanics for Engineering Students.

Edward Arnold, 300 North Charles Street, Baltimore, Maryland 21201, U.S.A.

Edward Arnold (Australia) Pty Ltd, 80 Waverley Road, Caulfield East, Victoria 3145, Australia

First published 1958
Reprinted 1960
Second Edition 1962
Reprinted 1965
Third edition 1969
Reprinted 1971
Fourth edition 1975
Reprinted 1976, 1979
Fifth edition 1984

British Library Cataloguing in Publication Data

Francis, J. R. D.
  Civil engineering hydraulics.
  1. Fluid mechanics
  I. Title II. Minton, P. III. Francis, J. R. D.
  Fluid mechanics for engineering students
  532′.002462 TA357

ISBN 0-7131-3514-X

90 04

Text set in Times New Roman
by Macmillan India Ltd., Bangalore
Printed and bound by
Thomson Litho Ltd, East Kilbride, Scotland

Preface

When John Francis died unexpectedly in 1979 the form of this fifth edition of the textbook had already been agreed. The objective of the revision was not only to up-date the material but also to move the text more into the field of Civil Engineering Hydraulics. As a colleague and admirer of John Francis' enthusiasm for, and achievements in, the field of hydraulics the task of revision fell to the present writer.

The major changes proposed for this edition were a revision of the chapter on hydraulic machines and the addition of more worked examples. This has been done, and in the case of the worked examples some use has been made of material from "Problems in Hydraulics and Fluid Mechanics", also by the authors of the present book. Due to the increased material it was decided, with great regret, no longer to include the chapter on compressible flow contributed by Professor G. Jackson, late of Brunel University. Luckily this change has not severed Professor Jackson's connections with the text and his comments and criticism have been welcomed and acted on by the writer. Thanks are also due to the writer's colleagues, Dr. J. D. Hardwick, Mr. M. J. Kenn, and Mr. A. Scott-Moncrieff whose general advice and criticism of the chapters on Open Channel Flow and Hydraulic Machinery have been particularly valuable.

In spite of the changes which have been made by the writer it is hoped that the original objectives described by John Francis in his preface to the previous edition will continue to be achieved.

'This book was originally written as an attempt to provide the least amount of instruction in the subject that can be accepted for a University degree in Engineering. No effort was made to write a complete text of all possible branches of the subject during a three-year course, but it is still believed that the necessary basis for at least the first two years' work is here. Third-year work often diverges in various ways according to the interests of the teacher and only some of these ways are reviewed in this text.

In an attempt to make this book simple and easy to follow, many rigid proofs of mathematical formulae are omitted, and simplifications made to others. Also, detailed tables of experimental data have been avoided as far as possible; the undergraduate should begin to acquaint himself with these at a later stage, from more detailed textbooks. On the other hand, considerable attention has been paid to explaining in full the limitations of any theoretically derived equations; to the engineer, the practical limitations of a theory are of more interest than a perfect proof.

The author of any textbook depends largely upon his predecessors, and I have gained much from the works of Addison, Jameson, Lewitt, Hunter Rouse and Hunsaker and Rightmire. In particular, the well-known volumes *Modern Developments in Fluid Mechanics*, edited by S. Goldstein, have been a constant source of information: this is undoubtedly the work to recommend to the graduate who is to practise in this field of engineering. The present book also owes much to the teachings of Emeritus Professor C. M. White, for many of the ideas and simplifications herein have been suggested by him in the course of his lectures.'

P. Minton

The pioneers

It is of considerable interest to know something about the men whose names are now regularly used in fluid mechanics. A very abbreviated list is given below. The original papers in which these pioneers have published their results are, in general, listed in the Royal Society Catalogue of Scientific Papers.

Archimedes	287–212 B.C.	Greek philosopher
H. Pitot	1695–1771	French inventor
D. Bernoulli	1700–1782	Swiss physicist
A. Chézy	1718–1798	French engineer
J. C. Borda	1733–1799	French mathematician
G. B. Venturi	1746–1822	Italian engineer
P. S. Laplace	1749–1827	French mathematician
J. L. Poiseuille	1799–1869	French physicist
H. Darcy	1803–1858	French engineer
W. E. Weber	1804–1891	German physicist
W. Froude	1810–1879	British naval architect
J. B. Francis	1815–1892	American engineer
R. Manning	1816–1897	Irish engineer
H. Bazin	1829–1917	French engineer
L. A. Pelton	1829–1908	American engineer
E. Mach	1838–1916	Austrian philosopher
O. Reynolds	1842–1912	British engineer
C. G. P. de Laval	1845–1913	Swedish engineer
L. Prandtl	1875–1953	German engineer
V. Kaplan	1876–1934	Czech engineer
T. von Kármán	1881–1963	Hungarian engineer
G. I. Taylor	1886–1975	British applied mathematician.

Contents

1

The properties of fluids

1.1 Definition of a fluid

Most people realize that the term *fluid* includes such different materials as water which is a *liquid* and air which is a *gas*. The essential property in common is that a volume of fluid cannot preserve its shape, unles it is constrained by surrounding surfaces. It is clear that if the sides of a barrel of water were suddenly removed, the cylinder of fluid within would collapse at once and spread out to a thin layer over a large area. The motion only stops if the water reaches another set of boundaries. The same thing would happen, though much more slowly, if the material within the barrel was oil or pitch; in these cases the spreading might take seconds or even weeks or months, depending on the chemical composition of the oil or pitch. In this respect, then, oil and pitch are fluids as well as water and air. The speed of spreading, and of losing the original shape under a deforming force, is governed by the property of a fluid known as *viscosity*. Viscosity essentially governs the speed of a fluid motion, there being no further change of shape when the force is removed, i.e. fluids behave plastically when sheared. Accordingly, the following comprehensive definition of a fluid may be based on this property:

A fluid is matter in a readily distortable form, so that the smallest unbalanced external force on it causes a continuous change of shape.

This definition clearly excludes solids, such as steel, concrete, wood or rubber, all of which distort only a certain amount when a shear force is applied. With a fluid, i.e. a liquid, gas, or vapour, the same shear force gives no definite amount of distortion, the change of shape being continuous as long as the force is applied.

1.2 Density and compressibility

The mass of a fluid is usually expressed in the terms of its *density*, ρ, defined as the mass per unit volume. Typical densities at 20°C, and atmospheric pressure are:

$$\text{water, } \rho = 1000 \text{ kg m}^{-3}$$
$$\text{air, } \rho = 1.25 \text{ kg m}^{-3}$$

1

Two other properties occasionally referred to are:–

a) *specific gravity*, s.g., which is the ratio of the density of a fluid to the density of water,
b) *specific weight*, w, the weight force of fluid per unit volume, whence

$$w = \rho g.$$

In the case of most solids in engineering the application of direct stresses, i.e. tension or compression, does produce a change in volume of the material, but this is so small that the change in density is rarely of importance. If a fluid is compressed it behaves in a similar manner to a solid, i.e. if the force per unit area or *pressure intensity*, p, acting on the fluid is increased by an amount δp, then the volume of the fluid V, will decrease by δV. For small changes the relationship is linear and hence:

$$\delta p = -K \frac{\delta V}{V}$$

where K is the *bulk modulus* or *compressibility* of the fluid. For a particular mass of the fluid $m = \rho V$, by definition, then the relationship can be expanded to:

$$\frac{\delta p}{K} = -\frac{\delta V}{V} = \frac{\delta \rho}{\rho}$$

With liquids the value of K changes little over the range of pressures met in hydraulics, but for gases or vapours the value of K will depend on their thermodynamic properties.

Example

A diving bell is lowered slowly from the water surface, where the pressure is $10^5 \, \text{N m}^{-2}$, to a depth of approximately 200 m where the pressure is $2.1 \, \text{MN m}^{-2}$. The initial density of air in the diving bell is $1.23 \, \text{kg m}^{-3}$ while the density of water at the surface is $1025 \, \text{kg m}^{-3}$. If the temperature of the air in the diving bell does not change and the water temperature remains constant with depth, compare the densities at depth with those at the surface. For water $K = 2.3 \, \text{kN mm}^{-2}$.

Solution

For the water $dp = K \dfrac{d\rho}{\rho}$.

$$\therefore d\rho = \frac{dp}{K} \rho = 2.1 \frac{\text{MN}}{\text{m}^2} \times \frac{\text{mm}^2}{2.3 \, \text{kN}} \times 1025 \frac{\text{kg}}{\text{m}^3} \times \frac{10^3 \, \text{kN}}{\text{MN}} \times \frac{\text{m}^2}{10^6 \, \text{mm}^2}$$

$$= 0.94 \, \text{kg m}^{-3}.$$

At a pressure of $2.1 \, \text{MNm}^{-2}$ the density of the water is $1025.9 \, \text{kg m}^{-3}$.

For the air, Boyle's Law gives $\dfrac{p_1}{\rho_1} = \dfrac{p_2}{\rho_2}$ as the temperature is constant.

Hence $\rho_2 = \rho_1 \dfrac{p_2}{p_1} = 1.23 \dfrac{\text{kg}}{\text{m}^3} \times \dfrac{2.1}{0.1} = 25.8 \dfrac{\text{kg}}{\text{m}^3}$

Note how the same change of pressure has increased the density of the air twenty times, but altered the density of the water by less than one part in a thousand.

In fluid dynamics the compressibility becomes important when the velocity is more than about 1/5 of the velocity of sound waves in the fluid. Under these conditions the pressure may rise suddenly across a very narrow region due to the compression of the fluid (a pressure wave), instead of varying smoothly from point to point.

The most convenient expressions for the speed of sound, a, are:

$$a = \sqrt{\frac{K}{\rho}} \text{ for liquids,}$$

and

$$a = \sqrt{\gamma R T} \text{ for gases,}$$

where γ is the adiabatic index and R the gas constant.

Example

Calculate the speed of sound in water where $K = 2.3 \text{ kN mm}^{-2}$ and in air at 20°C where $\gamma = 1.4$ and $R = 0.29 \text{ kJ kg}^{-1} \text{ K}^{-1}$

For water

$$a = \left[2.3 \frac{\text{kN}}{\text{mm}^2} \times \frac{\text{m}^3}{1000 \text{ kg}} \times \frac{10^3 \text{ kg m}}{\text{kN s}^2} \times \frac{10^6 \text{ mm}^2}{\text{m}^2} \right]^{1/2} = 1500 \frac{\text{m}}{\text{s}}.$$

For air $\quad a = \left[1.4 \times 0.29 \frac{\text{kJ}}{\text{kg K}} \times 293 \text{ K} \times \frac{10^3 \text{ kg m}^2}{\text{kJ s}^2} \right]^{1/2} = 345 \frac{\text{m}}{\text{s}}.$

The speed of sound in air is of the same order as the speed of aircraft and missiles and in these, and certain pipe flows, compressibility is of importance. The speed of sound in water is much higher than the speeds at which objects move through water, or water moves in pipes, for example even after falling 1000 m water could not be moving at greater than one tenth of the speed of sound. However in one case, the sudden deceleration of a liquid, the compressibility is of importance (Chapter 13).

1.3 Viscosity

Having calculated the effects of direct, compression, stresses on a fluid, the effect of shear stresses must be found. Consider a volume of moving fluid one view of which is the square ABCD shown in Fig. 1.1. The volume has unit length in a plane perpendicular to the paper. A shear stress τ (force per unit area) acts on top and bottom of ABCD in the directions shown, and from the definition of a fluid the top moves at a small speed δu relative to the bottom. In a short time δt, ABCD will distort to ABC'D', causing a change of shape or *shear strain* which can be expressed by the magnitude of the angle $\delta\theta$. The distances CC' and DD' are both given by the product of speed and time,

that is,
$$CC' = DD' = \delta u . \delta t.$$

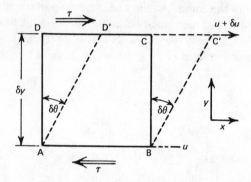

Fig. 1.1 Deformation of a cube of moving fluid ABCD due to a velocity gradient producing a higher velocity at the top than at the bottom. The rate of deformation $\delta\theta/\delta t$ is caused by a shear stress on opposite sides. The cube has deformed to the shape ABC'D' in a time δt.

But if $\delta\theta$ is small, $\delta\theta = \dfrac{CC'}{\delta y}$, where δy is the small distance between the top and bottom of ABCD

That is,
$$\delta\theta = \frac{\delta u . \delta t}{\delta y} = shear\ strain.$$

This shear strain must now be connected to the shear stress τ that has caused it. Since θ continues to increase with time, it is not possible to consider τ as dependent on θ as is the case with solid materials (this assumption produces the well-known coefficient of shear elasticity). Instead, τ is found to depend upon the *rate of change of* θ, $\delta\theta/\delta t$, thus using μ as the coefficient of proportionality,

$$\tau = \mu\frac{\delta\theta}{\delta t}$$

$$= \mu\frac{\delta u . \delta t}{\delta y . \delta t} = \mu\frac{\delta u}{\delta y}.$$

If the volume ABCD is now made infinitesimal τ is then the shear stress at one particular level, and

$$\tau = \mu \frac{du}{dy} \tag{1.1}$$

where du/dy is the gradient of velocity in the y direction, that is at right angles to the direction of the velocity itself.

Experimental data show that if a fluid is moving sufficiently slowly, within a tube of small cross-sectional area, then μ for that fluid depends neither on τ nor du/dy; it may decrease with the temperature of the fluid if a liquid, or increase if a gas. In this type of flow, which is called *laminar flow*, the shear stress is entirely balanced by intermolecular forces in the fluid, which try to prevent one layer moving over the next. Under these circumstances μ is called the *coefficient of molecular viscosity* and the fluids are called *Newtonian fluids* after Sir Isaac Newton who first observed their behaviour.

The molecular viscosity is a stress divided by a velocity gradient and has the dimensions

$$\frac{MLT^{-2}}{L^2} \div \frac{L}{TL} = ML^{-1}T^{-1}.$$

At 20°C the viscosities of water and air are 10^{-3} and 1.8×10^{-5} kg m^{-1} s^{-1}, (or N s m^{-2}) respectively. Another common unit used for the measurement of molecular viscosity is the 'poise', i.e. g cm^{-1} s^{-1}.

If, however, a fluid is moving more quickly or is flowing in a tube of larger cross-sectional area, then μ is far less simply defined. It depends now on τ, du/dy and on many other variables, and is always greater than the molecular value, perhaps as much as 10^8 times as great. The mechanism connecting τ with du/dy is quite different from the previous case since irregular motions known as *turbulence* have appeared in the fluid, and these create occasional rotatory motions called *eddies* (see Chapter 4). Under these conditions μ is called the *eddy viscosity* and is no longer constant for a given fluid and temperature. A discussion is given in Chapter 12 on the function of turbulence in the production of shear stresses.

Some materials appear on cursory inspection to be fluids, but prove after experiment to have variable values of μ, even when turbulence is absent. In these materials, non-Newtonian fluids, the intermolecular forces change with τ or du/dy and very complicated conditions of flow are set up. Mud, cream, and cheese are examples of these fluids, and are the concern of a separate branch of fluid mechanics, known as Rheology.

One further definition of viscosity is used in fluid mechanics. It is often convenient to use the ratio of the coefficient of molecular viscosity to the density of the fluid μ/ρ. This ratio is called the *kinematc viscosity*, ν, and has the dimensions

$$ML^{-1}T^{-1} \div ML^{-3} = L^2T^{-1}.$$

At 20°C and atmospheric pressure the kinematic viscosities of water and air are 10^{-6} and 1.5×10^{-5} m^2 s^{-1} respectively. The kinematic viscosity may be expressed in 'stokes', i.e. cm^2 s^{-1}.

While the production of shear stresses within a flowing fluid can be either due to laminar, viscous shear or turbulent shear there are particular flow regions where laminar, viscous shear is all important. At a solid boundary of a flow it can be argued, and confirmed by observation, that it is not possible for turbulent motions to exist. Hence shear stresses can only be transmitted from a boundary to a flow, and vice versa, by the presence of laminar, viscous shear stresses due to velocity gradients at the boundary. These velocity gradients exist because the fluid molecules adjacent to the boundary have no motion relative to the boundary. This observation, known as the *no slip condition* ensures that if there is any motion in a fluid relative to a boundary then shear stresses will occur.

By normal engineering standards shear stresses in fluids are small. Whereas a working stress in steel might be 10^8 Nm^{-2}, a pressure in a hydro-power pipe 10^7 Nm^{-2}, and a domestic water system pressure 10^5 Nm^{-2}, the shear stress on the walls of a water pipe would be of the order of 10^2 Nm^{-2}. However in spite of their small magnitudes these shear stresses can be of great importance in fluid flows, as described in Chapters 12 to 14.

Example

A long circular rod of 70 mm diameter slides concentrically in a 150 mm long fixed tube of 70.5 mm internal diameter. The annular space between the rod and the tube is filled with oil of viscosity 0.193 N s m^{-2}. What force is required to slide the rod through the tube with a velocity of 1 m s^{-1}?

Solution

To find the shear stress the velocity gradient must be known. As there is no relative motion between a fluid and the surface with which it is in contact the difference of velocity between the two sides of the oil film is 1 m s^{-1}. Making the simplest assumption that the velocity of the oil changes linearly between zero at the tube surface and 1 m s^{-1} at the rod surface

$$\frac{du}{dy} \simeq \frac{u}{\delta r} = \frac{1\,\text{m s}^{-1}}{(35.25 - 35)\,\text{mm}} = \frac{10^3\,\text{s}^{-1}}{0.25} = 4000\,\text{s}^{-1}.$$

Assuming laminar flow

$$\tau = \mu \frac{du}{dy} = 0.193\,\frac{\text{Ns}}{\text{m}^2} \times 4000\,\text{s}^{-1} = 772\,\frac{\text{N}}{\text{m}^2}.$$

Area of rod in contact with the oil $= \pi \times d \times l = \pi \times 70 \times 150$ mm^2
$$= 33 \times 10^3\,\text{mm}^2.$$

$$\text{Shear force} = \text{shear stress} \times \text{area} = 772\frac{N}{m^2} \times \frac{33}{10^3}m^2 = 25.5\,N.$$

If the rod does not slide lengthways but rotates, what would be the torque or moment required to turn it in the tube at 3500 revolutions per minute? Answer 11.4 Nm.

1.4 Vapour pressure

In Section 1.2 the effect of increasing the direct stress (pressure intensity) on a fluid was described. If the pressure in a gas is reduced the density–pressure relationship is applicable until the density becomes so low that the gas has to be considered in terms of its individual molecules. With a liquid the reduction of pressure will eventually result in the production of vapour and a liquid–vapour interface giving a sudden and non-uniform change in the density.

A body of liquid bounded partly by vapour or a gas–liquid interface continually sends off molecules of liquid in vapour form from the interface, until the pressure of vapour is such that no more evaporation occurs. If the pressure above the liquid is reduced bubbles of vapour are formed in the fluid and rise to the surface, and evaporation recommences until a new balance is reached. Attempts to lower the pressure still further simply result in more vigorous evaporation until finally no liquid remains.

In hydraulic engineering work the vapour pressure of a liquid is of importance, for there may be places of low local pressure, particularly when the liquid is flowing over a solid surface. If, in one of these places, the pressure is reduced until the liquid vaporizes, then bubbles are formed quite suddenly. When the liquid has moved on to a place of rather higher local pressure, the bubbles suddenly collapse. These very rapid collapsing motions cause high impact pressures if they occur against portions of the solid surface, and may eventually cause a local mechanical failure by fatigue of the solid surface. Severe pitting and damage of the surface may result. This effect is called *cavitation* and reduces the efficiency of hydraulic machines, even if damage is not done.

Somewhat similar effects occur when the gases of the air are dissolved by a liquid, for air bubbles may be released in the same way as vapour bubbles. Air bubbles usually occur at rather higher pressures than vapour cavities and as air bubbles collapse later they can decrease the damage caused by vapour cavity collapse.

1.5 Surface tension

The physical chemistry of every liquid is such that if one of its boundaries is a gas, the surface molecules of the liquid are always repelling each other. This gas–liquid interface is therefore in a state of tension, each molecule being kept in equilibrium by a tension on all sides. If, however, a solid surface intersects the interface, then molecules in contact with the solid still exert a tension on their companions, so that if there is a relative motion of the solid to the liquid, the interface is dragged along by this *surface tension* force. The force is proportional to the length of the intersection of the solid with the interface, and is usually small compared with the magnitude of the other forces found in engineering work, such as those due to viscosity, to pressure changes, or to the weight of the fluid. However, as will be described, small scale models of hydraulic engineering structures are often made and operated. In these models, surface tension forces on the fluid concerned, if there is a gas–liquid surface, may be relatively so much more important than those in the full size prototype that a simple scaling-up of total measured forces is misleading and errors will be introduced.

Surface tension is also of importance in some hydraulic measurements, when the height of a liquid–gas interface is required to be known accurately. The surface condition of the surrounding solid boundaries may be such as to deform the interface, as well as to cause the tension force (see Fig. 1.2). In some cases the deformation makes the interface nearly tangential to the solid surface. Consequently the surface tension force F is inclined to the major part of the interface, and will give a vertical component of force on the liquid, which will accordingly be raised or lowered somewhat. If the liquid is contained in a tube the vertical force will be exerted all round the periphery. This effect is sometimes called a *meniscus* or *capillarity* effect. As the deformation is critically dependent on the traces of impurities on the solid surface, the effect is very variable.

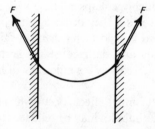

Fig. 1.2 Deformation of a gas–liquid interface by an intersecting solid surface. The tension force F is now inclined to the main part of the interface and a force is exerted tending to pull the solid surface into or out of the liquid.

Although surface tension forces are always small (for example, a clean water–air interface gives a force of only $0.073 \, \text{N} \, \text{m}^{-1}$, equivalent to a rise of water surface level of 3 mm in a 10 mm diameter tube), they can entirely stop a fluid motion if they happen to be larger than the other forces acting on the liquid. In this respect, they must be contrasted to the effect of viscosity, which slows a motion but never stops it entirely. The surface tension of an interface may be reduced by chemical means (a soap or detergent reducing the water–air force by about 50 per cent), and this is sometimes done in model testing work in order to reduce the errors due to surface tension.

1.6 The ideal fluid

The preceding paragraphs have reviewed the principal physical properties of fluids, but it is rare for a specific engineering problem to be solved taking into consideration all the properties concerned, for the mathematics soon becomes too complicated and a simplification is necessary. One simplification is to define an *ideal fluid* which complies with the definition of a fluid but its coefficient of viscosity is zero so that a velocity gradient cannot cause any shear stresses. Further, the ideal fluid is incompressible, has no surface tension and does not vaporize. Many problems of fluid motion can now be solved, although the results sometimes have an air of unreality about them, compared with the observed phenomena in real fluids such as air and water. For example, there can be no shear stresses in an ideal fluid, which would not be slowed down near a solid boundary by viscous effects and so would not conform to the no slip condition.

Experimental evidence must be used to convert the calculated result into a realistic solution to the problem, and the value of the empirical coefficients thus used are an indication of the accuracy of the assumption of an ideal fluid. It is fortunate that water and air are surprisingly near an ideal fluid in many respects, so that the above approach can be sufficiently successful as a first approximation to the solution of an engineering problem.

2

Pressure forces in fluids

2.1 Forces in static fluids

When a body of fluid is at rest relative to its boundaries there are no velocity gradients and so no shear forces. A great simplification is therefore made when calculating the forces exerted by the fluid, for only the components normal to the boundaries need be taken into account. Under these conditions, known as *hydrostatic*, exact solutions can be obtained to problems and no experimental evidence is needed: a rare occurrence in fluid mechanics.

The weight force of a static fluid produces at each and every point in the fluid a compression stress which is not necessarily the same everywhere. If an object, or a boundary, happens to be in a fluid, then this compression stress is felt on it as a *pressure intensity*, p, expressed as a force per unit area.* If the boundary area a is large, a pressure force F, which is a vector, will be felt on it so that

$$F = \int p \mathrm{d}a \qquad (2.1)$$

the integral being taken over the whole area a. In this way the engineer can find the fluid force, providing it is known how p varies from place to place over the area. The force F must always act in a direction *normal* to the boundary area a, otherwise F would have a component, a shear force, acting parallel to the boundary. A shear force, however, cannot exist in a static fluid.

2.2 Variation of pressure intensity with direction at one point

Consider a right-angled triangular prism ABC of fluid within a larger mass of static fluid, BC being horizontal, Fig. 2.1. The prism is of unit length. Let the average pressure intensities on the three sides be p_a, p_b, p_c respectively on BC, AC, and AB. The forces on the sides are therefore p_a. CB, p_b. AC, p_c. AB per unit length of prism all acting normally to the sides. The mass of the prism is $\frac{1}{2}\rho$ AC. BC per unit length, so that the downward weight force is $\frac{1}{2}\rho g$ AC. BC.

---

* Engineers may incorrectly use 'pressure' to mean both pressure intensity and pressure force. The context usually makes it clear which is meant.

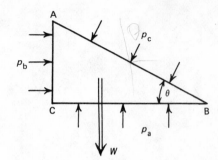

Fig. 2.1 End view of a triangular prism of hydrostatic fluid showing pressure intensities on every side due to the surrounding fluid, and the weight force.

Resolving the forces horizontally

$$p_b \cdot AC - p_c \cdot AB \sin \theta = 0$$

for the prism must be in equilibrium since it is in a static fluid. But from geometry,

$$AC = AB \sin \theta. \quad \text{So } p_b = p_c.$$

Now resolve the forces vertically.

$$p_a BC - p_c AB \cos \theta - \tfrac{1}{2}\rho g \, AC \cdot BC = 0 \text{ for equilibrium.}$$

But from geometry,

$$CB = AB \cos \theta. \quad \text{So } p_a = p_c + \tfrac{1}{2}\rho g AC.$$

Now consider the prism to be made smaller until in the limit, as the prism becomes of infinitesimal size, $AC \to 0$ so that the term $\tfrac{1}{2}\rho g AC$ tends to zero. Thus if A, B, C are coincident,

$$p_a = p_b = p_c \qquad (2.2)$$

So the pressure intensities at a point are the same in all directions, i.e. pressure is a scalar quantity.

This result is of fundamental importance in hydraulic engineering, for hydrostatic pressure forces, even though they are caused by the weight force of the fluid, are exerted undiminished in all directions, even vertically upwards.

2.3 Variation of pressure intensity with height in a static fluid

It is nearly axiomatic with most people that great pressures exist at the bottom of high columns of fluid. The following proofs are given to place this common knowledge on a quantitative basis.

Consider a small vertical cylinder of fluid in equilibrium in a larger mass of static fluid (Fig. 2.2). Its cross-sectional area is a and its length δz. The bottom of the cylinder is at a height z above a purely arbitrary datum level. The fluid

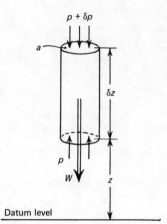

Fig. 2.2 Perspective view of a vertical
cylinder of hydrostatic fluid, showing
pressure and weight forces acting.

density is ρ so that the weight force W acting downwards is $W = \rho ag\delta z$.
Because the fluid is static, this weight force must be balanced to preserve
vertical equilibrium: there can be no shear forces on the curved surface so there
only remain the pressure intensity forces acting on the ends of the cylinder.

Let the pressure intensity on the bottom acting upwards be p, and that on
the top acting downwards be $p + \delta p$. (Notice that since z is taken as increasing
upwards, so must all other variables, including p. If in fact p *decreases* upwards,
a minus sign will appear.)

Thus the balancing force $= pa - (p + \delta p)a = -a\delta p$
So $\rho ag\delta z = -a\delta p$

or $\dfrac{\delta p}{\delta z} = -\rho g.$

As δz approaches zero, the gradient of pressure becomes

$$\frac{\mathrm{d}p}{\mathrm{d}z} = -\rho g. \tag{2.3}$$

This is the *hydrostatic equation* for the pressure gradient at a point: it is only
of direct use if it can be integrated to find p which will now be done for two
particular cases.

2.4 Solutions of the hydrostatic equation

(*a*) For an incompressible fluid, whose density ρ is independent of pressure
intensity and therefore constant, the hydrostatic equation is easily solved.
Water, and most other liquids, are sufficiently incompressible for the following
analysis to be valid (see Chapter 1).

Writing $\mathrm{d}p = -\rho g \, \mathrm{d}z$

and integrating between the limits z_1 and z_2, p_1 and p_2,

$$p_1 - p_2 = -\rho g(z_1 - z_2),$$

or in words, the change of pressure intensity between two levels is proportional to the difference of height between the levels. In the common case of a volume of liquid at rest with a free surface exposed to the atmosphere, where the pressure is p_a, the equation gives

$$p_a - p_1 = -\rho g(z_a - z_1)$$

where z_a is the height of the surface above datum level.

But if
$$z_a - z_1 = h,$$

where h is the depth below the surface of the point where the pressure intensity p_1 has been developed,

thus
$$p_a - p_1 = \rho g h$$

or
$$p_1 = p_a + \rho g h \tag{2.4}$$

From this equation it can be seen that the pressure intensity at any point is partly due to (i) to the atmospheric pressure on the free surface and (ii) to the density. (i) may thus be regarded as being transmitted without diminution throughout the fluid. The pressure intensity p_1 is reckoned from the same datum pressure as is p_a, that is, above a perfect vacuum: therefore p_1 is called the *absolute pressure*.

Engineers, however, often prefer to measure pressure intensity above a datum pressure at atmospheric pressure. They call $\rho g h$ the *gauge pressure*, and imply that if absolute pressure is required, p_a should be added. This is justified since it is often desired to find the difference of pressure intensity between two points. If the atmospheric pressure is the same at both points, the difference of gauge pressures is therefore the same as the difference of absolute pressures. The term gauge pressure also shows one use of the hydrostatic equation. When a pressure intensity has been measured, it is convenient to express it as h, the height or *head* of the fluid which, if static, produces this pressure. A great variety of pressure measuring instruments, called manometers, use the principle of balancing a static column of fluid (or several fluids) against the pressure, and then measuring h. Some of these are described in an Appendix, p. 26.

(*b*) For a compressible fluid, the law connecting ρ with p must be used. For instance, many gases obey the perfect gas law over a range of pressures and temperatures. That is $pV = RT$, where V is the volume of unit mass of gas at pressure p and absolute temperature T. Since ρ (mass per unit volume) is V^{-1} the gas law can be rewritten

$$\frac{p}{\rho} = RT \quad \text{or} \quad \rho = \frac{p}{RT}$$

and substituted into the hydrostatic equation to give

$$\frac{dp}{dz} = -p\frac{g}{RT}$$

or

$$\frac{dp}{p} = -dz\frac{g}{RT}$$

If the temperature is constant at all heights, an integration can be made between the pressure limits p_1 and p_2, and height limits z_1 and z_2 to give

$$\log_e\frac{p_2}{p_1} = -\frac{g}{RT}(z_2 - z_1)$$

or

$$\log_{10}\frac{p_2}{p_1} = -0.434\frac{g}{RT}h \qquad (2.5)$$

where $h = z_2 - z_1$ is the vertical height between the two places where the pressure intensities are p_1 and p_2 respectively.

The logarithmic decrease of pressure with height in an isothermal gas is an approximation to the actual pressure distribution in the lower part of the atmosphere where in fact T varies considerably with z. On the average, there is a temperature gradient of about 6.5 °C per 1000 m, which can be inserted into the preceding integration to give a more accurate expression for the change of pressure with height. However, a close approximation can always be made with the simpler relationship, Eqn 2.5, using the mean temperature over the height range z_1 to z_2. The errors so introduced are usually smaller than those caused by neglecting the water vapour content of the air, which can also vary from layer to layer, causing different values of the gas constant R.

In any case, for engineering purposes, it is only necessary to assume that air is compressible if the height difference $z_2 - z_1 = h$ is greater than a kilometre. For small values of h a sufficiently accurate estimate is usually made with the incompressible formula, $p = p_a + \rho gh$, Eqn 2.4, assuming a mean value of the density ρ.

2.5 The forces due to hydrostatic pressure

If the variations of the pressure intensity p within a fluid are known, an integration $F = \int p\,da$ can be made to find the force F due to that pressure on a certain area a of the vessel walls containing the fluid. The integrations for finding F will be shown first when the pressure intensity remains constant everywhere over the area, and second when the pressure intensity varies linearly with depth. As force and area are vectors the integral will have to be of components in a particular direction. The first case is approximated when the change of p between top and bottom of the area concerned is small compared to the mean pressure intensity. The second case is required for the calculations of a multitude of engineering works where water is restrained by a solid wall or structure. The case of the pressure intensity varying logarithmically with depth (i.e. a compressible gas with variable ρ) is not dealt with, for engineering works are not large enough to make this variation significant. Any variation of pressure intensity with position can be treated by similar methods to those shown here, and the force found: for example, when the fluid is moving relative to a solid object, the pressure intensity may vary in a much more complicated fashion than the simple hydrostatic way. (See Chapter 7 for the reasons why the pressure should so vary.)

Having found the magnitude of the fluid force spread all over a surface, the engineer often wishes to balance it by a single force of the same magnitude acting at some point on the surface. This point is called the *centre of pressure* and is not to be confused with the centre of area (centroid) of the surface, though in exceptional cases these centres coincide.

Case 1

Pressure intensity constant everywhere: surface area plane.

In Fig. 2.3, AB is the cross section of a plane area a subject to a constant pressure intensity p all over it to produce a total force F. On an elementary area δa the force will be $\delta F = p\delta a$, and since all these elementary forces act along parallel directions, normal to AB, they may be added arithmetically to give

$$F = pa.$$

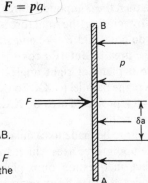

Fig. 2.3 End view of a plane surface AB, on one side of which the pressure intensity is p greater than on the other. F is the single balancing force, acting at the centre of pressure.

Also, since the elementary forces are uniform all over AB, then their resultant acts at the centroid of the surface area concerned. It is at this point, therefore, that a balancing force must be applied to preserve equilibrium. So in this exceptional case, the centre of pressure and centroid of the area are coincident.

Case 2

Pressure intensity constant everywhere: surface area *not* plane (Fig. 2.4).

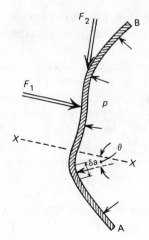

Fig. 2.4 End view of a curved surface AB on one side of which the pressure intensity is p greater than on the other. F_1 and F_2 are component forces at right angles which just balance the pressure force. Note how the elementary pressure forces are always normal to the surface where they act.

AB is now the cross section of a non-plane area with elementary pressure force components $\delta F = \rho\,\delta a$ acting normally to every element of area δa. Thus δF is now no longer always in the same direction, and to find the resultant force on the surface, the elements must be added vectorially. It is convenient to find components of the total force in directions parallel to an arbitrary axis such as XX, and at right angles to XX. Thus if θ is the angle between one typical force element and XX, then the component force along XX is $p\delta a \cos\theta$, and normal to it is $p\delta a \sin\theta$. Integrating, the total forces along and normal to the direction XX are $F_1 = \int p\,da \cos\theta$ and $F_2 = \int p\,da \sin\theta$ respectively, the integral being taken all over the area a. But $\int da \cos\theta$ is the area of the projection of the surface **AB** onto a plane at right angles to XX, and $\int da \sin\theta$ is the area projected onto a plane parallel to XX. So that $F_1 = p \times$ projected area normal to XX, and $F_2 = p \times$ projected area on XX. These force components can then be combined vectorially.

A rule can therefore be made to calculate the hydrostatic pressure force for this case: 'Decide upon three axes which are mutually at right angles; project the curved surface in question onto planes at right angles to these axes; multiply the projected areas by the constant pressure intensity to find the pressure force components along the axes; and combine the components vectorially to find the total pressure force.' If there is an axis of symmetry to the curved surface the projected areas on to this axis from each side of it are equal, so that the pressures forces normal to the axis on each half of the surface are equal and opposite (Fig. 2.5). There will therefore be no resultant pressure force in the direction normal to the axis of symmetry.

The centre of pressure of each component force F_1, F_2, F_3 is at the centroid of each projected area so that each component is parallel to its axis and acts through the centroid of its projected area. The centre of pressure of the total

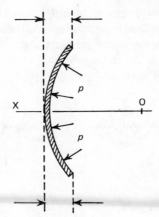

Fig. 2.5 Side view of a curved cylindrical surface with an axis of symmetry OX. The projection of the area onto OX is symmetrical above and below OX so that there is no resultant pressure force normal to OX.

force will therefore be at the intersection of the lines of action of the component forces.

Case 3

Pressure intensity increases with depth in fluid: surface plane.

Because so many engineering problems are concerned with liquids, which if static produce pressure intensities uniformly increasing with depth, this case of finding the resulting pressure forces is important. For example, the forces on lock gates, valves, walls and floors of engineering structures can all be estimated accurately.

Consider a vertical plane surface such as ACBD in Fig 2.6 which is subject to

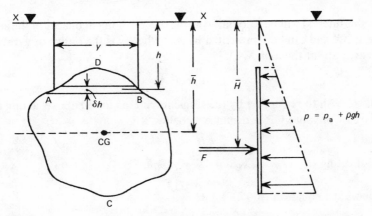

Fig. 2.6 Front and side elevations of a plane surface subject to a pressure intensity which increases with the depth h below a free surface of the fluid XX. Side elevation shows the pressure diagram and the balancing force F which acts at the centre of pressure.

a hydrostatic pressure intensity on one side, and atmospheric pressure on the other. If the fluid is incompressible, then along any narrow horizontal strip AB, the resultant pressure intensity above atmosphere is $p = \rho g h$, where h is the depth below the free surface XX of the fluid. Thus the pressure force on AB which is δh wide and y long is

$$\delta F = \rho g h \, y \delta h$$

So that the total pressure force will be

$$F = \int_C^D \rho g h \, y \mathrm{d}h$$

$$= \rho g \int_C^D y h \mathrm{d}h$$

But $\int y h \mathrm{d}h$ will be recognized as the first moment of the area about the surface XX and is equivalent to $A\bar{h}$, where A is the toal area ABCD and $\bar{h}$ is the depth below XX of the centroid of the area.

Therefore $F = \rho g A \bar{h} = A \rho g \bar{h}$ (2.6)

or in words, 'The pressure force is the product of the area of the surface, and the pressure intensity at the centroid of the surface'.

To find the centre of pressure of the area, the moment δM of the elementary force about XX is found.

Thus $\delta M = \rho g h \, y \delta h \, h$

and the total moment is $M = \rho g \int_C^D y h^2 \, \mathrm{d}h.$

But this integral will be recognized as the second moment of area of the surface about XX and can be represented by Ak^2 where k is the radius of gyration of the area about the axis XX.

So $M = \rho g A k^2.$

But we wish to replace M by the single force F as found above, acting at the centre of pressure CP, a distance $\bar{H}$ below XX.

That is, $F \bar{H} = \rho g A k^2$

Substituting $F = A \rho g \bar{h},$

reduces the equation to $\bar{H} = \dfrac{k^2}{\bar{h}}$

This equation for $\bar{H}$ is inconvenient to use because k for an area varies according to the distance of the area from the axis XX. However, by the theorem of parallel axes, we can express k^2 in terms of I_{CG}, the second moment of the area about an axis parallel to XX running through its centroid.

That is
$$Ak^2 = A\bar{h}^2 + I_{CG}$$

Substituting for k^2

$$\bar{H} = \bar{h} + \frac{I_{CG}}{A\bar{h}} \qquad (2.7)$$

This is an important result, for since the term $I_{CG}/A\bar{h}$ is always positive, then $\bar{H}$ must always be greater than $\bar{h}$: that is, the centre of pressure invariably lies below the centroid. Further, since this term varies inversely with $\bar{h}$, the deeper a given surface is immersed, then the nearer $\bar{H}$ gets to $\bar{h}$: at great depths the centre of pressure is close to the centroid and the difference may in some cases be ignored.

Some confusion is caused because though the magnitude of the force F depends on the magnitude of $\bar{h}$, the balancing force F does *not* act at this depth: it acts at $\bar{H}$ which is always below $\bar{h}$. A further confusion arises because a special application of the above formula gives an easily remembered answer. If the surface is a rectangle of depth b, with the edge a lying in the surface XX, Fig. 2.7,

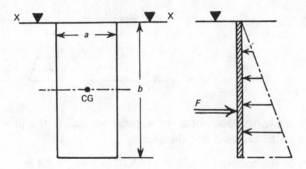

Fig. 2.7 The special case of the force on a rectangular surface with one edge in the surface. The pressure diagram is a triangle and the balancing force acts at $\frac{2}{3}$ of the depth of the rectangle below the surface.

then
$$I_{CG} = \frac{ab^3}{12} \quad \text{and} \quad \bar{h} = \frac{b}{2}$$

so that
$$\bar{H} = \frac{b}{2} + \frac{ab^3}{12ab\,b/2} = \frac{2}{3}b$$

for the special case.

This result is sometimes used quite erroneously for other shapes or for rectangles which do not have one side in the surface. The special case of a rectangle with one side in the surface is a common one, however, in engineering work, for the underwater parts of walls, gates and other structures are often equivalent to such a rectangle.

Example

A 3 m diameter circular sewer is to be closed by a watertight bulkhead supported by two horizontal beams. The sewage, density 1150 kg m^{-3}, completely fills the sewer on one side of the bulkhead and stands to a level 0.9 m above the roof level. If one beam is 0.3 m from the bottom of the sewer find the position for the other so that it bears the same load. Find the load.

If the sewer is completely dry on both sides of the bulkhead, but the air pressure on one side is 20 kN m^{-2} above the other, what are now the forces on the beams?

Solution

It is good practice to sketch a pressure distribution diagram, hydrostatic pressure as at (*a*), air (constant) pressure as at (*b*).

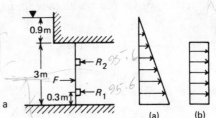

Fig. 2.8 Pressure distributions on a bulkhead in a sewer.

The total hydrostatic force on a plane surface is the product of the pressure at the centroid and the area.

Pressure at the centroid $= \rho g \bar{h} = 1150 \dfrac{kg}{m^3} \times 9.8 \dfrac{m}{s^2} \times 2.4\,m$

$$= 27.0\,kN\,m^{-2}.$$

Total force $= 27.0\,kN\,m^{-2} \times \pi \times 1.5^2\,m^2 = 191\,kN.$
Load on each beam $= 95.5\,kN.$

Depth of the centre of pressure below centroid $= k_g^2/\bar{h}$ where k_g is the radius of gyration of the area about a horizontal axis through the centroid. For the circle

$$k_g = r/2.$$

Depth of C of P below sewerage surface $= \bar{h} + \dfrac{k_g^2}{\bar{h}}$

$$= 2.4\,m + \dfrac{2.25}{4 \times 2.4}\,m = 2.63\,m.$$

To bear the same loads the beams must be equal distances from C of P. Position of C of P above bottom is $(3.9 - 2.63)\,\text{m} = 1.27\,\text{m}$. Second beam must be $1.27 + 1.27 - 0.3 = 2.24\,\text{m}$ from bottom.

With air the pressure will be uniform, hence the force is

$$F = p \times A = 20\,\text{kN}\,\text{m}^{-2} \times \pi \times 1.5^2\,\text{m}^2 = \underline{141\,\text{kN}}.$$

The resultant force will now act at the centroid of the area as the pressure is uniform. By resolution and moments for the reactions R

$$R_1 + R_2 = 141\,\text{kN}$$

and $\qquad\qquad 141 \times 1.5\,\text{kN} = 0.3\,R_1 + 2.24\,R_2.$

Hence $R_1 = 53.8\,\text{kN}; R_2 = 87.2\,\text{kN}.$

For plane surfaces which are not vertical it can be demonstrated that the resultant pressure force is still given by Equation 2.6. The expression (2.7) may be used to calculate the depth of the centre of pressure provided the distances $\bar{h}$ and $\bar{H}$ are measured to the liquid surface along the plane of the surface on which the force is being computed.

Case 4

Pressure intensity increases with depth in the fluid: surface not plane.

This case occurs when assessing the forces on non-plane surfaces such as valves, gates, etc. It has a great similarity with Case 2. The technique, Case 2, of finding component forces in arbitrary directions is again used. The surface in question is projected in directions at right angles, and the forces due to the fluid on the imaginary plane surfaces of these projections are found, together with their centres of pressure, by the methods of Case 3. These component forces are then combined to find the total force and its line of action. It is nearly always convenient to take one of the arbitrary directions vertically downwards, for then the component force vertically downwards is merely the weight of fluid supported above the surface in question. For example, in Fig. 2.9, AB is a side view of a non-plane surface in a fluid whose free surface is XX. The downward component of force on AB, F_1, is merely the weight of the prism of fluid ABCD above it, for no other forces are applied to this part of the fluid to keep it static. The horizontal component of force F_2 is the product of the projected area ZZ and the mean pressure on ZZ assessed as in Case 3. The total force on AB will be the combination of the two components F_1 and F_2.

If the surface is curved in one direction only (for example a part of the surface of a cylinder) it is convenient to choose one axis in the direction along which there will be no projected area (the direction of the cylinder's axis). There will only be two force components now to consider.

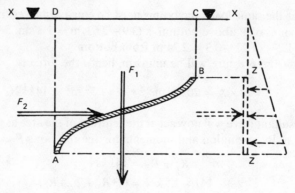

Fig. 2.9 Side elevation of a non-plane surface AB subject to a pressure intensity increasing with depth. Vertical component of force F_1 is the weight force of the fluid above AB: horizontal force F_2 is that on the projected area ZZ.

Example

The figure shows a radial gate of 10 m radius on a reservoir spillway. Specify completely the resultant hydrostatic thrust per metre length of the closed gate.

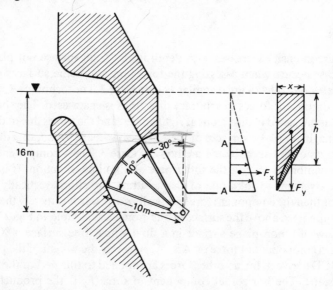

Fig. 2.10 Section through a radial gate on a dam spillway.

Solution

The upper surface of the gate is subjected to hydrostatic pressures so it is most convenient to calculate the horizontal and vertical components of the resultant force separately.

The horizontal component of the thrust F_x is equal to the thrust on the projected plane surface A-A (see Fig. 2.10)

$$F_x = \rho g a \overline{h}$$

where
$$a = 10\,\text{m}\,(\cos 30° - \cos 70°) = 5.24\,\text{m}^2 \text{ per m length}$$
$$\overline{h} = 6\,\text{m} + 10\,\text{m}\,(1 - \cos 30°) + 5\,\text{m}\,(\cos 30° - \cos 70°)$$
$$= 9.96\,\text{m}$$
$$F_x = 10^3\,\text{kg}\,\text{m}^{-3} \times 9.8\,\text{m}\,\text{s}^{-2} \times 5.24\,\text{m}^2 \times 9.96\,\text{m}$$
$$= 0.51\,\text{MN per m length.}$$

The vertical component of thrust F_Y is the weight of water supported vertically above the surface

Volume supported

$$= x\overline{h} - \text{area between chord and arc}$$
$$= 9.96\,\text{m} \times 10\,\text{m}\,(\sin 70° - \sin 30°) - 100\,\text{m}^2 \left(\frac{\theta}{2} - \sin\frac{\theta}{2}\cos\frac{\theta}{2}\right)$$
$$= 41.0\,\text{m}^3 \text{ per m length}$$

So $F_Y = 10^3\,\text{kg}\,\text{m}^{-3} \times 9.8\,\text{m}\,\text{s}^{-2} \times 41.0\,\text{m}^3 = 0.40\,\text{MN per m length.}$

Now $F = \sqrt{(F_x{}^2 + F_Y{}^2)} = 0.65\,\text{MN per m length.}$

Direction of force to the horizontal $= \tan^{-1} F_x/F_Y = 51.9°$

Now as the pressure acts normal to the surface, i.e. radially, the resultant pressure force must act radially. It must pass therefore through the centre of curvature of the gate.

2.6 Pressure forces on surfaces in moving fluids

Although the theory in the previous parts of this chapter has been developed to calculate the resultant force caused by a hydrostatic pressure distribution and its line of action, it is just a special case of the general method of dealing with distributed forces or stresses. For a surface with fluid moving past it there will be stresses on the surface both due to hydrostatic pressure forces and due to the motion. The stresses caused by the motion of the fluid are, by convention, resolved into normal stresses (pressures), and shear stresses due to viscosity. The total pressure at all the points on the surface can be found either by calculation or experiment using pressure measuring instruments such as the manometers described in the Appendix. Numerical or graphical methods of integration can then be used to solve the equations for the force, F, and its moment, M, about some axis X:

$$F = \int p\,\mathrm{d}a$$

$$M = \int p\,x\,\mathrm{d}a$$

It is possible to find the resultant shear force in a similar way but this is rarely necessary in Civil Engineering Hydraulics.

Example

A wind-tunnel test of a thin plane aeroplane wing (aerofoil) 25 cm wide showed that the following pressures relative to the undisturbed pressure of the oncoming stream of air existed on the upper and lower surfaces. Positive signs show pressures, measured with a water manometer, above undisturbed pressure. What is the total lift force on a 1 m length of the wing, and where should a single force of this magnitude act so as to balance the lift?

Distance from leading edge	x	0	2.5	5	7.5	10	cm
Pressure on upper surface	h_1	-7.5	-3.5	-2.5	-2.0	-1.75	cm of water
Pressure on lower surface	h_2	$+2.5$	$+1.25$	$+0.75$	$+0.5$	$+0.25$	cm of water

Distance from leading edge	x	12.5	15	17.5	20	22.5	25 cm
Pressure on upper surface	h_1	-1.25	-0.75	-0.5	-0.25	0	0 cm of water
Pressure on lower surface	h_2	$+0.25$	$+0.25$	$+0.25$	0	0	0 cm of water

Solution

Difference of pressure intensity $(p_1 - p_2)$ between top and bottom of wing at any value of x is the resultant vertical pressure intensity which causes a vertical pressure force $(p_1 - p_2)\mathrm{d}x$, where $\mathrm{d}x$ is the element of area of a unit length of the wing on which this pressure acts (see Fig. 2.11). The total lift is

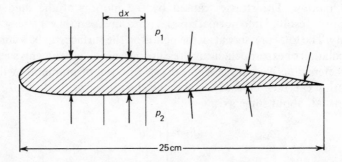

Fig. 2.11 Section of a thin aerofoil.

then $\int (p_1 - p_2)\mathrm{d}x$. But $p = \rho g h$ where h is the height of the static fluid in the manometer.

The total force is therefore given by the area

$$\rho g \int (h_2 - h_1)\mathrm{d}x = \frac{1000 \times 9.81}{100 \times 100} \int (h_2 - h_1)\mathrm{d}x \text{ newtons per metre length of}$$

wing if h is measured in cm and x in cm. Evaluate the integral graphically by the mid-ordinate rule as the area under the $(h_2 - h_1)$ curve (Fig. 2.12).

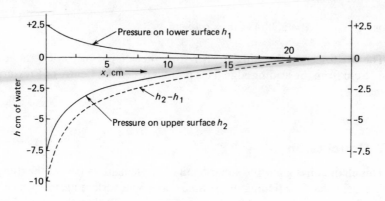

Fig. 2.12 Pressure distribution on an aerofoil.

$$\text{Area} = 49.4 \text{ (cm of water} \times \text{cm)}$$

$$\text{Total lift force} = 49.4 \times \frac{1000 \times 9.81}{10\,000} \text{ N per m length of wing}$$

$$= 48.6 \text{ N m}^{-1}$$

To find the point of application of the balancing force, it is necessary to find the moment of the elementary forces about a fixed point on the wing section. It is convenient to take the leading edge for this purpose. Then the moment of the pressure forces is

$$\int (p_2 - p_1)x\,\mathrm{d}x$$

or, converting pressures to the corresponding head of water,

$$\frac{1000 \times 9.81}{100 \times 100 \times 100} \int (h_2 - h_1)x\,\mathrm{d}x.$$

The integral is the area under the $(h_2 - h_1)x$ curve, and this is found by the mid-ordinate rule to be 304 cm of water $\times$ cm².

Thus the total moment of the pressure forces about the leading edge is

$$\frac{304}{100} \times \frac{1000 \times 9.81}{100 \times 100} \, \text{N m per m length of wing}$$

$$= 2.98 \, \text{N m/m}$$

But this moment also equals the balancing force, $48.6 \, \text{N m}^{-1}$ acting at a distance $\bar{x}$ from the leading edge.

That is, $2.98 = 48.6\bar{x}$

or $\qquad \bar{x} = 2.98/48.6 \, \text{m} = 6.15 \, \text{cm}$

The lift force on the wing is, therefore, $48.6 \, \text{N}$ per metre length, acting $6.15 \, \text{cm}$ from the leading edge. This point is called the Centre of Pressure of the wing.

2.7 Conclusion

This chapter has gone into some detail of an elementary part of the study of fluid mechanics. The finding of the fluid force on a surface against which is a static fluid is required so often in engineering work that the student must acquire complete facility with the methods used. Indeed, some engineers do no other calculations in fluid mechanics, for in many cases the hydrostatic pressure forces are overwhelmingly important, and the forces due to motion are insignificant.

It should not be thought, however, that these methods can be used for static fluids only. Even if the fluid is in motion the normal forces on a surface caused by the pressure distribution thereon can be found by essentially the same methods. This force is sometimes called the 'hydrostatic' force even if the fluid is not static. There will in general be other forces acting as well as this force: for example, there are likely to be shear forces acting parallel to the surface. The shear forces must be assessed by quite different methods, often involving experimental data, and may be of the same order of magnitude as the hydrostatic force. The two forces interact upon each other so that their combined effect is far more difficult to assess than is the single effect of the normal force in truly hydrostatic conditions.

Appendix: The measurement of pressure intensity

There are three essentially different ways of measuring the pressure intensity at a point in a fluid, whether static or moving. The fluid pressure may be applied to a movable diaphragm or piston which is balanced by weights, the pressure intensity being then the applied force divided by the diaphragm area: or the fluid pressure may be balanced by a hydrostatic column of fluid of height h (either the same fluid as that in which the pressure has been generated, or another), when the pressure intensity p is then given by the hydrostatic equation for incompressible fluids, $p = \rho g h$. The third alternative is to apply the fluid pressure to a device which either deflects or alters some other property which can be measured mechanically or electrically and the output calibrated to evaluate the pressure.

A deadweight piston gauge (Fig. 2.13) is an example of the first type of measuring instrument. The pressure is transmitted from the point in question by a narrow tube A and applied to a piston P inside a closely fitting cylinder C. Weights are then applied to P until the pressure force is balanced. Such gauges are mainly used to calibrate gauges of the third type which are less bulky and easier to read, such as the *Bourdon gauge*. In this gauge the pressure is applied to the inside of a bent and flattened tube whose end is blocked. The tube tends to straighten due to the internal pressure and the straightening is limited by the elasticity of the tube. A pointer connected to the blocked end of the tube then shows the equilibrium position of the tube on a scale which is directly calibrated in pressure intensity. Bourdon gauges are convenient for engineering use because they can be made to suit a wide variety of pressures and are compact. They invariably require calibration at frequent intervals against some other sort of pressure gauge if they are to be relied upon.

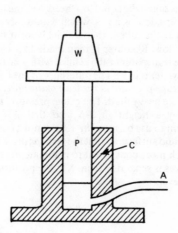

Fig. 2.13 A dead-weight piston gauge.

A more convenient method of measuring the deflection of the Bourdon tube may be to use electrical resistance strain gauges which give the possibility of transmitting the reading of the gauge to a remote station. More sophisticated pressure gauges make use of the piezoelectric properties of crystals whose resistance changes with pressure. These may be built into pressure transducers whose output may be measured electrically and used for automatic control systems.

The second kind of measuring instrument ('manometer') has more variation in its possible arrangement. The simplest device shown in Fig. 2.14(a) is a vertical transparent tube (a *piezometer*) from the pipe or container P, inside of which is the liquid whose

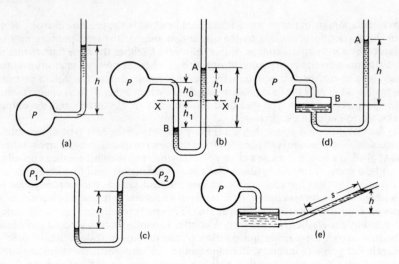

Fig. 2.14 Simple manometers.

pressure p is to be measured. The liquid, water for example, rises to a height h, when the pressure p is said to be equal to a 'head' h of the liquid. This simple device cannot be used if the fluid inside P is a gas, which would escape: nor is it suitable if h is large (above about 2 m) as the tube is then unwieldy; or if h is small (below about 7 cm), when the accuracy is low. Recourse is then made to a U-tube manometer, Fig. 2.14(b), where a bend in the transparent tube is filled with a heavier fluid (water or alcohol if air is in P; mercury if water is in P). The difference of level, h, of the two surfaces of the heavier fluid gives the gauge pressure *at the level of the lower surface B*, that is, $\rho_1 gh$, where ρ_1 is the density of the heavy fluid. The gauge pressure, p, at P is smaller than $\rho_1 gh$ because P is above B by a height $(h_1 + h_0)$ so that $p = \rho_1 gh - \rho g(h_1 + h_0)$. This equation is inconvenient to use because $(h_1 + h_0)$ is not a constant; it changes with every position of the heavy fluid surface, so that the term $\rho g(h_1 + h_0)$ must be computed for each value of h. It is much more convenient to eliminate h_1, the amount by which each of the heavy fluid surfaces is separated from XX, the position of the surfaces when they are both at the same level.

Since now $\qquad\qquad h = 2h_1$

$$p = \rho_1 gh - \rho g \left(\frac{h}{2} + h_0 \right)$$

or $\qquad\qquad p = gh \left(\rho_1 - \frac{\rho}{2} \right) - \rho g h_0.$

Both $(\rho_1 - \rho/2)$ and $\rho g h_0$ are now constants, and h is the only variable. Note that for a mercury-filled manometer used for finding the pressure of water $\rho_1 = 13.56 \times 10^3$, $\rho = 1.0 \times 10^3$ kg m^{-3}, so that

$$p = (13.06 \times 10^3 \times 9.81h - 9.81 \times 10^3 h_0) \, \text{N m}^{-3}$$

in the SI system of units, where p is measured in $N\,m^{-2}$ and h in metres.

The above scheme of measurement is also used when a U-tube manometer has each side connected to pressure tappings in a piece of apparatus so that it measures the *difference* between the pressures at the tappings, Fig. 2.14(*c*). It will be then found that the pressure difference is $p_1 - p_2 = gh(\rho_1 - \rho) - \rho gh_0$, where h_0 is now the difference of height between the tapping points, the one at which the pressure is p_1 being assumed lower than the other.

Two modifications of the common U-tube manometer are shown in Fig. 2.14(*d*) and (*e*). In (*d*) one limb of the manometer has been widened so that its cross-sectional area is much larger (100 times or more) than the other. For all practical purposes the movement of the surface of the heavy fluid in the widened side is now negligible, compared with the movement on the narrow side. The level B can therefore be regarded as constant so that only the height of the surface at A need be measured. Only one reading of the height of a meniscus is therefore required to find h in the narrow tube, and only this tube need be transparent.

The second modification, in Fig. 2.14(*e*), is used when the pressure to be measured is small. By sloping the transparent tube of a widened limb manometer, there is a magnification of the distance that the meniscus moves along the tube, for a given pressure. If the slope is θ, then the distance s along the tube is $s = \dfrac{h}{\sin\theta}$. By making θ sufficiently small, s can be made large and it can therefore be measured more accurately than if the tube were vertical. It is not usual for the slope to be less than $\tan^{-1}\frac{1}{25}$, for then small changes of surface tension forces due to greasy patches in the bore of the tube may cause the meniscus to stick at some places, and so give inaccurate readings.

There are also other ways of measuring differences of pressure (differential pressures), particularly between pairs of tapping points in pipe systems. The simplest method is to arrange two piezometers alongside, each being connected to its tapping point, Fig. 2.15(*a*). The difference of the heights of the fluid surfaces in them is the differential pressure. If this difference is large then each tapping point is led to one side of a U-tube manometer containing a heavy fluid as has been described. If the pressure

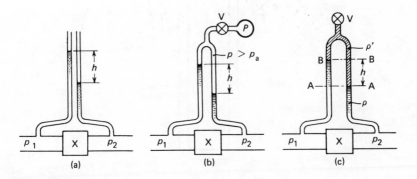

Fig. 2.15 Methods of measuring a pressure difference caused by an apparatus X in a pipe.

difference is small enough to use the piezometer method, but both pressures are high compared to the atmospheric pressure, the tubes will be inordinately long: but if the tops of the tubes are connected together and to a small air pump, a false 'atmospheric' pressure may be applied to both fluid surfaces which are then equally depressed without affecting the difference, Fig. 2.15(*b*). Thus much shorter tubes can be used. If the pressure difference is small, the tubes may be sloped to increase the distance traversed by the meniscus along the tube. Another method of magnifying the movement of the meniscus is to join the tops of a pair of piezometers and to fill the space above the surface of the fluid with another fluid of a slightly lower density, ρ'. As shown in Fig. 2.15(*c*), the difference of pressure at the level AA is due to the difference between the pressures at the bases of columns of fluid of height *h*, and densities ρ and ρ' respectively. Thus $p_1 - p_2 = gh(\rho - \rho')$. Theoretically a great magnification of the difference of level *h* may be obtained by making $\rho - \rho'$ small enough: but a limit is soon reached because the meniscus between fluids of near densities becomes very sensitive to changes in the surface tension within the tube, due to traces of grease in the bore. The meniscus may then adhere to parts of the bore more than others and be deformed so that it is not horizontal and its level cannot be determined with accuracy.

The above descriptions are for a few basic types of manometer only. There are many other types each for its own range of duty but certain precautions must be taken when measuring pressures with any sort of manometer. First, the connecting tubes from the tapping points must be full of the fluid whose pressure is being measured, and there must be neither bubbles of liquid in the tubes of gas system nor gas bubbles in a liquid system: the density ρ in the connecting tubes must be uniform. Secondly, it is desirable to be able to confirm, at any time, that if the pressures are equalized, then the readings of the surfaces of the manometer fluid are the same. A manometer system ought always to have valves in it to flush the connecting tubes, and to have an equalizing valve to ensure that both tubes can be brought to the same pressure as shown in Fig. 2.16.

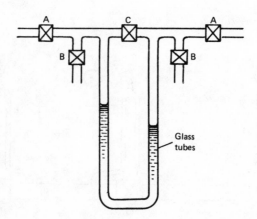

Fig. 2.16 Valves A are to isolate the manometer from tapping points. Valves B are for flushing, C is the equalizer valve.

Problems

1. An 8 m long, 2 m diameter cylindrical tank is mounted on a lorry and is full of petrol (specific gravity 0.7). When the lorry decelerates at 4 m s^{-2} what will be the total force on the front end of the tank and where will it act? Sketch the pressure distribution along the bottom of the tank. Find also the inclination of the free surface if the tank were only partly filled.

   Ans. 92 kN at 1.06 m from top to tank: 22.2°.

2. A radial gate on the crest of a weir 10 m wide is an arc of 7 m radius, arranged to lift by pivoting about its centre of curvature which is 4.2 m above and downstream of the crest. Find (*a*) the position, (*b*) the magnitude, and (*c*) the direction of the resultant force when the upstream water-level is 6.3 m above the crest and the gate is both closed and dry downstream.

   Ans. (*a*) through centre, (*b*) 2055 kN, (*c*) $19\frac{1}{2}°$ upward.

3. A closed rectangular tank, 3.3 m high can be filled to a depth of 3.0 m with a volatile liquid (s.g 1.60). A safety valve in the roof is set to blow at 7 kN m^{-2} gauge. (*a*) What is the maximum loading per horizontal metre run on each side? (*b*) Where should a tie bar be placed to take this load? (*c*) What is the maximum load when there is no liquid in the tank?

   Ans. 93.7 kN m^{-1}: 1.16 m from bottom: 23.1 kN m^{-1}.

4. Vertical boards are being used to shore up completely waterlogged soil at the side of an excavation 6 m deep. Where should two horizontal beams be placed, one above the other, so that they are equally loaded? What is the load in each per metre run of boarding in the horizontal direction?

   Ans. Equi-spaced about C.P but minimum bending if 2.84 m and 5·17 m from top: 88.5 kN m^{-1}.

5. A model of a 1.2 m pipe submerged in a river is tested in a stream and discloses the following pressures on the full-size pipe surface. Plot the pressures and find the horizontal and vertical components of the loading per metre run. Angles measured from the horizontal.

θ degrees	p:kN m^{-2}	θ	p	θ	p	θ	p
0	20.7	80	6.2	180	8.8	280	1.4
20	17.2	100	8.4	200	8.1	300	1.7
40	10.6	120	9.2	220	7.0	320	8.0
60	6.1	140	9.2	240	5.7	340	17.1
		160	9.1	260	3.7		

Ans. Horizontal 8810: Vertical 6270 kN m^{-1}.

6. A spherical container is made up of two hemispheres, one resting on the other with the interface horizontal. The sphere is completely filled through a small hole in the top by a weight W of liquid. What is the minimum weight of the upper hemisphere in order to prevent it from lifting?   *Ans. W/4.*

3
Buoyancy forces on immersed objects

3.1 Vertical forces in a fluid at rest

The preceding chapter has dealt generally with the forces on surfaces which are subject to the pressure of a fluid. This chapter will examine the particular case of the resultant vertical (buoyancy) force on a body enclosing a finite volume surrounded by the fluid. Fig. 3.1 shows the side view of a solid body totally immersed in a fluid whose free surface is XX. Consider a vertical column of the body such as AA whose cross-sectional area in a horizontal plane is a and whose length is x. Then the downward hydrostatic force on the upper end of the column is $\rho g h a$, and the upward force on the lower end is $\rho g(h + x)a$, so that the resultant force on the column is $\rho g x a$, in an upward direction. But xa is the volume of the column, V, so that the upward force F on the whole body, which is made up of many columns, is

$$F = \Sigma \rho g V = \rho g V_i \qquad (3.1)$$

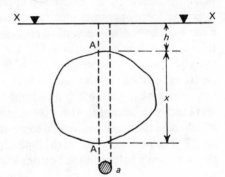

Fig. 3.1 Side view of an object totally submerged in a fluid whose free surface is XX. A vertical cylindrical surface (cross-sectional area a) is shown intersecting the object.

where V_i is the total immersed volume of the body. F is the *buoyancy force* and, as there is no resultant horizontal force is the total pressure force acting on the immersed volume.

Since the upward hydrostatic force on the bottom area of the cylinder, and also the downward force on the top, both act through the centroid of the cross-sectional area, therefore the resultant force $\rho g x a$ acts through the centre of the

33

cylinder. The total force $\rho g V_i$ therefore acts through the centre of mass of fluid displaced by the total immersed volume V_i.

The relationship $F = \rho g V_i$, attributed to Archimedes, is often expressed in words as 'The upward buoyancy force on a body is the weight force of the fluid displaced by the body', and it is used in all calculations where a body is wholly or partially surrounded by fluid. If the weight force of the body Mg (where M is the mass of the body) exceeds the buoyancy force, then the body will sink, or alternatively an upward force $(Mg - \rho g V_i)$ must be applied in some other way to preserve equilibrium. Correspondingly, if the weight force is smaller than the buoyancy force, then the body rises through the surface until the immersed volume has so decreased that $\rho g V_i = Mg$, where V_i is now not the total volume of the body, but merely the volume of the portion below the fluid surface. At this stage the body will float indefinitely if the conditions are preserved.

Archimedes' principle determines, then, whether a particular body will sink or float; or alternatively, what volume of the body must be immersed in order to balance its weight and therefore allow it to float. But engineers are not only concerned with this principle as they also have to consider whether a particular body is stable in any one position. In other words, they are concerned to know whether the body will float right way up or whether it will capsize.

3.2 Stability of floating bodies

An object, acted upon by any set of forces (which may include buoyancy forces as above), is said to be in stable equilibrium if a small change of its position caused by an externally applied force or couple gives rise to an opposing force or couple which just balances the applied force. Thus consider a rectangular box whose vertical cross section is ABCD in Fig. 3.2(a), which is floating in a fluid with surface XX. The buoyancy force F is, by Archimedes' principle, equal to the weight force of the fluid displaced and acts through the centre of mass of the displaced volume of fluid PQCD, that is at H. The only other force acting upon it is its weight force Mg acting vertically downwards and exactly balancing F. Mg acts through the centre of gravity G of the body, which must therefore be situated somewhere on the vertical line through H, i.e. on ZZ. The position of G on this line depends on the distribution of weight in the box, heavy ballast in the bottom lowering G, but weights on the deck AB raising G.

Now consider what happens if a couple of value Wgx is applied to the box; Fig. 3.2(b), which causes it to tilt or roll through an angle θ. Assuming that the weights in the box are secured and do not move as the box rolls, G will not be affected and will remain somewhere on ZZ. The effect of the change of attitude is to change the shape of the cross-section of the immersed volume to P'Q'CD, a trapezoidal shape, Fig. 3.2(b). The area of P'Q'CD is still the same as PQCD, for the downward forces on the box are unaltered and the balancing buoyancy

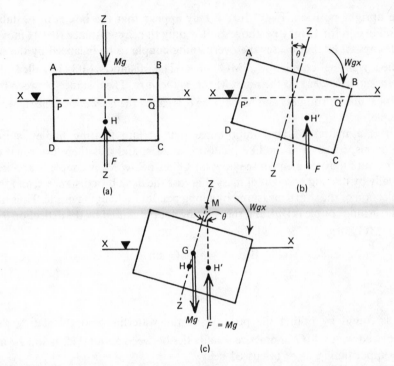

Fig. 3.2 The forces on an object floating in a fluid, surface, XX. (a) The object at rest with an axis of symmetry ZZ vertical. The vertical buoyancy force F acts through H, the centre of the displaced mass represented by PQCD. (b) If the object is subjected to a torque Wgx, it heels over to an angle θ. In doing so, the immersed volume changes to a trapezium P'Q'CD with the centre of displaced mass at H', through which F now acts. (c) If the centre of gravity G of the mass of the body lies below M, then there is a restoring couple Mg. $\overline{MG} \sin\theta$ which just balances the torque Wxg, and the object is stable.

force therefore stays constant. But the position through which the buoyancy force acts has changed, for the centre of mass H' of the displaced fluid whose cross-section is the trapezium P'Q'CD is nearer to the long edge CQ' than the previous centre of mass H. The buoyancy force acts therefore on the *vertical* line H'M, M being the place where the buoyancy force cuts the original vertical centre line ZZ. [See Fig. 3.2(c).]

Providing the centre of gravity G lies *below* M, the lines of action of the weight force and the buoyancy force are so separated that they cause a couple Mg. $\overline{MG} \sin\theta$ tending to bring the box upright. If this couple just equals the overturning couple Wgx the box will be in equilibrium at the angle θ. Thus the condition for the box to be in stable equilibrium is that G shall lie somewhere below M. There is no restriction in this respect about the position of G relative to H, the original position through which the buoyancy force acted. Thus in the upright position, Fig. 3.2(a), it may appear that the box is in unstable equilibrium for G may be above H: it is only the circumstance that H moves sideways to H' that causes the overturning couple to be balanced by the so-called righting couple, $Mg . \overline{\text{MG}} \sin \theta$. The distance $\overline{\text{MG}}$ is called the *metacentric height* and the point M the *metacentre*. The distance between the vertical lines through G and M, $\overline{\text{MG}} \sin \theta$, is called the *lever arm* of the righting couple.

It is a matter of vital importance that certain floating bodies, ships, pontoons, barges and the like, shall not capsize, and it is therefore usual to carry out an experiment to see how far G lies below M. A couple is applied, usually by moving a weight of mass W across the deck by a distance x, and the deflection of the ship θ measured by a long pendulum hanging inside. Thus the overturning couple is Wgx and the righting couple caused by the deflection θ is $Mg . \overline{\text{MG}} \sin \theta$.

So
$$Wgx = Mg . \overline{\text{MG}} \sin \theta$$

$$\overline{\text{MG}} = \frac{W}{M} \frac{x}{\sin \theta} \tag{3.2}$$

M is found by noting the position of the waterline and calculating the displaced water. $\overline{\text{MG}}$ in practice usually lies between about 0.15 m and 1.3 m; the upper limit will be discussed below.

The experimental method of finding the metacentric height is of course only possible if the box or ship is already afloat and ballasted. It is often necessary to determine in advance what $\overline{\text{MG}}$ will be for a proposed loading system. For example, when the sections of a bridge are being put into position from barges, the stability is of paramount importance as the sections are jacked up. It is clear that a unit volume of displaced water creates a larger righting couple if it is far from the pivoting axis of a box than if it is near the axis: a wide box, i.e. AB large, gives a larger couple and a larger $\overline{\text{MG}}$ than a narrow box of the same volume. $\overline{\text{MG}}$ thus depends on the way in which the waterline area is distributed, the waterline area being the area of the body intersected by the waterline XX. (In the case of a box this area is a rectangle: for a ship the area is more of an elliptical shape with the ship rolling about the long axis, Fig. 3.3.) In fact, it can be shown that for angles of roll, θ, small enough for the deck of the ship not to become immersed and for the ship's sides being parallel in a vertical cross section as Fig. 3.2, the height $\overline{\text{MH}}$ is given by $\overline{\text{MH}} = I_{\text{WLA}}(1 + \frac{1}{2}\tan^2 \theta)$. I_{WLA} is the second moment of the waterline area about the axis of roll and V_i is, as before, the immersed volume. Notice that $\overline{\text{MH}}$ is not the metacentric height directly, but that $\overline{\text{MH}} = \overline{\text{MG}} + \overline{\text{GH}}$. $\overline{\text{GH}}$ must be found from other evidence such as the distribution of weights in the ship which fix the position of G relative to H.

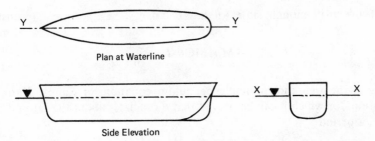

Plan at Waterline

Side Elevation

Fig. 3.3 Plan and elevation of a ship to show the waterline area (on the plan) of which the second moment of area I_{WLA} is required about the axis of roll YY, to obtain the height of the metacentre above the centre of buoyancy.

All these terms can be calculated and $\overline{MG}$ found before a ship is constructed, though the determination of G is a long job for such a complicated structure. The predicted value of $\overline{MG}$ is usually checked by the experimental method after the ship is launched. To ensure that $\overline{MG}$ is suitable and is not so small as to endanger the ship, it is common practice to stipulate limitations to the loading of ships so that G does not rise so high that $\overline{MG}$ vanishes.

3.3 The period of rolling of floating bodies

Though the civil engineering problems associated with the stability of floating bodies are usually concerned with the static aspects described above, it is of interest in the field of ship design to examine what happens if an overturning couple is suddenly removed from a body which has been inclined to an angle θ. If the overturning couple is suddenly removed, Fig. 3.4, the only force acting on the body is the righting couple as above, $Mg \cdot \overline{MG} \sin \theta$, so by Newton's Second Law of Motion a rate of change of angular momentum is caused, denoted by $I_s(d^2\theta/dt^2)$, where $d^2\theta/dt^2$ is the angular acceleration of the ship and I_s is the moment of inertia of the weights and the ship about the axis of roll. Thus $Mg \cdot \overline{MG} \sin \theta = -I_s(d^2\theta/dt^2)$, the minus sign indicating that the couple is acting in a direction such that it tends to decrease θ.

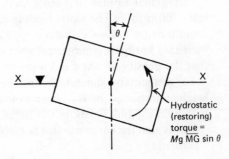

Hydrostatic (restoring) torque = $Mg \overline{MG} \sin \theta$

Fig. 3.4 If the overturning torque *Wxg* is removed, the restoring torque causes the object to rotate about its axis of roll.

If θ is small enough, $\sin \theta \to \theta$, $\tan^2 \theta \to 0$

or
$$-Mg \cdot \overline{MG}\theta / I_s = \frac{d^2\theta}{dt^2}.$$

This is the well-known simple pendulum equation, or Simple Harmonic Motion, for which it can be proved that θ oscillates about the zero position with a period

$$t = 2\pi \sqrt{(I_g / Mg \cdot \overline{MG})}.$$

Thus t is the period of rolling of the ship. In the absence of fluid friction, or damping, the oscillation will continue undiminished with this period. Fortunately, the relative motion between ship and water causes fluid forces which oppose the motion in whatever direction it occurs, so that rolling motions die out fairly quickly, unless of course a new overturning couple is applied.

In the above expression it will be seen that a large metacentric height $\overline{MG}$ causes a *small* period of rolling with correspondingly large accelerations: such a ship will roll rapidly from one side to the other and will not only be uncomfortable to sail in but may damage its own structure. A small $\overline{MG}$, though giving a slower roll, is undesirable because a small error in loading cargo, or a small amount of water on deck in a gale, may change G so much that the ship will capsize. It is usual to get $\overline{MG}$ in the range 0.15 to 1.3 m, and a ship with only a small amount of cargo on board sometimes stows it on a deck so as to have G in the required position to give a comfortably slow roll.

Example

A barge is to be designed to carry loads of up to 500 tonnes to a construction site on a river. When laden, the depth of the bottom of the barge below the water surface (the draught), d, is to be limited to 3 m and the breadth, b, to 10 m. For initial calculations it may be assumed that the barge is rectangular in plan and cross-section and constructed of steel plate of mass $160 \, \text{kg m}^{-2}$. The worst loading case is expected to be when two long 2.5 m diameter hollow cylinders, each of mass 250 tonnes are placed with their axes 3 m apart symmetrically about the centre line of the barge. Assume that the density of water is 1.0 tonne m^{-3}.

Suggest suitable dimensions for the barge, and estimate its metacentric height and period of roll when laden. Neglect the mass of any internal structure, e.g. bulkheads, in the barge.

Check that the barge will still be stable when one of the cylinders has been unloaded.

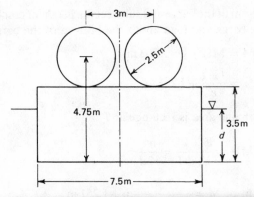

Fig. 3.5 Cross-section of a laden barge.

Solution

To support the load a volume of water V must be displaced such that:

$$\rho g V = 500 \text{ tonne} \times g$$

and as the density of water is 1 tonne m^{-3}

$$V = 500 \text{ m}^3$$

If the load causes a draught of 2 m, leaving 1 m of draught available to support the weight of the barge, then the water plane area must be 250 m^2. Assuming a beam of 7.5 m, which gives plenty of room for two 2.5 m cylinders, the length of the barge would be $250/7.5 = 33.33$ m, which is not unreasonable.

If x is the unladen draught of the barge,

$$\rho g \times \text{immersed volume} = g \times \text{mass of barge}$$
$$= g \times \text{mass steel per unit area} \times \text{area}$$

$$1 \text{ tonne m}^{-3} \times 250 \text{ m}^2 x = \frac{160 \text{ tonne}}{1000 \text{ m}^2} \times \text{area of material}$$

The area of material in the barge is the sum of the areas of the deck, bottom, sides and ends. Assuming the deck of the loaded barge is 1 m above water level then area of material A:

$$A = 2 \times 250 \text{ m}^2 + 2 \times 7.5 \text{ m } (3 \text{ m} + x) + 2 \times 33.3 \text{ m}(3 \text{ m} + x)$$
$$= 745 \text{ m}^2 + 81.7 \, x \text{ m}$$

hence $\qquad 250x = 0.16(745 \text{ m} + 81.7 \, x)$

and $\qquad x = 0.503$ m.

Suggested dimensions for barge:–

length, l 33.33 m, breadth, b 7.5 m, height 3.5 m.
mass = 125.7 tonne, unladen draught = 0.503 m.

To calculate the metacentric height, calculate the height of metacentre above the bottom of the barge and subtract the height of combined centre of gravity of the barge and load above the bottom of the barge.

$$\text{Height of M} = \text{MH} + \tfrac{1}{2} \times 2.503 \text{ m}$$

$$= \frac{I_{WLA}}{V} + 1.251 \text{ m}$$

As the water line area is rectangular $I = \dfrac{lb^3}{12}$

and hence $\text{MH} = \dfrac{lb^3}{12} \dfrac{1}{lbd} = \dfrac{b^2}{12d}$

Hence height of $\text{M} = \dfrac{7.5^2}{12 \times 2.503} \text{ m} + 1.251 \text{ m} = 3.124 \text{ m}$

If y is height of combined c.g. above bottom, then taking moments about the bottom of the barge:

$$(500 + 125.7) \text{ tonne } y = 500 \text{ tonne} \times 4.75 \text{ m} + 125.7 \text{ tonne} \times 1.75 \text{ m}$$
$$\text{whence } y = 4.147 \text{ m}$$

So the metacentric height is $(3.124 - 4.147)$ m and is negative. Therefore the barge would be unstable and the suggested design is unsuitable. If the breadth of the barge is increased it will become more stable. It would be prudent to see if the barge can be made stable at the specified maximum breadth of 10 m, but keeping the same draught approximately. Using the same method as previously the dimensions of the barge become:

length 25 m, breadth 10 m, height 3.5 m,
mass = 119.2 tonne   unladen draught = 0.476 m,

$$\text{height of metacentre} = \frac{10^2}{12 \times 2.476} \text{ m} + \tfrac{1}{2} \times 2.476 \text{ m} = 4.604 \text{ m},$$

height of centre of gravity = 4.172 m,
hence metacentric height = $4.604 - 4.172 = 0.432$ m

which looks reasonable.

To calculate the period of roll it is necessary to assume that the barge and load roll about their combined centre of gravity and to calculate the moment of inertia of the barge and load about this axis. This may be done by using the theorem of parallel axes.

The barge may be split unto separate sides and the moment of inertia of each of these found about the axis of the barge and then all components added.

For the deck $I = \dfrac{mb^2}{12} + m\left(\dfrac{h}{2}\right)^2$

where h is the height of the barge and m the mass of the deck.

Whence I deck $= I$ bottom $= 250 \times 0.16 \left[\dfrac{100}{12} + \left(\dfrac{3.5}{2} \right)^2 \right]$ tonne m^2

$$= 455.83 \text{ tonne m}^2$$

Similarly for one side, I side $= 364.29$ tonne m^2

For one end $I = m \dfrac{(b^2 + h^2)}{12} = 52.38$ tonne m^2

Total moment of inertia about axis $= 1745$ tonne m^2
Moment of inertia of barge about c.g. of
  combination $= 1745$ tonne m$^2 + 119.2$ tonne $(4.172 - 1.75)^2$ m^2

$$= 2444 \text{ tonne m}^2$$

Considering the load, assume that the radius of gyration of each cylinder
is 1.2 m

then I for cylinder about axis $= 250 \times 1.2^2 = 360$ tonne m^2
I about combined c.g. $= 360 + 250[1.5^2 + (4.75 - 4.172)^2]$ tonne m^2
$$= 1006 \text{ tonne m}^2$$
Total moment of inertia of barge and load $= 4456$ tonne m^2

Period of roll $= 2\pi \sqrt{\dfrac{I_s}{Mg\,\overline{MG}}}$

$$= 2\pi \sqrt{\dfrac{4456 \text{ tonne m}^2}{619.2 \text{ tonne } 9.81 \text{ ms}^{-2} \ 0.432 \text{ m}}}$$

$$= 8.2 \text{ s}$$

Consider the case when one cylinder is removed. This is most easily
calculated by considering the mass of the barge and remaining load
combined on the centreline plus a moment due to the load being off centre.

Draught of barge with one cylinder $= 1.476$ m
Hence height of metacentre $= 5.646(1 + \frac{1}{2}\tan^2 \theta) + 0.738$ m
$$= 6.38 + 2.82 \tan^2 \theta \text{ m}$$
Height of centre of gravity $= 3.78$ m
So metacentric height $= 2.6 + 2.82 \tan^2 \theta$ m
The moment causing the barge to roll $= 250$ tonne $\times g \times 1.5$ m $\cos \theta$
$$= \rho g V \times \text{metacentric height } \sin \theta$$
i.e. $250 \times 1.5 \cos \theta = 369.2(2.6 + 2.82 \tan^2 \theta) \sin \theta$
$$1.016 = 2.6 \tan \theta + 2.82 \tan^3 \theta$$
by successive approximation,
first approximation as $\tan \theta$ is small for stability

$$\tan \theta = \frac{1.016}{2.6} = 0.390$$

If $\tan\theta = 0.35$, right hand side = 1.03.
Finally if $\tan\theta = 0.346$, right hand side = 1.016,
i.e. angle of heel = 19°.
Check to see that deck is not submerged.
Freeboard at half load = $3.5 - 1.476 = 2.0$ m.
Half width = 5 m so $\tan\hat{\theta} = 0.4$ $\hat{\theta} = 21.8°$.

The final design will not be much narrower than the 10 m of the second approximation as the metacentric height at full load is small and the freeboard at half load is also small.

Problem

A floating rectangular pontoon of mass 50 tonnes with C of G at the waterline rolls 10° when an overturning torque of 59.8 kN m is applied. How high may a 5 tonne mass be raised on a light scaffolding?

Ans. GM 0.7 m; 7.7 m above W.L.

4
Definitions concerning fluids in motion

4.1 The vocabulary of fluid motion

The phenomena exhibited by fluids in motion are more complex than those which have already been described for static fluids. The motions of a fluid may vary from place to place, or from time to time, or both: they may appear simple at first but may later appear much more complicated as attention is concentrated on certain aspects of the flow. It is accordingly necessary to make certain definitions so that a flow is described accurately.

The first division of the types of flow is made by considering the 'steadiness' or way in which the flow changes with *time*. A flow whose properties, at one place, do not change with time is called *steady*: if any property of the flow changes with time it is called *non-steady*. Many fluid motions of interest to engineers are steady, for example, the flow through a pipeline when a short time has elapsed after the controlling valves have been opened. A great deal of experimental and theoretical work has been carried out on steady flow, which is usually far simpler to analyse than non-steady flow. However, considerable engineering problems are associated with non-steady flow: for instance, the conditions occurring in a long pipeline when a valve is suddenly shut may give rise to stresses which can wreck the pipe. An introduction to the solution of these problems is given in Section 13.12.

A further division of types of flow is concerned with 'uniformity' or way in which the properties change with *position*. If, for example, the velocity does not change within a particular zone, then the velocity is said to be *uniform* there: if the velocity changes from place to place, then it is *non-uniform*. Two distinct cases of non-uniformity can be distinguished. The velocity may vary over a cross section of a flow (for example, in a river where the water is slower at the sides and bottom than at the middle); or it may vary along the flow direction (for example, at the entrance to a river from a lake, where the water accelerates from a standstill to a relatively high velocity). Both sorts of non-uniformity often coexist, as in the latter case. However, it is observed that changes of the first sort of non-uniformity are connected with changes of the second sort. A flow which is non-uniform along the direction of motion may either suppress or enhance the non-uniformity across it. In a converging pipeline, for instance, the flow is made more uniform as the cross-sectional area is decreased, the slower fluid near the walls of the pipe being speeded up more than the fluid in the centre. In a diverging pipe the reverse is the case, and the non-uniformity across the flow is accentuated (Fig. 4.1). Complete uniformity in the direction of a flow is common in engineering problems (for example, flow in a constant-bore pipe), but complete uniformity across the flow is rare. Shear stresses are present, making the velocity lower near solid surfaces in a part of the flow called the *boundary layer*. The flow may be nearly uniform over limited areas of the cross section far removed from such surfaces.

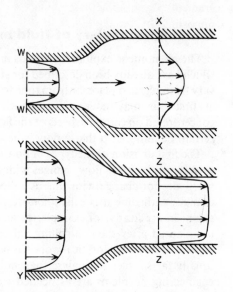

Fig. 4.1 Non-uniformity of the cross-section of the stream along the direction of flow affects the uniformity of the velocity distribution. (Above) A diverging flow makes the velocity distribution less uniform (Below) A convergence tends to make a uniform velocity across the flow— a property used in the design of inlets for hydraulic machines.

The next division of types of flow concerns the small irregular motions which are often superimposed upon the main motion. In the atmosphere, for example, the wind always blows in gusts with lulls between, and the direction of the wind constantly alters. In this respect the wind is air in non-steady motion, for the velocity is changing with time. But if average velocities are computed, each over several successive periods, and with each period fairly long compared with the time of a gust, it will be found that these averages will not change; the average wind is therefore steady though it has the non-steady motions of the gusts superimposed on it. This type of motion is mentioned in Section 1.3 in connection with the Shear stresses and is called *turbulent*, and the relative magnitude of the superimposed non-steady motions can be used to express the degree of *turbulence* in the motion. The irregular motions do not affect all the fluid at the same instant, so that the general effect is of a continually changing map showing patches of fluid moving differently from the long-term average speed. At the boundaries of the patches there must exist fairly abrupt changes

Fig. 4.2 Laminar and turbulent flow expressed as vectors. (Above) Steady laminar flow at a point is represented by a single vector of length *u*. (Below) Steady turbulent flow of the same magnitude is represented by a mean flow vector *ū*, with the addition of fluctuating velocities *u'*, *v'*, *w'*, which average out to zero over a sufficiently long period.

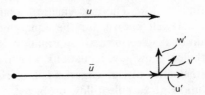

of speed, i.e. there are large velocity gradients. The presence of velocity gradients implies shear stresses and these may be large locally. A force opposing a movement of matter (whether fluid or solid) requires energy to be expended in doing so, proportional to the product of force and velocity, and these stresses and irregular motions are no exceptions. Mechanical energy (potential or kinetic) is continually being expended by turbulence and is converted into low-grade thermal energy. This constant degradation of energy is always present in turbulent flow, and is most noticeable at places where the irregular motions (sometimes called 'turbulent eddies') are greatest.*

Occasionally, however, there is no turbulence present in a flow, so that the mean velocity and the actual instantaneous velocity are exactly the same. In this case the flow is called *laminar*. It has been found that laminar flow tends to occur if velocities are low, or viscosity high, or if the boundaries of the flow are close together. Such conditions are rare in engineering problems, where a low

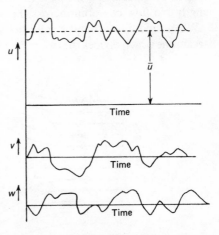

Fig. 4.3 Another way of expressing steady turbulent flow. The velocity at a point is plotted against time, all three components of motion *u*, *v* and *w* (in the *x*, *y* and *z* directions respectively) being shown. Components *v* and *w* average out to zero over a long period. Component *u* averages out to *ū* over the same period. The frequency of the turbulent motion is of the order of a thousand or more cycles per second.

*The degradation of mechanical energy into thermal energy causes a small increase of temperature of the fluid. In turn, the rise of temperature may cause chemical changes to take place, or the viscosity to change, or cause a local expansion and therefore reduction of density to occur. The first change is only of importance if chemically active fluids are being used: the second and third changes are usually assumed not to affect the conditions of flow although in certain meteorological problems the buoyancy forces due to the changed density may be comparable to other forces acting in the atmosphere.

viscosity fluid, air or water, is usually moving at a sufficiently high speed to give turbulent flow. The mathematical analysis of laminar flow problems is in general well known, for the equations concerned can be solved with no additional experimental information, but the corresponding analyses for turbulent flow give equations which cannot be directly solved with the mathematics at our command. Experimental data are required to obtain solutions to turbulent flow problems.

4.2 Flow patterns

It is sometimes desirable to have a complete diagram of the direction of motion at a number of points in a fluid motion (Fig. 4.4). This can be done by drawing *streamlines* on a plan or elevation of a flow. These lines are drawn so that they are tangential to the direction of flow at any point on them. Streamlines are therefore often curved and an infinite number of them can be drawn in any particular part of a flow. For clarity a few only are shown on any one diagram. No two streamlines can cross unless the fluid is at rest at that point, for if they did, the particle of fluid at the intersection would have two directions of motion, one tangential to each streamline. Chapter 5 will describe how streamlines can be plotted and Section 5.2 describes methods of making them visible in fluid flows. A streamline has no width and so has no cross-sectional area. If it is desired to consider a finite portion of a flow, a *streamtube* is often postulated. This is a prism of fluid bounded by streamlines along its length. No flow can occur across the walls of a streamtube.

When fluid flows round a solid body it divides to pass on either side. The streamline which divides the two parts of the flow meets the surface of the body

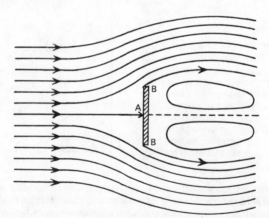

Fig. 4.4 Streamline pattern around a badly streamlined shape (a flat plate) in an extensive stream. 'A' marks the stagnation point where the stagnation streamline meets. the body. Notice the breakaway points BB and the large, permanent eddies behind the shape.

at the *stagnation point* (A on Fig. 4.4), so called because of the low flow velocities in the region. At the stagnation point the stagnation streamline divides and may coincide with the surface of the body. It is sometimes found that a streamline which was following the surface at one place is no longer in contact with the surface further downstream. The flow is then said to *breakaway* or *separate* from the surface. A mass of fluid called the *wake* which does not take part in the main flow remains between the separated streamline and the surface. This mass may rotate and move slowly forming eddies, as shown in Fig. 4.4. Such eddies which are large and slow moving should not be confused with the irregular motions in turbulent flow, which can also make rotary, though non-permanent, eddies. Some shapes, notably aircraft wings and well-designed ship hulls, have no breakaway on them at all in their normal operating condition, so that their shape is that of a streamline. Their shapes are therefore called *streamlined*—a word which has been badly misused.

4.3 The calculation of the kinetic properties of fluid flows

If it were possible to define a small element of fluid and follow its motion along a streamtube the analysis of fluid motion could be performed by the techniques of solid body mechanics. The motion of the element would be followed, the kinematics and dynamics of the motion calculated by what is sometimes referred to as *Lagrange's method*, and a separate technology of fluid mechanics would not be needed. However if a small element of fluid is defined by adding dye to a flow it is a common experience that in a very short time the dye is diffused, by turbulence or molecular motion, and the element ceases to exist as originally defined.

To overcome the limitations of Lagrange's method there is an alternative means of analysis named after the Swiss mathematician Leonhard Euler. The basis of *Euler's method* is that the motion of the fluid is considered as the flow across an imaginary surface fixed relative to the observer called a *control surface*. To find the details of a flow, control surfaces may be defined at those cross-sections where information is required. Fig. 4.1 shows four control surfaces (WW, XX, YY, ZZ) drawn to illustrate the effects of an expansion and contraction on a flow. One advantage of Euler's method is that if the flow is steady, diagrams such as Fig. 4.1 define the flow without having time as a variable. However, if Lagrange's method is used, as the element of fluid is followed both its position and the time have to be considered and these may complicate the analysis.

If a control surface is drawn across a flow it is possible to calculate the mechanical properties of the flow across the control surface. For a non-uniform flow it is necessary to have information on how the velocity varies with position across the control surface. Such information is usually in the form of a *velocity distribution* curve and Fig. 4.5 gives a typical velocity distribution curve

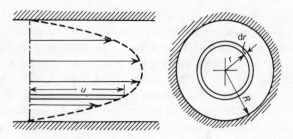

Fig. 4.5 Non-uniform flow across a control surface in a circular pipe. The velocity distribution curve has a maximum at the centre.

for a control surface drawn across a circular pipe. If the flow is steady the properties calculated are unique, but if the flow is unsteady the properties of the flow may be calculated for each time for which the velocity distribution is known.

The total quantity of fluid passing a control surface is usually defined by the *mass flow rate m*. In general, m is found by an integration of the form $m = \int \rho u \, da$, where ρ is the fluid density and u the velocity at right angles to the element of area of the control surface, da. In the case of a fluid where the density is constant across the cross-section it is usual to consider the *volumetric flow rate* or *discharge*, Q, where $m = \rho Q = \rho \int u \, da$.

In the case of a river the process of measuring the flow is known as *gauging* and is usually performed by measuring the flow at a number of points on the control surface with a *current meter* (see Fig. 4.6) and then having to perform a numerical integration of the results due to the two dimensional non-uniformity of the flow. For the circular pipe, as shown in Fig. 4.5 the flow is usually axially symmetrical and the element of area is the circular ring of radius r, on all points of which the velocity is uniform, and of width dr. Thus the mass flow rate in the pipe is

$$m = \int_0^R \rho u \, 2\pi r dr \tag{4.1}$$

where R is the radius of the bore of the pipe. If the density of the fluid is uniform across the control surface then the discharge may be calculated and the mass flow rate derived as

$$Q = \int_0^R u \, 2\pi r dr \quad \text{and} \quad m = \rho Q. \tag{4.2}$$

The discharge through a pipe or along a river is often expressed by the engineer in terms of the mean velocity $\bar{u}$ across the section where $\bar{u} = Q/A$, A being the total cross-sectional area (πR^2 for the circular pipe). The mean velocity cannot be measured directly although there will be a number of points on the control surface where $u = \bar{u}$. The position of these points will not

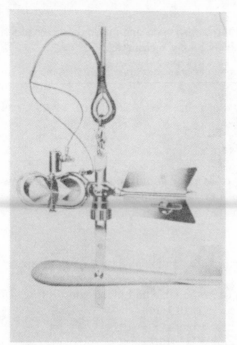

Fig. 4.6 A current meter used to find water speeds in rivers. It is suspended on a wire cable kept nearly vertical by the streamlined weight below. The rotor is revolved by the fluid forces on the conical buckets and an electrical pulse is sent once per revolution to the operator above. (Photo by Hilger and Watts Ltd.)

usually be the same for all flow rates and it is not prudent to assume that $\bar{u}$ can be measured accurately by placing a velocity-measuring instrument at a particular place in a flow.

It is possible to calculate not only the mass flow rate but also the flow rates of momentum and kinetic energy across a control surface if the velocity distribution is known. As in solid body mechanics each unit mass of fluid has a momentum u and a kinetic energy $u^2/2$ then across a control surface,

$$\text{momentum flow rate} = \int \rho u^2 \, da \text{ and,} \tag{4.3}$$

$$\text{kinetic energy flow rate} = \int \tfrac{1}{2} \rho u^3 \, da. \tag{4.4}$$

If the flow is non uniform the momentum flow rate is not equal to the product of the mass flow rate and the mean velocity nor is the kinetic energy flow rate equal to the product of the mass flow rate and $\tfrac{1}{2} \bar{u}^2$. In each case the actual flow rate of momentum or kinetic energy is greater than that calculated from the mean velocity, although in many calculations the latter is used for convenience.

Example

At a control surface across a one metre diameter pipe full of water the velocities in Table 4.1 were measured.

Table 4.1 Calculation of the kinetic properties of a flow through a circular pipe.

Given Values	Radius r/(m)	0	0.1	0.2	0.3	0.4	0.45	0.475	0.5
	Velocity u/(ms^{-1})	4.99	4.95	4.88	4.73	4.29	3.65	2.75	0
Calculated from given values	ur/(m^{2}s^{-1})	0	0.495	0.976	1.42	1.72	1.64	1.31	0
	$u^2 r$/(m^{3}s^{-2})	0	2.45	4.76	6.71	7.36	6.00	3.59	0
	$u^3 r$/(m^{4}s^{-3})	0	12.1	23.2	31.8	31.6	21.9	9.9	0

What are the mass, momentum, and kinetic energy flow rates across the control surface and how would their values compare with the same quantities calculated from the mean velocities?

Solution

a) Mass flow rate $m = \displaystyle\int_0^R 2\pi r \rho u \, dr = 2\pi\rho \int_0^R ru \, dr$

Calculate ru as in the table and plot against r (Fig. 4.7).
Area under curve is 10.7 units.
1 unit $= 0.1 \, \text{m} \times 0.5 \, \text{m}^2\,\text{s}^{-1} = 0.05 \, \text{m}^3\,\text{s}^{-1}$.
So $m = 2\pi\rho \times 0.05 \times 10.7 \, \text{m}^3\,\text{s}^{-1}$

$\qquad = 10^3 \, \text{kg m}^{-3} \times 3.36 \, \text{m}^3\,\text{s}^{-1}$

$\qquad = 3360 \, \text{kg s}^{-1}$,

$Q = 3.36 \, \text{m}^3\,\text{s}^{-1}$,

and $\bar{u} = Q/A = 3.36/0.25\,\pi = 4.28 \, \text{m s}^{-1}$.

b) Momentum flow rate $= \displaystyle\int_0^R 2\pi r \rho u^2 \, dr = 2\pi\rho \int_0^R ru^2 \, dr$.

Calculate ru^2 as in the table and plot against r (Fig. 4.7).
Area under curve is 9.2 units.
1 unit $= 0.1 \, \text{m} \times 2.5 \, \text{m}^3\,\text{s}^{-2} = 0.25 \, \text{m}^4\,\text{s}^{-2}$.
So momentum flow rate $= 2\pi \times 10^3 \, \text{kg m}^{-3} \times 9.2 \times 0.25 \, \text{m}^4\,\text{s}^{-2}$

$\qquad\qquad = 14.45 \, \text{kN}$.

c) Kinetic energy flow rate $= \displaystyle\int_0^R 2\pi r \rho \frac{u^3}{2} \, dr = \pi\rho \int_0^R ru^3 \, dr$

Calculate ru^3 as in the table and plot against r (Fig. 4.7).
Area under curve is 10.8 units.
1 unit $= 0.1 \, \text{m} \times 10 \, \text{m}^4\,\text{s}^{-3} = 1 \, \text{m}^5\,\text{s}^{-3}$.
So kinetic energy flow rate $= \pi \times 10^3 \, \text{kg m}^{-3} \times 10.8 \, \text{m}^5\,\text{s}^{-3}$

$\qquad\qquad = 33.9 \, \text{kW}$.

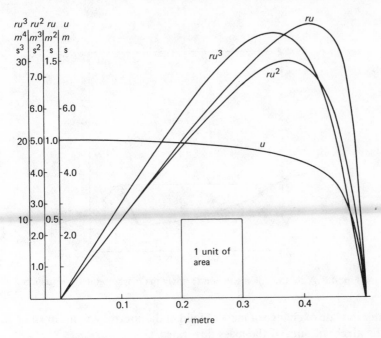

Fig. 4.7 The properties of a flow through a circular pipe.

Using the mean velocity the mass flow rate would be identical, but the momentum and kinetic energy flow rates would be approximated as:

$$\text{momentum flow rate } (\bar{u}) = m\bar{u} = 3360 \text{ kg s}^{-1} \times 4.28 \text{ m s}^{-1} = 14.38 \text{ kN}$$
$$\text{kinetic energy flow rate } (\bar{u}) = \tfrac{1}{2} m\bar{u}^2 = \tfrac{1}{2} \times 3360 \text{ kg s}^{-1} \times 4.28^2 \text{ m}^2 \text{ s}^{-2}$$
$$= 30.77 \text{ kW}$$

Note how the actual momentum and kinetic energy flow rates are greater than those calculated from the mean velocity. The values are typical of those in a turbulent pipe flow.

4.4 The conservation of matter in a fluid flow

Control surfaces can be used to enclose a volume in space fixed relative to the observer which is known as a *control volume*. Having defined a control volume it is possible to apply the conservation equations (of mass, momentum, and energy) to the flows into and out of the control volume.

Consider a control volume as shown in Fig. 4.8 into and out of which there are a number of mass flow rates $m_1, m_2, \ldots m_n$. By the law of conservation of

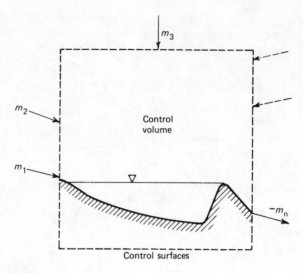

Fig. 4.8 A control volume drawn to apply the mass-continuity equation.

matter the rate of change of mass of fluid in the control volume must be equal to the algebraic sum of the mass flow rates, i.e.

$$\sum_1^n m_x = \text{rate of change of mass in control volume.} \qquad (4.5)$$

This equation is often referred to as the *mass-continuity equation*, but for an incompressible fluid such as water, whose density is constant, the volume continuity equation may be used, i.e.

$$\sum_1^n Q_x = \text{rate of change of volume of liquid stored.} \qquad (4.6)$$

In steady flow there is no change of mass or volume of liquid stored so the continuity equations become

$$\sum_1^n m_x = \sum_1^n Q_x = 0 \qquad (4.7)$$

for compressible and incompressible flow respectively. For the case of a control volume with only one inflow and one outflow the mass continuity equation reduces to

$$\rho_1 a_1 \bar{u}_1 = \rho_2 a_2 \bar{u}_2 \qquad (4.8)$$

while for incompressible flow the volumetric continuity equation is

$$a_1 \bar{u}_1 = a_2 \bar{u}_2 \qquad (4.9)$$

An important application of the continuity equation is to the filling of water storage reservoirs. Reservoirs are frequently made in valleys by building a dam across a river and extending the embankment to both sides. If the reservoir is full, and water continues to be added, an overflow device such as a spillway must be provided so that the surplus is safely passed over the dam. If the water-level rises to a height h above the spillway crest, then there is a discharge, Q_{out}, which in general depends on h and so is not steady. The inflow to the reservoir Q_{in} is usually non-steady. Rainstorms on the catchment area of the river cause a fluctuation of Q_{in}, and a common shape of the inflow curve (a *hydrograph*) is shown in Fig. 4.9 as a graph of Q_{in} against time. The difference, $Q_{in} - Q_{out}$, is retained in the reservoir, increasing the volume of water there, and raising the water-level. The relation between the small change of level dh in a small time dt is therefore

$$Q_{in}\,dt = Q_{out}\,dt + A\,dh,$$

where A is the plan area of the water surface at a height h.

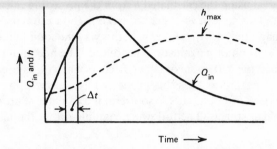

Fig. 4.9 An *inflow hydrograph*, which is a graph of inflow to a reservoir against time. Most rainstorms give hydrographs of this shape. The dotted line shows the height of water in the reservoir, which has a maximum much later than the maximum of the inflow.

It is not usually possible to solve this differential equation by direct algebraical integration to find h at every t. Every term of the equation changes in a complicated way: Q_{in} changes with t so that it cannot be expressed as an easy algebraic function: Q_{out} can change with h in several different ways (though $Q_{out} \propto h^{3/2}$ for simple weirs, more complicated ways of spilling water are often used): and A commonly increases somewhat with h again in no simple fashion. Thus approximate (that is, numerical) methods must be used to produce a curve of h against t.

One way of carrying out such an integration is to divide the inlet hydrograph into arbitrary and finite time intervals Δt, as in Fig. 4.9 and so to obtain $Q_{in}\Delta t$, which is the area under the hydrograph in this time interval, and is also the volume of water entering the reservoir. A tentative value of h is then estimated for the water-level at the end of the time interval, so that the *mean* water-level and thus the mean Q_{out} in this interval can also be estimated if an expression is available for the variation of Q_{out} with h. From Q_{out}, a tentative value of $Q_{out}\Delta t$ is calculated, the volume allowed to escape from the reservoir. The difference,

$Q_{in}\Delta t - Q_{out}\Delta t$ is then the tentative volume of water retained in the reservoir, $A \Delta h$. The area of the reservoir can then be found, if it changes with h, by using the *mean* value of h. Using this, the calculated change in height Δh is

$$\Delta h = (Q_{in}\Delta t - Q_{out}\Delta t)/A,$$

which may be added to the height of the water surface at the beginning of the interval to produce the calculated height at the end. This is now compared with the estimated height used at the beginning of the calculation: if the estimate was a good one, then the calculated value of h will be the same as the estimated one. Usually, of course, there will be a discrepancy so that the calculation must be repeated with a new h, based on the result of the previous calculation. Eventually, fair agreement will be reached, so the computation proceeds to the next time interval using the calculated h from the preceding calculation as the starting-point. This time it is easier to make a good estimate of h because the order of magnitude of Δh is known. In this way the computation is built up in layers and a graph of h against t is produced. The maximum value of h is always later than the maximum of Q_{in} and occurs when Q_{in} is equal to Q_{out}, i.e. when Q_{out} is a maximum. It is this property of a reservoir in reducing and delaying the peaks of floods that is so useful if a catchment area of a river is subject to sudden storms. The following example shows the method.

Example

A reservoir has the following water surface areas:

at 100 m O.D., $2.5 \times 10^6 \, \text{m}^2$
    101 m O.D., $2.7 \times 10^6 \, \text{m}^2$
    102 m O.D., $3.0 \times 10^6 \, \text{m}^2$
    103 m O.D., $3.5 \times 10^6 \, \text{m}^2$

A rainstorm on the catchment gives an inflow to the reservoir which consists of the flow increasing from zero to $10^3 \, \text{m}^3 \, \text{s}^{-1}$ linearly in 2 hours followed by a linear decrease to zero flow again in a further 8 hours. The only outlet from the reservoir is a weir whose crest is at 100 m O.D., the flow over the weir being given by $Q = 220 \, \text{m}^{3/2} \, \text{s}^{-1} \, h^{3/2}$ where h is the height of water above the weir crest.

If the storm inflow starts when the water level is at 100 m O.D., find the maximum water level and the outflow hydrograph.

Solution.

Consider a control volume which encloses the reservoir. There is then only one inflow, from the storm, and there is only one outflow, over the weir. Hence the method of the previous section can be applied directly.

First sketch the inflow hydrograph as in Fig. 4.9 and then consider time intervals of 1 hour. As an approximation to the change Δh over the first hour assume that no water flows over the weir. This will provide an estimate greater than the actual value of Δh.

At start of inflow, $t = 0$, inflow rate $= 0$.
when $t = 1$ hour, inflow rate, $Q_{in} = 500\,\mathrm{m^3\,s^{-1}}$.
Hence average inflow rate $= 250\,\mathrm{m^3\,s^{-1}}$ and in the time
$\Delta t = 1$ hour, volume inflow $= 250 \times 3600 = 9 \times 10^5\,\mathrm{m^3}$.

Taking the surface area to be $2.5 \times 10^6\,\mathrm{m^2}$ as constant for this approximation, then the resulting rise in the water surface would be

$$\Delta h = \frac{9 \times 10^5\,\mathrm{m^3}}{2.5 \times 10^6\,\mathrm{m^2}} = 0.36\,\mathrm{m}.$$

As this is larger than the real value a better first estimate might be $\Delta h = 0.3$ m.

Then average h during $\Delta t = 0.15$ m.
Average outflow rate $= 220 \times 0.15^{3/2} = 12.8\,\mathrm{m^3\,s^{-1}}$.
Volume outflow $\qquad Q_{out}\Delta t = 12.8 \times 3600 = 46 \times 10^3\,\mathrm{m^3}$.
Volume stored $= (900 - 46) \times 10^3 = 854 \times 10^3\,\mathrm{m^3}$.
Mean surface area $= \frac{1}{2}(2.5 + 2.5 + 0.3 \times 0.2) \times 10^6 = 2.53 \times 10^6\,\mathrm{m^2}$.

$$\text{Calculated } \Delta h = \frac{854 \times 10^3\,\mathrm{m^3}}{2.53 \times 10^6\,\mathrm{m^2}} = 0.338\,\mathrm{m}.$$

Hence the first estimate was too small so a second estimate, say $\Delta h = 0.33$ m, is necessary.

The calculation is carried out most economically by a tabular method, using a programmable calculator if one is available, Table 4.2.

If the variation of h with time is plotted as the calculation continues, the extrapolation of the curve acts as a useful guide to the next estimate. Fig. 4.10 shows the results of the calculations. Note how the maximum of the outflow hydrograph is less than the maximum inflow rate and occurs after it. This property is the basis of flood control by storage in reservoirs. The maximum outflow and hence maximum water level in the reservoir occur where the two hydrographs intersect.

Table 4.2 Calculation of an outflow hydrograph for a reservoir.

Estimated height above weir at end of period h (m)	Time from start of flood t (hours)	Time period considered Δt (hours)	Mean inflow rate from hydrograph during period Q_{in} (m^3s^{-1})	Volume inflow during period V_{in} ($10^6 m^3$)	Mean head over weir from estimated h, $\bar{h}$ (m)	Mean outflow based on $\bar{h}$, Q_{out} (m^3s^{-1})	Estimated volume outflow during period V_{out} ($10^6 m^3$)	Volume stored in reservoir $V_s = V_{in} - V_{out}$ ($10^6 m^3$)	Mean area of water surface at $\bar{h}$, A ($10^6 m^2$)	Calculated rise in water level Δh, $\Delta h = \frac{V_s}{A}$ (m)	Computed water level (m O.D.)	Comments
0	0										100	
0.33	1	1	250	0.9	0.165	14.75	0.053	0.847	2.53	0.334	100.334	Agreement with estimated value satisfactory
					For period 1–2 hours start with water-level at 100.33 m and estimate $h = 1.0$ m							
0.33	1	1									100.33	
1.0	2	1	750	2.7	0.665	119.3	0.429	2.27	2.62	0.866	101.196	Estimated h too small Try $h = 1.16$ m
1.16	2	1	750	2.7	0.745	141.5	0.509	2.19	2.64	0.830	101.16	Agreement

For period 2–3 hours start with water-level at 101.16 m and estimate h = 2.0 m

1.16	2		937.5	3.38	1.58	437	1.57	1.80	2.87	0.627	101.16	
2.0	3	1	937.5	3.38	1.48	396	1.43	1.95	2.84	0.686	101.79	Overestimate Try h = 1.8 m
1.8	3	1	937.5	3.38	1.495	402	1.45	1.93	2.85	0.677	101.85	Nearer but not sufficiently accurate Try h = 1.83 m
1.83	3	1	937.5	3.38							101.837	Near enough to carry on

For the remaining periods only the final values are given

2.08	4	1	812.5	2.93	1.995	601	2.16	0.76	2.99	0.254	102.08	
2.11	5	1	687.5	2.48	2.10	669.5	2.41	0.73	3.05	0.024	102.10	
2.01	6	1	562.5	2.03	2.06	650	2.34	−0.32	3.03	−0.104	102.00	
1.83	7	1	437.5	1.58	1.92	585	2.11	−0.53	2.98	−0.179	101.82	
1.60	8	1	312.5	1.13	1.72	494	1.78	−0.65	2.91	−0.224	101.60	
1.34	9	1	187.5	0.675	1.47	392	1.41	−0.74	2.84	−0.259	101.34	
1.05	10	1	62.5	0.225	1.20	287	1.03	−0.81	2.76	−0.294	101.05	

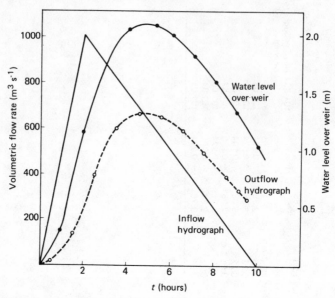

Fig. 4.10 Inflow hydrograph, outflow hydrograph and water level for a reservoir. The maximum height of the water level above the weir is 2.1 m when $Q_{in} = Q_{out}$ = 670 m^3 s^{-1}.

Problems

1. In steady laminar flow in a circular pipe the velocity distribution across a diameter is parabolic with the maximum velocity V on the pipe axis and zero velocity on the pipe wall, i.e. if u is the velocity at radius r and R is the pipe radius

$$u = V\left(1 - \left(\frac{r}{R}\right)^2\right)$$

Show that the average velocity $\bar{u}$ is equal to half the maximum velocity.

2. The flow entering a lake on a certain day is

Time	Flow m^3 s^{-1}	
until noon	28	
1 p.m.	170	With sinusoidal increase
2 p.m.	310	and decrease.
7 p.m.	170	
midnight and after	28	

The lake discharges over a simple weir for which $Q = 49.5\,h^{3/2}$ m$^{3/2}$ s^{-1}. The surface area of the lake is 3.7×10^6 m^2 at crest level, plus 3×10^5 m^2 per metre of rise. Find the maximum level in the lake and the time at which it occurs.   *Ans.* 1.7 m at 8.40 p.m.

5

Plotting streamlines—A problem in surveying

5.1 Streamlines and the velocity of flow

Chapter 4 described how a fluid flow can be represented by its streamlines. These are lines drawn in the fluid so that tangents drawn to them are in the direction of flow at the tangent points. Streamlines cannot cross, for if they did so, then the fluid at the intersection would have two velocities, one along the tangent to each streamline.

In the same way that a country is not known until a map has been made of it, so the knowledge of a fluid flow is not complete until the streamlines are known. For a particular configuration of the solid boundaries, and of the inlets and outlets of the flow, a map can be drawn up showing a selection of the infinite number of streamlines existing in the stream, remembering that solid boundaries are streamlines.

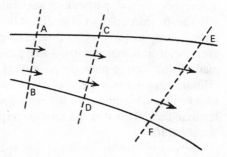

Fig. 5.1 Portion of a fluid flow with only two streamlines ACE, BDF shown. No fluid can pass across either streamline, so that if in a steady incompressible two-dimensional flow, a volume Q passes across the control surface AB in unit time, then Q also passes across CD and EF. Since EF > AB, velocity at AB > velocity at EF.

In steady flow if two control surfaces are drawn across the same streamtube the mass flow rates across them will be the same. Furthermore if the flow is incompressible the volumetric flow rates across the two control surfaces will be the same. In a two-dimensional incompressible flow*, i.e. where the depth of the flow is constant in a direction at right angles to the plane of the streamline

---

*In this brief summary of streamline plotting, steady two-dimensional incompressible flow is assumed throughout. In the far more complicated case of three-dimensional flow, i.e. when the depth of the stream varies, and the pattern changes from one level to another, the convergencies and divergencies of streamlines in all three directions must be considered.

pattern and the patterns at all levels are the same, the velocity of flow will vary inversely with the spacing of the streamlines. Hence the importance of streamline patterns; they enable not only the direction of the flow but its speed to be calculated. It will be shown in Chapter 7 that in certain circumstances it is possible to deduce the pressures in the flow if the velocities are known.

5.2 The experimental method of streamline plotting

If the direction of flow is known at a number of successive points, then the streamlines can be drawn as tangents to these directions. This technique, though tedious, can be used in a fluid stream, with suitable vanes, or flags to show the directions. A more direct method, often employed, uses particles suspended in or floating on, the fluid as in Fig. 5.2(b). This shows the motion of water under a train of waves. Particles of aluminium, strongly illuminated, were photographed with a time exposure equal to the period of the wave, thus showing the orbital motions near the surface, which become elliptical and finally straight lines at the bottom. The wave on the surface was not photographed because at any one place the surface had moved through the whole height of the wave in the time of exposure, and was therefore fogged and indistinguishable on the photographic plate. If the flow is steady the *particle paths* follow the *streamlines*. Streamlines have been plotted in this way in large-scale fluid flows such as rivers, sometimes from photographs of floats taken at short intervals.

Another method, particularly applicable to laboratory studies, is to release into the stream a small jet of smoke (if the fluid is a gas) or dye (if a liquid) of the same density as the flowing fluid. The particles liberated successively at the same point trace out the *streak line* passing through the point: the trail, now made visible, can be photographed. Very slow, laminar flow around a circular cylinder between two sheets of glass 1 mm apart is shown in Fig. 5.2(c). The streak lines were made visible with dye which only diffuses slowly in the laminar flow. This flow pattern corresponds closely with that predicted in Section 5.10.

The use of dye or smoke is less successful where there is mixing as in Fig. 5.2(a) where the dye is passing through a sand bed in a laboratory tank. The flow is from right to left, the water being taken off from the left hand surface of the sand. While turbulent flows cause dye and smoke to mix rapidly the technique can still be of use in particular flow areas as seen in Fig. 10.1.

All the above methods, and variations of them, are used in engineering fluid mechanics where often the shape of the boundaries of the flow is so complicated that the theoretical methods of finding streamlines, described later, are tedious and therefore expensive. Furthermore, as these theoretical methods often break down, giving a quite unreal map of the streamline in certain areas of a flow, the experimental methods are always needed as a check.

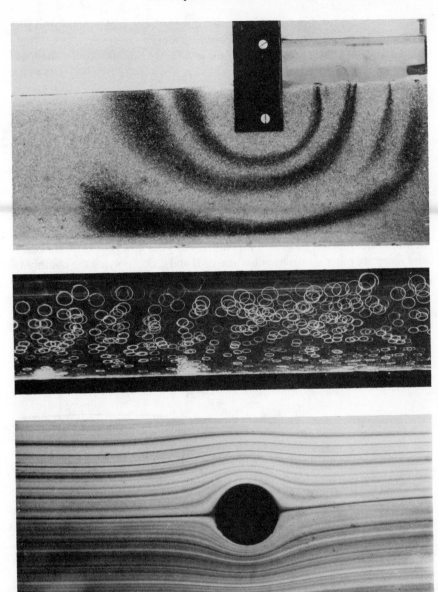

Fig. 5.2 Examples of flow visualization (a) Flow of water in a bed of sand, under an impermeable vertical wall. (b) Water motions under a train of waves in a glass-sided laboratory channel. (Photo by F. Ruellan and M. Wallet; reproduced by permission of SOGREAH, Grenoble, France.) (c) Streamlines of the very slow flow of water around a cylinder sandwiched between the two glass sides of a tank only 1 mm apart

In general, the experimental methods give a less precise map than the theoretical methods, but this lack of precision must be accepted in areas where the latter give a misleading answer anyway.

5.3 Theoretical properties of streamlines

Although experimental methods of finding streamlines are the only final evidence of a flow pattern, it is frequently desirable or necessary to predict a flow pattern at the design stage of an engineering job.

By using the essential property of streamlines, that they do not cross, it is possible to produce a method of plotting the lines on the drawing-board in advance of experiment. The limitations to the method are reviewed in Section 5.11.

Consider any two streamlines, AB and CD, not necessarily parallel, of a fluid in motion (Fig. 5.3). The fluid is incompressible, the flow is of constant depth and there are other streamlines between AB and CD. Applying the continuity equation to the control volume ABCD, since there is no flow across streamlines, the same quantity of fluid per unit time passes AD as passes BC.

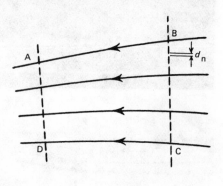

Fig. 5.3 A control volume in a streamline pattern.

That is
$$\int_A^D u \, \mathrm{d}n = \int_C^B u \, \mathrm{d}n = \text{Constant} = \psi,$$

where u is the velocity normal to the element of area of the control surfaces AD or BC. The quantity ψ is expressed as a volume per unit time, or, in other words, streamline CD is always a 'discharge' of ψ distant from AB even if the length it is distant from AB changes. Consequently CD can be labelled by its discharge from AB, and it can be called streamline ψ ($= \int_A^D u \, \mathrm{d}n$) relative to AB. Another streamline such as EF, further from AB than CD, will clearly have a larger value of ψ than CD. ψ is called the *stream function* of a streamline and is a method of labelling streamlines, to give them a quantitative meaning, depending on their position within the stream.

5.4 The potential function

A quite different method of marking a flow which eventually proves useful is by the plotting of 'potential' lines on it. Along any one streamline values of a quantity $\phi = \int u\,ds$, called the *potential*, can be marked off, where s is the distance measured along the streamline. The potential, though having this precise meaning, is a fictitious quantity and it cannot be measured directly with instruments. Referring to Fig. 5.3, if A is taken as the zero point for potential on AB, then

$$\phi \text{ at } B = \int_A^B u\,ds.$$

(Note the family likenesses and differences with ψ: n is measured across streamlines, s along them.) Points on other streamlines can also be marked off with their values of ϕ, and those with the same value on different streamlines linked to give contours of the same ϕ, called *equipotential lines*. Since the potential is by definition a property that increases only along the direction of a streamline, and never has a component across it, the equipotential lines are *always* normal to streamlines.

There is also another important link between the streamlines and equipotential lines. It has already been shown in Section 5.1 that if the velocity u increases along a streamline, then adjacent streamlines come closer and converge: and if u decreases they diverge. But if u increases, then a given potential $\phi (= \int u\,ds)$ is developed in a *smaller* distance s than it did for a smaller u. If streamlines converge, then the equipotential lines become closer together also.

The two properties, stream function ψ and potential ϕ taken together, are used to draw streamline maps or patterns by a method of successive approximations.

5.5 Numbering convention

It is conventional to label a streamline by the value of its stream function ψ, that is by the quantity of the flow passing between the streamline and an arbitrary, reference streamline. It is also conventional to regard the value as increasing positively to the *left* of the reference line, while looking *downstream* (see Fig. 5.6). The unit of stream function used for such numbering is entirely optional, but it is often convenient to use $m^3\,s^{-1}$, passing in a stream of 1 m depth.

5.6 Drawing streamlines by a method of successive approximations

Consider a streamline map drawn perfectly correctly for the boundaries of the flow concerned (Fig. 5.4). The streamlines are drawn at equal intervals of ψ. At one arbitrarily chosen point draw a line normal to the streamline there and continue the line (curving it as necessary) to cross all other streamlines at 90°, e.g. ABCD. This line will be an equipotential line in the flow and can be given a value of ϕ = zero, say. Any number of other equipotential lines can also be drawn, each of which will have a value of ϕ, but consider only EFGH which has been drawn so that its value of ϕ is $\int_A^B u\,dn$.

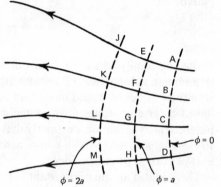

Fig. 5.4 A perfectly correct streamline pattern with equipotential lines also drawn in. If ABFE is chosen to be a 'square', then so is FBCG, GCDH.

That is
$$\phi = \int_A^E u\,ds = \int_A^B u\,dn.$$

If the streamlines have been drawn fairly closely together, then u is nearly uniform at all places in the area ABFE, and in the limit, when there are an infinite number of streamlines drawn, u will be quite uniform. Thus to a first approximation, u is independent of both s and n if the streamline interval is fairly small, and
$$\int_A^E ds = \int_A^B dn$$
or
$$EA = AB.$$

So EABF is a shape bounded by curves which intersect at 90° and which has two sides EA and AB of the same length. The shape is conveniently called a 'square'. Similarly, FBCG and GCDH are 'squares', though not the same size as EABF: FB = GC = EA only if AB = BC = CD, the condition of uniform velocity. If the streamlines and equipotential lines are close enough, then shapes such as ABFE, GCDH will be true squares.

Again, another equipotential line JKLM may be drawn at the same interval of ϕ, e.g.

$$\int_E^J u\,ds = \int_E^F u\,dn$$

so that,
$$JE = EF.$$

Thus another set of 'squares' are produced, again not all of the same size, getting larger as the streamlines diverge. However, there is one common property: if the interval of ϕ between equipotential lines is kept the same, and was chosen to give 'squares' at one part of the map, then 'squares' are produced in *every* part of the map if the streamline pattern is already correct. Even if the streamlines are drawn well apart the 'squares' will still be approximately formed.

This important property can be used by the engineer to draw streamline patterns for flows around complicated shapes. The boundaries of the flow are drawn to scale, and a 'guessed' set of streamlines put in, using the boundaries as streamlines, for there is no flow across them. At one part of this 'guessed' pattern, preferably at a place where the velocity is uniform, commence drawing smooth curved lines at 90° to the guessed streamlines, spacing these lines so that at one point they produce 'squares'. It will soon be found that if 'squares' are drawn in the space between two streamlines then the continuation of the lines at 90° will *not* result in 'squares' between other pairs of streamlines. Clearly this is due to errors in the original guessed streamlines. Revisions can therefore be made to the streamlines at the places where the 'squares' are most in error, but this process will be found to make the original '90°' lines now no longer correct, so that revisions are again necessary to the '90°' lines. In this way, successive adjustments to the streamlines and to the lines at 90° to them make a closer and closer approximation to the correct streamline and equipotential line pattern. In places where there are rapid changes of streamline spacing, additional streamlines and equipotential lines can be drawn between the original set of both. If the pattern is correct, then the smaller 'squares' so formed are even better approximations to true squares than the big ones.

As a final check on the accuracy of the squares, diagonals can be drawn across them. These should be smooth curves over the whole pattern. The amount of correction necessary will depend, of course, on the experience, judgement and intuition of the draughtsman in producing the first, guessed, pattern, but comparatively unskilled people can eventually produce correct patterns. Large scale diagrams are advisable: soft pencils and plenty of indiarubber are essential. The method has no limitations on it so far as boundaries are concerned provided the flow does not break away but the more complicated configurations demand more adjustments. Patience and accuracy will invariably succeed.* An example is shown in Fig. 5.5.

* A slightly different method of drawing streamlines by an approximate method is given by S. Leliavsky in *The Engineer*, vol. 185, pp. 464—5 and 488—90. Some valuable hints which simplify the drawing of the first guessed pattern are given by H. A. Foster in *Trans. Amer. Soc. Civ. Eng*, vol. 110, pp. 1237—51.

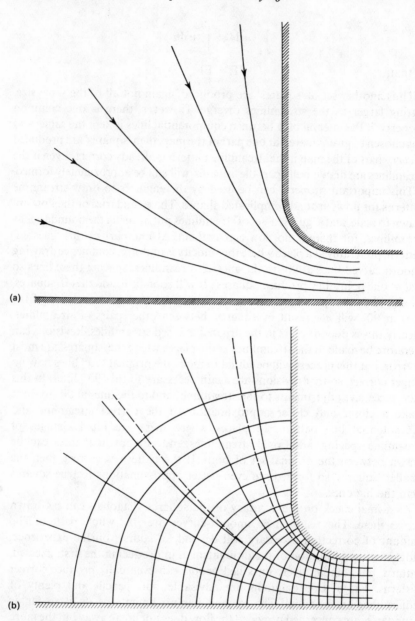

(a)

(b)

Fig. 5.5 The approximate trial-and-error method of drawing streamlines applied to the flow into the well-rounded entry to a pipe. (a) The boundaries and a set of guessed streamlines are drawn. (b) On a guessed pattern (full lines) are drawn lines at 90° to the streamlines, starting at the constant velocity part of the map, where the spacing of the 90° lines is such as to produce 'squares'. In many places the streamlines are clearly in error because the 90° lines make elongated rectangles instead of 'squares'. The dotted

(c)

(d)

lines are revisions to the streamlines to improve the 'squares'. But these revisions put the original 90° lines now in error. (c) The revised streamlines of (b) are now transferred, and a new set of 90° lines drawn. There are now errors only in the regions marked X. (d) Small revisions are now made to the streamlines of (c), and yet another set of 90° lines inserted. The errors have now nearly disappeared, leaving a network of 'squares' only. A useful check is to draw the diagonals, dotted, which must be smooth curves.

5.7 Simple streamline patterns

Although the trial-and-error method above will eventually prduce a correct streamline pattern for any given boundaries, a more accurate way is to combine two or more simple patterns together, until one of the streamlines coincides with the shape of the boundaries concerned. Some of the simple patterns are shown in Fig. 5.6. These are:

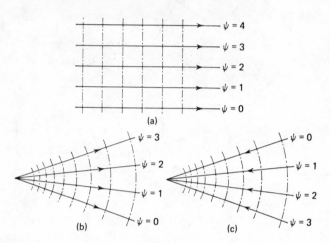

Fig. 5.6 Three simple streamline patterns, all drawn with the streamlines at the same interval of stream function. Potential lines are dotted. Note the numbering convention.

(a) Uniform straight line flow. Both the streamlines and equipotential lines are uniformly spaced forming a rectangular grid.

(b) A point source of fluid. The flow is radially outward so that the streamlines diverge. The spacing of the circular equipotential lines will therefore increase outward from the centre. The streamline of value $\psi = 0$ is a radius, quite arbitrarily chosen, and the remaining radial streamlines are successively numbered as shown in accordance with the numbering convention.

(c) A point 'sink' or outlet of fluid. The pattern is radial, precisely the same as the source, but with the direction changed. The numbering is thus reversed, according to the numbering convention.

5.8 Combination of streamline patterns by a graphical method

Consider two superimposed simple patterns, A and B, each with its streamlines drawn at the same interval of ψ. At a certain point x, y two streamlines intersect, one of pattern A whose stream function is p, called $_A\psi_p$:

the other form B whose stream function is q, called $_{B}\psi_{q}$. Since 'discharge' (flow per unit time) is a scalar quantity, the total discharge at x, y, due to both patterns, is simply $p + q$: so that the stream function of the combined pattern at that point is the algebraic sum $p + q$. The streamline of the combined pattern through this point is called $_{A+B}\psi_{p+q}$.

All other intersections of streamlines can thus be labelled with the sum of the stream functions, and smooth curves drawn between points of the same value. These curves are the streamlines of the combined pattern. In this graphical manner any number of simple patterns, centred or oriented differently, can be successively combined to form patterns of considerable complexity. Sometimes the combination will give one streamline that is a closed curve: in this case the streamlines outside the curve are those which would result from the fluid flowing around a solid object of the shape of the curve (see Fig. 5.7).

5.9 Combination of streamline patterns by an algebraical method

Graphical methods to produce complicated combinations of more than two or three simple patterns tend to be tedious and slight errors of plotting gradually accumulate to give inadmissible errors. An alternative method, always to be preferred if more than three simple patterns are involved, is to add algebraically the equations of ψ for the patterns and to plot the resultant pattern. Any number of simple patterns can thus be combined, and although the resultant equation for ψ may be complicated, it can always be evaluated at given points $x_1, y_1; x_2, y_2 \ldots$, etc. With ψ known at a number of places, lines of constant ψ can be drawn which are the streamlines of the combined pattern. This type of solution is most easily performed by a computer.

The combined streamline pattern for a uniform straight line flow of speed U and a point source of Q cusecs is obtained as follows: First, take the x-axis along the direction of the straight-line flow, and the y-axis normal to it. The equation to *any* streamline of the uniform flow is $_{A}\psi = Uy$ since the flow between the streamline concerned and the x-axis is independent of x. Secondly, it is convenient to take the point source on the co-ordinate origin and to number the streamlines due to the source from the x-axis as a reference streamline. Due to the numbering convention, the value of ψ at P (Fig. 5.8) will be positive. Thus the equation to any streamline of the source such as OP is

$$_{B}\psi = \theta Q/2\pi.$$

Alternatively, if a point x, y, is considered, whose radius to O makes an angle of $\tan^{-1} y/x$ with the x-axis, the stream function of a source streamline passing through that point is

$$_{B}\psi = \frac{Q}{2\pi} \tan^{-1} \frac{y}{x}.$$

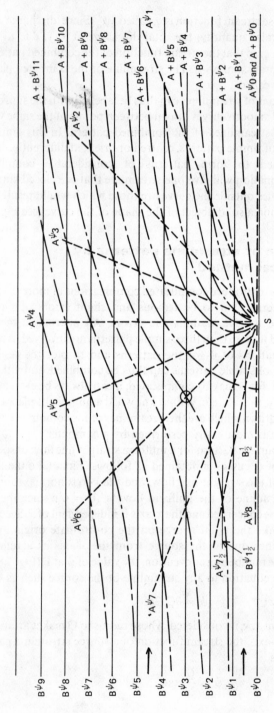

Fig. 5.7 Graphical combination of 2 streamline patterns. A source S of 16 m³ s⁻¹ (pattern Aψ) is combined with a uniform stream of 1 m s⁻¹ flowing 1 m deep (Bψ). Note that the streamline intervals for both Aψ and Bψ are the same, namely 1 m³ s⁻¹, and that only half the pattern is shown. The pattern is symmetrical about the axis Bψ₀. The numbering conforms to the convention. The encircled point lies on both Aψ₆ and also on Bψ₃ so that the combined streamline passing through there has the value A+Bψ₃₊₆ = A+Bψ₉. Additional streamlines Aψ₇½ and Bψ½, Bψ₁½ are drawn to improve the accuracy of the points of inflexion on the upstream edge of the combined pattern.

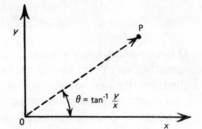

Fig. 5.8 System of axes for combining a uniform straight line flow parallel to the x-axis, with a point source at 0.

Thirdly, the two equations for $_A\psi$ and $_B\psi$ are added to give the stream function of streamlines at the point x, y, due to the combined pattern, so that

$$_{A+B}\psi = Uy + \frac{Q}{2\pi} \tan^{-1} \frac{y}{x}.$$

This is the equation to the streamlines of the combined pattern. The plotting of this equation can be done in two ways. The first, as already mentioned, is to find values of $_{A+B}\psi$ at a number of points, $x_1, y_1; x_2, y_2$, etc., and to interpolate the streamlines: the second way is to decide on a value of $_{A+B}\psi$ for which a streamline is desired, and then find pairs of values of x and y which satisfy the equation. A number of co-ordinates are thus found for one streamline. The process must be repeated for each streamline drawn on the diagram. The second method of plotting is more tedious than the first but does not involve interpolation. The streamlines drawn to the above equation will be found to be precisely the same as those drawn in Fig. 5.7 by the graphical method.

5.10 An important combination of streamlines

The case of the streamlines given by the combination of a point source and a point sink leads to a most important pattern. The combining may be done by the graphical method, but the algebraical method will be demonstrated as it is more convenient later.

Consider a source S, with a flow Q coming from it situated a distance SK from a sink K with the same flow going into it (Fig. 5.9). Let the reference direction for a polar co-ordinate system be SK. Then at a point P the stream function ψ due to S is $_A\psi = \frac{\theta_1}{2\pi}Q$, and that due to K is $_B\psi = \frac{\theta'_2}{2\pi}Q$. (Note the influence of the numbering convention on the directions in which θ_1 and θ'_2 are measured.) The combined stream function at P is therefore

$$_{A+B}\psi = \frac{Q}{2\pi}(\theta_1 + \theta'_2)$$

but

$$\theta'_2 = -\theta_2.$$

So
$$_{A+B}\psi = \frac{Q}{2\pi}(\theta_1 - \theta_2).$$

Since Q is a constant, every value of $_{A+B}\psi$ has its value of the angle $(\theta_1 - \theta_2)$. Consequently, every point having the same $_{A+B}\psi$, that is every point on that streamline, has the same value of $\theta_1 - \theta_2$ subtended at it by the length SK. By simple geometry, all these points must be on a circle, so that the streamlines must all be circles passing through S and K (see Fig. 5.9).

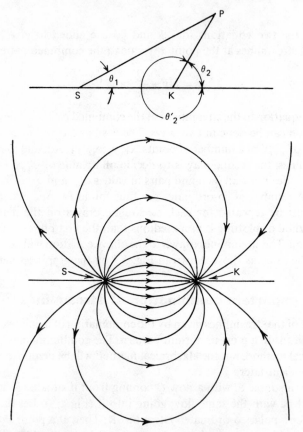

Fig. 5.9 (Above) Definition sketch for the algebraical combination of a source at S and a sink at K. (Below) The combined streamline pattern for a source at S and a sink at K.

It is now convenient to work in terms of the radial distances r_1 and r_2 of P from S and K respectively. For example,

$$\frac{SK}{\sin(\theta_2 - \theta_1)} = \frac{r_2}{\sin\theta_1}.$$

So the stream function equation can be rewritten

$$_{A+B}\psi = \frac{Q}{2\pi}(\theta_1 - \theta_2)\frac{SK \sin \theta_1}{r_2 \sin (\theta_2 - \theta_1)}.$$

The importance of the source and sink combination lies in the streamline pattern which results if SK tends to zero, that is, as the sink and source become coincident. To avoid an indeterminacy it is assumed that Q is increased so that the product of Q and SK remains constant. In this case

$$\theta_1 \to \theta_2 \to \theta; \ \sin (\theta_2 - \theta_1) \to \theta_2 - \theta_1 \to -(\theta_1 - \theta_2);$$

and $$r_1 \to r_2 \to r.$$

So that $$_{A+B}\psi = -\frac{Q}{2\pi}\frac{SK}{r} \sin \theta$$

or since $$Q.SK = \text{Constant} = C$$

$$_{A+B}\psi = -\frac{C}{2\pi}\frac{\sin \theta}{r},$$

r now being the radial distance of the point P from the source-sink position.

This pattern is again a series of circles (see Fig. 5.10) all passing through the common point where both source and sink are assumed to be. The pattern is called a *doublet*. It has no existence in a real fluid for it is impossible for a source and sink to coexist at the same point. It is studied for the sole reason that when it is combined again with a straight uniform flow a pattern of great significance is produced. For example, add a doublet pattern to the pattern $_C\psi = Uy$, where U is a constant and y is a distance normal to the line SK of the source-sink

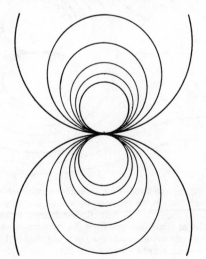

Fig. 5.10 Streamline pattern for a doublet.

pattern. The combined pattern is now

$$_{A+B+C}\psi = Uy - \frac{C}{2\pi}\frac{\sin\theta}{r}.$$

Converting polar to rectangular co-ordinates, taking the origin at the source-sink position, then

$$\sin\theta = \frac{y}{r} \text{ and } r = \sqrt{(x^2+y^2)}$$

so that

$$_{A+B+C}\psi = Uy - \frac{C}{2\pi}\frac{y}{x^2+y^2}.$$

For convenience only, write $\frac{C}{2\pi} = Ua^2$, where a is another suitable constant.

Then

$$_{A+B+C}\psi = Uy\left(1 - \frac{a^2}{x^2+y^2}\right).$$

At values of x and y which are large compared with a it will be seen that $_{ABC}\psi \rightarrow Uy$: that is, the uniform straight-line flow is unaffected by the doublet flow at large distances from the doublet point, as might be expected. For points where $x^2 + y^2 = a^2$, $_{ABC}\psi = 0$, so that a streamline passes through these points. Further, since $x^2 + y^2 = a^2$ is the equation to a circle, the streamline is a circle with the source-sink point as centre (see Fig. 5.11).

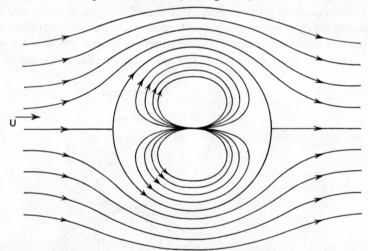

Fig. 5.11 Streamline pattern for a doublet combined with a uniform stream. The single circular streamline can be replaced by a solid surface, in which case the remaining streamlines are those of a uniform flow deflected around a circular cylinder. Observe how the flow at some distance from the cylinder is less affected than the flow near the cylinder. The streamlines inside the circle are those of the doublet flow, which, though deformed, do not lose their identity.

It has already been explained that any streamline can be replaced with a solid surface without changing any of the other streamlines. The circle $_{A+B+C}\psi = 0$ can so be replaced, and the remaining pattern outside the circle then becomes the flow pattern around a cylinder in a straight uniform flow. This pattern is a most important one in fluid mechanics where frequently cylindrical structures are exposed to fluid flows.

Still more complicated patterns may be built up by distributing doublets of varying strengths along a line, and combining them with a straight-line flow. There is often a closed streamline in the combined pattern and a well chosen set of doublets can give shapes similar to aerofoils. The flow around, and pressure against, aerofoils can be predicted by this sort of calculation.

5.11 Limitations of the theoretical streamline patterns

The methods (just described) for drawing streamlines give patterns which are confirmed by experiment only if certain additional conditions are observed. These conditions refer to the ways in which forces and rotations are applied to the fluid. Some ways produce streamline patterns having equipotential lines not at right angles to streamlines; or, if lines are drawn at right angles to the streamlines, then they are not equipotential lines. Thus the theoretical methods, which depend on the property that *streamlines are normal to potential lines*, sometimes fail, and produce a quite misleading pattern. Streamlines still exist, but they do not obey the laws of construction which have been given above.

Consider a fluid flow, part of which is shown in Fig. 5.12, where the streamlines AB, CD, have been constrained to be straight and parallel, such as occurs when a fluid flows in a straight pipe. Suppose an external shear force is applied to the fluid in such a way that the velocity along AB, u_1, is greater than that along CD, u_2: this situation arises if the flow is near a solid boundary XY so that the shear stress on XY slows the nearer fluid more than the further. Attempt now to draw equipotential lines at an interval of ϕ. Along AB this increment of potential requires a distance ϕ/u_1, and along CD, ϕ/u_2.

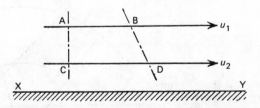

Fig. 5.12 A straight-line flow near a solid surface XY. Due to the shear stresses the velocity near the surface is less than that further away. Lines drawn at an equal potential are not normal to the streamlines.

Since $u_1 > u_2$, $\phi/u_1 < \phi/u_2$ and   AB < CD.

But the line equipotential BD is no longer normal to the streamlines, even if AC was normal to them. Alternatively, if lines are drawn normal to the streamlines, then they are not always equipotential lines. The rules for drawing streamline patterns have broken down.

Another possible type of flow is shown in Fig. 5.13, where AB and CD are streamlines of a fluid which are bent to circular arcs about a centre Z. The fluid is further constrained to move so that it has a constant angular velocity ω everywhere. The velocities along AB and CD are therefore ωr_1 and ωr_2 respectively. Such a situation occurs when a paddlewheel rotates without slipping in a fluid, and makes all the fluid move in the same flywheel manner. As before, an increment of potential ϕ is measured out along each streamline, and occupies along AB a distance $\phi/\omega r_1$ and CD $\phi/\omega r_2$. Since $r_1 < r_2$, the distance $\phi/\omega r_1 = $ A′B′ is greater than $\phi/\omega r_2 = $ C′D′ so that the equipotential lines A′C′ and B′D′ for the potential interval ϕ are not everywhere normal to the streamlines. The rules have again broken down.

The theoretical methods of drawing are applicable, however, if there are no shear stresses on the fluid (either an ideal fluid, or a real fluid some distance from a solid surface), or if a type of rotation is present which allows equipotential lines to be normal to the streamlines. For example, consider Fig. 5.13 again, but now the rotation is such that $u_1 r_1 = u_2 r_2 = K$, i.e. the velocity *decreases* outward. The increment of ϕ is now achieved in distances $\phi r_1/K$ and $\phi r_2/K$ respectively along AB and CD. These distances increase proportionately with the radius, so that the equipotential lines are radii, normal to the streamlines. The theoretical methods of plotting are thus applicable to such a flow, which is called a *potential* or *free vortex*, in contrast to a *forced* or *flywheel vortex* for which the theoretical methods do not apply. Alternatively, if the plotting methods are used on a set of boundaries which involve curved flow, then a free vortex will be predicted, even if a forced vortex is actually present. Other properties of free and forced vortices are discussed in Chapter 11.

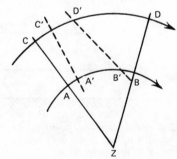

Fig. 5.13 A rotary flow about a centre Z. If the velocity increases outward proportionately to the radius, equipotential lines are not normal to the streamlines. They are only normal if the velocity is inversely proportional to the radius.

It must not, however, be thought that all forces applied to a fluid cause the plotting methods to be vitiated. If the restraining forces are applied uniformly all over the fluid instead of being applied only at a boundary, then the theoretical methods give a pattern which is correct. Take the instance of a fluid moving through a bed of sand or other small particles, the resistance of each grain contributing to the total resistance to the fluid. The fluid at no one streamline is retarded more than any other, contrary to frictional forces at boundaries which affect nearer fluid more than that further away. Experiment, moreover, shows that a pressure p is necessary to drive the fluid through the bed, and that $p \propto \int u \, ds$ (u = velocity, s = distance along streamlines). The physical quantity p thus has precisely the same properties as the potential ϕ, so that the equipotential lines as plotted are also lines of equal pressure in the sand bed. Streamline plotting methods are thus of considerable importance and value in finding the pressures and flows inside sand and gravel beds, which often pose important problems to the engineer.

Example
A concrete dam is built on a sand stratum 20 m thick, of permeability 2×10^{-5} m s^{-1}, overlying rock. The sand surface is at 190 m O.D. and the rock is at 170 m O.D. The dam is of triangular cross section 14 m high and 12 m base with its crest at 200 m O.D. Maximum water level is 198 m O.D.

Plot by an approximate method, the streamlines of flow in the sand and from the diagram deduce the leakage per m length of dam. Downstream of the dam, is there any likelihood of the sand rising due to upward pressures if sand only fills half of any volume, the remainder being filled with water?

Solution
The streamline pattern may be approximated by the 'squares' method as in Fig. 5.14. (see also Fig. 5.2)

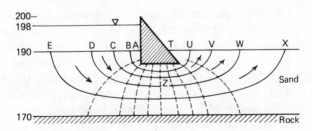

Fig. 5.14 The stream and potential lines under a dam.

The gap between any two streamlines contains 10 'squares' made by potential lines, between upstream and downstream surfaces of sand; and these squares represent a head decrease of 8 m of water.

So one square represents a drop of head of $\frac{8}{10}$ m.
Let a square such as Z have a side x m to scale.

Then pressure gradient $= \dfrac{8}{10x}$ m of water per metre.

So velocity in direction of motion $= 2 \times 10^{-5}\,\mathrm{m\,s^{-1}} \times \dfrac{8}{10x} = u$, and discharge between streamlines CV and DW $= ux$ per metre width

$$= 2 \times 10^{-5} \times \frac{8}{10x} x \;\mathrm{m^2\,s^{-1}}$$

The discharge between every pair of streamlines is the same and since there are 5 pairs of lines

Total leakage discharge per metre of dam $= 5 \times 2 \times 10^{-5} \times \dfrac{8}{10}\,\mathrm{m^2\,s^{-1}}$

$$= 8 \times 10^{-5}\,\mathrm{m^2\,s^{-1}}$$
$$= 6.9\,\mathrm{m^2} \text{ per day}$$

For uplift pressures on sand:
Near UV, the uppermost 'square' averages about 4 m (to scale) high.
Thus for sand of sp. gr. 2.65 (in air), the submerged weight of sand in that square is $(2.65 - 1.0) \times 10^3 \times 9.81 \times 4\,\mathrm{N\,m^{-2}}$ of plan area $= 64.75\,\mathrm{kN\,m^{-2}}$ if sand completely fills the volume
As sand only fills half the volume the submerged weight is $32.4\,\mathrm{kN\,m^{-2}}$
Upthrust by water on one square is, however, $\frac{8}{10}$ m of water, equivalent to

$$\tfrac{8}{10} \times 10^3 \times 9.81\,\mathrm{N\,m^{-2}} = 7.95\,\mathrm{kN\,m^{-2}}$$

Thus the risk of sand rising is very small, there being a factor of safety of $\frac{32.4}{7.85} = $ about $4{:}1$

A striking example of the limitations of the theoretical streamline plotting method is given by the case of the flow round a circular cylinder. The theoretical pattern was described and shown in Fig. 5.11 and it will be seen that the pattern is symmetrical, with the leading half-cylinder having a pattern near it which is exactly the same as that near the trailing (rear) half. If the pattern is found experimentally it will be found that the theoretical pattern is fairly well reproduced at low speeds of the fluid, or with small cylinders or with great kinematic viscosities v of the fluid. (In fact, the combination is such that $Ud/v < 1.0$ for this condition, see Fig. 5.2.) At higher speeds, however, the streamlines appear to be those given by the theoretical pattern only near the leading edge: around the trailing half-cylinder the pattern is widely different, there being a large *wake*, see Section 4.2. The discrepancy can be attributed to the shear forces exerted on the fluid by the friction with the cylinder's surface.

At low speeds they cause little modification to the theoretical pattern: at higher speeds they lead to separation, thus causing the actual flow pattern to be greatly different from that predicted by theory see Fig. 12.20.

The flow of real fluids around circular cylinders are illustrated in Fig. 10.1, page 143.

5.12 Summing up

Methods of plotting streamlines using the theoretical properties of ideal fluids can be usefully employed if there is no boundary friction and if the rotations are of the free vortex type. Such a restriction must appear onerous to engineers accustomed to real fluids. However, it has been found that conditions sufficiently approximating to those restrictions are found near the leading edge of solid objects in a fluid stream, for instance, aeroplane wings and ship hulls. Further back, nearer the trailing edge of such bodies, the frictional forces may modify the flow sufficiently to cause serious changes from the theoretical pattern. Streamline patterns calculated or plotted from the theoretical considerations described may therefore be used, with caution, for many real fluid flows, and a great deal of useful information about velocity changes may be obtained from them.

Appendix

Stream and potential functions have been extensively studied, and there are available other more elaborate proofs of their properties. In particular, it can be shown that ψ is a number which obeys the laws

$$\partial\psi/\partial y = u \quad \text{and} \quad \partial\psi/\partial x = -v$$

with respect to rectangular co-ordinates x and y. Here u and v are velocity components at a point in the x and y directions respectively. It is assumed that there are no velocity components in the z direction, i.e. that the flow is 2-dimensional.

Now, as described in Section 5.11, the effect of boundary friction or of forced vortex motion is to cause elements of fluid to rotate about their own centres. This rotation, or *vorticity*, can be written

$$R = \frac{\partial v}{\partial x} - \frac{\partial u}{\partial y}$$

Substitution between the above equations gives Poisson's equation

$$\frac{\partial^2\psi}{\partial x^2} + \frac{\partial^2\psi}{\partial y^2} = R$$

This equation can be used to determine streamlines in a flow near boundaries. Each boundary is assigned a value of ψ and elsewhere each point has its value ψ_1 which must be related to the values at neighbouring points so that Poisson's equation is correct everywhere. When ψ is known at a large number of points, lines of equal values of ψ can be drawn which are, of course, streamlines. R is not necessarily a constant, but may vary with x or y.

By a similar set of proofs it can be shown that the potential ϕ is defined as

$$\partial\phi/\partial x = u \quad \text{and} \quad \partial\phi/\partial y = v$$

Also, by assessing the discharges of fluid entering and leaving a given control volume the nett increase or decrease of fluid within the volume (if the flow is steady) can be shown to be

$$V = \frac{\partial u}{\partial x} + \frac{\partial v}{\partial y}$$

Again, substitution between these properties gives a Poisson equation, this time in ϕ,

$$\frac{\partial^2\phi}{\partial x^2} + \frac{\partial^2\phi}{\partial y^2} = V$$

and this may be used to find values of ϕ throughout a flow.

In the special case of a two dimensional incompressible flow which is continuous everywhere (so that the same volumetric discharge leaves a given control volume as enters it), and which is irrotational everywhere, then

$$R = 0 \quad \text{and} \quad V = 0$$

so that $\qquad \dfrac{\partial^2\psi}{\partial x^2} + \dfrac{\partial^2\psi}{\partial y^2} = 0 \quad \text{and} \quad \dfrac{\partial^2\phi}{\partial x^2} + \dfrac{\partial^2\phi}{\partial y^2} = 0$

These are the celebrated Laplace equations. Their form is such that it can be shown that lines of equal ψ must be at right angles to lines of equal ϕ: this is the basis of the graphical 'squares' method of drawing streamlines (Section 5.6). The equations can be used to determine ψ and ϕ arithmetically at every one of a number of points in a flow, from which lines of equal ψ and ϕ can be interpolated. Notice, however, that such use of Laplace equations is tantamount to assuming an incompressible inviscid fluid in 2-dimensional flow with no boundary friction, and with no rotations other than free vortex motions. A flow pattern of streamlines so obtained, graphically or by computation, may therefore be quite different in some respects from a pattern found experimentally in a real fluid.

An excellent book on the more advanced aspects of potential flow is *Advanced Fluid Mechanics* by A. J. Raudkivi and R. A. Callander (Arnold).

Problems

1. Draw the streamline and potential patterns for
   (a) A uniform flow of 18 m s^{-1} parallel to the x-axis.
   (b) A source of strength 3.6 m^3 s^{-1} at the origin.
   (c) A source of strength 3.6 m^3 s^{-1} in a uniform flow of 18 m s^{-1}.
   (d) A doublet of strength $\frac{9}{4}\pi$ at the origin.

   In each case specify the scale of the diagram, and the interval between streamlines and potential lines.

2. The stream function ψ for the flow round a circular cylinder of radius a in a stream of velocity U is

$$\psi = U\left(-y + \frac{a^2 y}{x^2 + y^2}\right) \quad \text{in rectangular co-ordinates}$$

or $\qquad \psi = -U\left(r - \dfrac{a^2}{r}\right)$ sin θ in polar co-ordinates.

Compute and plot the streamlines $\psi = 0$, $\psi = 2a$, $\psi = 3a$.

3. The ground levels on the two sides of a vertical retaining wall are 210 m and 200 m above datum respectively, and the 2 m thick wall has its base at 197 m above datum. The sandy soil rests on an impervious clay bed at level 190 m. Sketch the flow pattern by the approximate 'squares' method, if the soil is everywhere waterlogged but there is no free water standing above either ground level. Use the sketch to estimate if there is a risk of the ground lifting on the lower side.

Hint. Will the weight of the soil be everywhere greater than the hydrostatic force tending to lift it?

4. Sketch the streamlines at the surface of a canal through which a boat is progressing steadily in a straight line (*a*) as seen from a road bridge, (*b*) as seen from the boat. In each case give two patterns, one for an ideal, frictionless fluid, the other for the real fluid in which the boat leaves a wake far astern.

5. To reduce the leakage of water beneath the dam in the example on page 77 it is decided to build a concrete slab 50 m long downstream of the dam. What will be the new seepage rate and the upthrust on the slab?

$\qquad\qquad$ *Ans.* $3.1 \, \mathrm{m^2}$ per day, $1.6 \, \mathrm{MN \, m^{-1}}$.

6. Determine the stream function for a free vortex whose tangential velocity V at radius r is

$$V = -\frac{10}{r} \, \mathrm{cm \, s^{-1}}.$$

If a free stream defined by $u = 20 \, \mathrm{cm \, s^{-1}}$ is superimposed on the vortex, determine the stream function for the combined flow pattern. On a diagram locate a stagnation point and plot the streamline $\psi = 5$.

$\qquad$ *Ans.* $\psi = 10 \log r$; $\psi = 20y + 10 \log r$; $(0, -0.5)$;

$\log r = 0.5 - 2y$—so choose values of r and solve for y.

7. The stream function $\psi = \dfrac{\omega R^2}{2}\left[\dfrac{\theta}{2} - \left(\dfrac{r}{R}\right)^2 + \dfrac{1}{3}\left(\dfrac{r}{R}\right)^3\right]$ represents in polar coordinates the absolute motion of ideal fluid through the impeller of a pump where the impeller has a radius R and constant angular velocity ω. Determine the radial and tangential velocity components, v_r and v_θ respectively. Show that $\psi_r = \dfrac{\omega R^2}{4}\left[\theta + \dfrac{2}{3}\left(\dfrac{r}{R}\right)^3\right]$ is the flow pattern relative to the rotating impeller, and plot a few points on one of the streamlines of ψ_r in the range $0.2 \leqslant r/R \leqslant 1.0$.

6

Forces due to fluids in motion

6.1 Newton's second law of motion and fluid flows

In general, the force system on a control volume in a fluid is complicated, each force due to pressure, shear, or weight, requiring separate data to enable it to be estimated in size and direction. Problems of motion involving all the possible forces, and confining the fluid between complicated boundaries therefore need complex algebra for their solution; indeed, our knowledge of mathematics is usually inadequate for general solutions. Some problems can be solved by numerical methods using computers, but even then the process may be time consuming and expensive.

In many problems of importance to the engineer, the force system is less complicated, and in a few it is found that the effect of only one force predominates; these simpler problems can often be solved in a general way. Among these simpler problems are those where a force caused by a pressure difference just balances the reaction of an acceleration, which in turn produces the observed velocities; or those where a shear force just balances a pressure force, and so causes the observed velocities. Much of this book is concerned with these two simplified families of problems, but it should not be forgotten that, in addition, more complicated families exist (e.g. shear, pressure and accelerations) and these often are the important ones in engineering practice. Solutions of these complicated families can often only be made by direct experiment, and the methods of presenting such information are reviewed in Chapter 10. In addition to hydrodynamic forces, there are of course always hydrostatic forces, due to the weight of the fluid. The latter must always be added to the former to obtain, in a specific case, the total force acting at a point. The methods of Chapter 2 are available for estimating 'hydrostatic' forces in moving fluids when the streamlines are straight. The more complicated pressure distributions in curved flows are considered in Chapter 11.

Estimating the hydrodynamic forces when there are accelerations present *always* involves using Newton's Second Law of Motion.

If Lagrange's method is used the force, δF, on each particle of fluid mass, δm, velocity, u, can be calculated from the equation.

$$\delta F = \frac{\mathrm{d}}{\mathrm{d}t}(\delta m u)$$

82

where the means that the properties are vectors and the force in a particular direction is due to the change of motion in that direction. To find the total force the forces δF have to be summed (integrated) for each particle over the whole flow field.

An alternative approach is to use Euler's method and apply Newton's Second Law of Motion to a control volume in the fluid. If the control volume chosen is a streamtube, Fig. 6.1, then the control surfaces which are streamlines, AD, BC, have no flow of fluid mass or momentum across them. The only changes of momentum in the control volume are due to the flows across AB, DC, which can be calculated by the method given on page 49 if the normal component of velocity, u, is known for each element of area of the control surface δa. For a control surface across one streamtube the velocity may be assumed to be uniform and hence the flow rate of momentum, δM, across AB *into* the control volume would be given by

$$\delta M_1 = \rho_1 u_1{}^2 \delta a_1, \quad \text{normal to AB,}$$

while the flow rate of momentum across CD *out* of the control volume is

$$\delta M_2 = \rho_2 u_2{}^2 \delta a_2 \quad \text{normal to CD.}$$

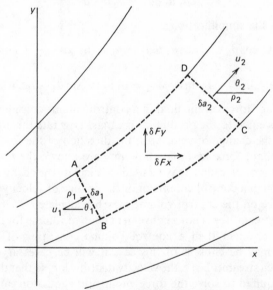

Fig. 6.1 A control volume chosen as part of a streamtube in a flow pattern. Broken lines show the control surfaces

Remembering that momentum flow rate is a vector quantity the components may be written:

$$\delta M_{1x} = \rho_1 u_1{}^2 \delta a_1 \cos \theta_1, \qquad \delta M_{1y} = \rho_1 u_1{}^2 \delta a_1 \sin \theta_1,$$
$$\delta M_{2x} = \rho_2 u_2{}^2 \delta a_2 \cos \theta_2 \qquad \delta M_{2y} = \rho_2 u_2{}^2 \delta a_2 \sin \theta_2$$

For steady flow the total rate of change of momentum in each direction in the control volume must be zero. Newton's Second Law of Motion states that a force will cause a rate of change of momentum and similarly any difference between the momentum inflow and outflow rates will cause a rate of change of momentum in the control volume. Hence if δF_x is the x component of the force δF in the flow direction, for the x direction momentum in the control volume to remain constant; then

$$\delta F_x + \delta M_{1x} - \delta M_{2x} = 0$$

or

$$\delta F_x = \delta M_{2x} - \delta M_{1x}$$
$$= \rho_2 u_2{}^2 \cos \theta_2 \, \delta a_2 - \rho_1 u_1{}^2 \cos \theta_1 \, \delta a_1.$$

The reader might like to show by a similar argument that

$$\delta F_y = \rho_2 u_2{}^2 \sin \theta_2 \, \delta a_2 - \rho_1 u_1{}^2 \sin \theta_1 \, \delta a_1.$$

As the flow is steady the mass continuity equation states that the mass inflow rate into the control volume must equal the mass outflow rate (see equation 4.8) whence

$$\rho_1 u_1 \, \delta a_1 = \rho_2 u_2 \, \delta a_2,$$

which allows the simplification

$$\delta F_x = \rho_2 u_2 (u_2 \cos \theta_2 - u_1 \cos \theta_1)\delta a_2 = \rho_1 u_1 (u_2 \cos \theta_2 - u_1 \cos \theta_1)\delta a_1 \quad (6.1)$$

$$\delta F_y = \rho_2 u_2 (u_2 \sin \theta_2 - u_1 \sin \theta_1)\delta a_2 = \rho_1 u_1 (u_2 \sin \theta_2 - u_1 \sin \theta_1)\delta a_1 \quad (6.2)$$

i.e. the resultant force on the fluid in a control volume considered as a single streamtube is equal to the product of the mass flow rate through the control volume and the change of velocity in the direction of the force.

If the resultant force and mass flow rate are known the *force–momentum equation* as it is often called, may be used to calculate the change in velocity of the flow through a control volume or, if the change of velocity is known, the resultant force on the control volume may be calculated.

If a single streamtube is not a convenient control volume for the solution of a particular problem then a control volume consisting of a number of streamtubes may be chosen. In this case it will be necessary to know the velocity in each streamtube i.e. the velocity distribution, so that the momentum flow rates required to solve the force–momentum equation can be found by integration (see page 49). However, it is often possible to simplify a problem by assuming the velocity is uniformly distributed across a control surface and by choosing the control volume such that it may be treated as a single streamtube. Aligning one of the axes with the flow direction e.g. the x-axis, will make $\sin \theta_1 = 0$ and $\cos \theta_1 = 1.0$ and if the flow is only in this direction $F_y = 0$, so

$$F_x = \rho_1 u_1 a_1 (u_2 - u_1) = \rho_2 u_2 a_2 (u_2 - u_1) \quad (6.3)$$

To summarize, the force–momentum equation states that the change of momentum flow rate through a control volume in a particular direction is equal to the resultant force on the control volume in that direction.

6.2 The forces on a control volume in a fluid

To use the force–momentum equation successfully the forces which make up the resultant force on a control volume must be understood. For example consider the bend in a ventilating system duct or wind-tunnel which contains turning vanes as shown in Fig. 6.2(a). The flow has to be represented by a large number of streamtubes, only a few of which are drawn, due to the presence of the turning vanes and their wakes. Fig. 6.2(b) shows one possible control

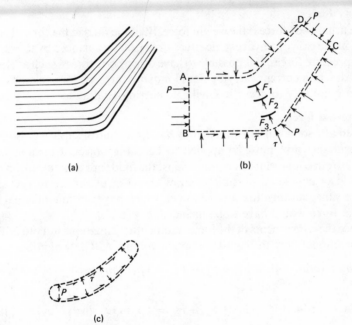

(a)

(b)

(c)

Fig. 6.2 The flow in a converging duct with turning vanes (a) streamlines of the flow in the duct (b) pressure and shear stresses exerted by the duct wall on the flow and total forces exerted by turning vanes (c) enlarged diagram of pressure and shear stresses whose resultant is F exerted on the flow by one turning vane.

volume which includes all the streamtubes of the flow and the turning vanes but none of the surfaces of the duct. The two control surfaces AB, CD which are not replacing solid boundaries are drawn perpendicular to the flow directions. There are then three types of force acting on the fluid in the control volume:

1) Surface forces

a) The resultant pressure force acting on the control surfaces has components, F_{px}, F_{py}, which are usually calculated by integration using projected areas as described in Chapter 2.

$$F_{px} = \int p \, da_x, \quad F_{py} = \int p \, da_y$$

b) The shear forces acting on the surfaces AD, BC, are tackled in the same way giving components F_{sx} and F_{sy}. Notice that if the convention is used that all forces are positive in the direction of motion the shear forces have to be drawn in the flow direction, but will have negative numerical values. The choice of control surfaces AB, CD, normal to the flow direction means there will be no shear stresses acting on them.

2) Body forces

The usual body force is the weight force, W, of the fluid in the control volume, which in the case of a hydrostatic situation is exactly balanced by the resultant of the pressure forces. It is possible to have other body forces, such as electrical forces due to a current carrying liquid moving through a magnetic field, but these are unlikely to occur in Civil Engineering Hydraulics.

3) External forces

When any solid object is isolated by having a control surface pass through it a resultant force must be applied to keep the object, a turning vane in Fig. 6.2(b), in equilibrium by acting against the fluid forces on the turning vane. It would of course be possible to draw the control surface to exclude the turning vane, but then the surface forces, which have a resultant equal to the external force would have to be included, Fig. 6.2(c).

Hence the components of the force–momentum equations (6.1), (6.2), for the two directional flow through the control volume ABCD would be:

$$\int p \, da_x + \int \tau \, da_x + W_x + F_{1x} + F_{2x} + F_{3x} = \int \rho_1 u_1 (u_2 \cos \theta_2 - u_1 \cos \theta_1) \, da_1 \tag{6.4}$$

$$\int \rho \, da_y + \int \tau \, da_y + W_y + F_{1y} + F_{2y} + F_{3y} = \int \rho_1 u_1 (u_2 \sin \theta_2 - u_1 \sin \theta_1) \, da_1 \tag{6.5}$$

where the subscripts x and y indicate the directions of the components. Fortunately the equations can often be simplified either because some of the forces are zero, or by judicious choice of the control surfaces or the direction of the axes as the following example will show.

Example

A symmetrical nozzle through which water is flowing vertically downwards at a rate of $0.034 \text{ m}^3 \text{ s}^{-1}$ is fitted on the end of a circular pipe. The area of the nozzle at inlet is $1.85 \times 10^{-2} \text{ m}^2$ and at outlet $4.64 \times 10^{-3} \text{ m}^2$. The gauge pressure measured at the nozzle inlet is 2.55×10^4 Nm^{-2} and the discharge is at atmospheric pressure, i.e. zero gauge pressure. If the volume of water contained in the nozzle is $1.66 \times 10^{-3} \mathrm{m}^3$ estimate the force the flow exerts on the nozzle.

Solution

The first decision to be taken is which control volume to use and which way the forces should be expressed. Fig. 6.3 shows four alternatives, and there could be more. Control volume (a) has the disadvantage that information would be required on the weight of the nozzle, forces in the bolts, pressures on the nozzle flange etc. While this information might be appropriate for a stress analysis of the nozzle it does not simplify the force–momentum equation. Control volume (b) simplifies the forces on the surfaces normal to the flow but considers complicated unknown pressure and shear distributions on the other surfaces. Control volume (c) has the disadvantages of both (a) and (b), but if the pressures and shear forces on the curved control surfaces are combined and considered as external forces the simple arrangement of Fig. 6.3(d) results. Note that there are no pressure forces on the outlet control surface as the problem is to be solved using gauge pressures.

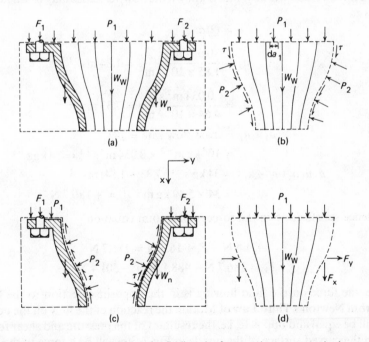

Fig. 6.3 Alternative control volumes for finding the force acting on a nozzle due to the flow of a fluid.

If the x axis is considered to lie along the centreline of the nozzle and the force momentum equation is applied to control volume 6.3(d) the equations 6.4 and 6.5 become

$$\int p_1 \, da_1 + F_x + W_w = \int \rho_1 u_1 (u_2 - u_1) da_1 \qquad (6.6)$$

$$F_y = 0.$$

As an approximation the velocity and pressure distributions may be assumed uniform and equation 6.6 reduces to

$$P_1 a_1 + F_x + W_w = \rho_1 u_1 a_1 (u_2 - u_1)$$

where a_1 is the inlet area of the nozzle.

Evaluating the terms independently

$$P_1 a_1 = 2.55 \times 10^4 \, \text{Nm}^{-2} \times 1.85 \times 10^{-2} \, \text{m}^2 = +471.8 \, \text{N}$$

$$W_w = \rho g \times \text{volume of water}$$

$$= 1000 \, \text{kgm}^{-3} \times 9.81 \, \text{ms}^{-2} \times 1.66 \times 10^{-3} \, \text{m}^3 = +16.3 \, \text{N}$$

the plus signs show both forces are in the flow direction.

As the volumetric flow rate is known the flow velocities may be calculated from

$$u = Q/a.$$

$$u_1 = \frac{0.034 \, \text{m}^3 \, \text{s}^{-1}}{1.85 \times 10^{-2} \, \text{m}^2} = 1.84 \, \text{ms}^{-1}$$

$$u_2 = \frac{0.034 \, \text{m}^2 \, \text{s}^{-1}}{4.64 \times 10^{-3} \, \text{m}^2} = 7.33 \, \text{ms}^{-1}$$

$$\rho_1 u_1 a_1 = \text{mass flow rate} = \rho Q$$

$$= 10^3 \, \text{kg m}^{-3} \times 0.034 \, \text{m}^3 \, \text{s}^{-1} = 34 \, \text{kg s}^{-1}$$

$$\rho_1 u_1 a_1 (u_2 - u_1) = 34 \, \text{kg s}^{-1} \, (7.33 - 1.84) \, \text{ms}^{-1}$$

$$= 34 \times 5.49 \, \text{kg m s}^{-1} = +186.7 \, \text{N}$$

Hence substituting in the force–momentum equation

$$471.8 \, \text{N} + F_x + 16.3 \, \text{N} = 186.7 \, \text{N}$$

$$F_x = 186.7 \, \text{N} - 488.1 \, \text{N} = -301.4 \, \text{N}$$

i.e. the force on the fluid flowing is in the opposite direction to the flow. From Newton's Third Law of Motion the reaction of the flow on the nozzle will be equal and opposite, i.e. the resultant of the pressure and shear forces on the curved surface of the nozzle in Fig. 6.3(c) will be a force in the flow direction on the nozzle of 301.4 N.

6.3 The force of a liquid jet on a normal surface

If the jet from the nozzle described in the previous example strikes a large plate held normal to the direction of flow it will exert a force on it which can be estimated using the force momentum equation. A control volume can be drawn around the jet and plate, but before calculations can commence it is necessary to know how the fluid flows on striking the plate. It can be observed, for example in the flow of water from a tap into a large empty sink, that water striking a solid surface flows along that surface and does not rebound. Hence the flow pattern and control volume can be drawn as in Fig. 6.4.

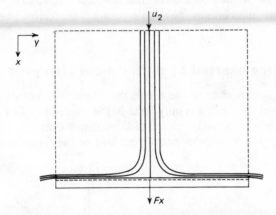

Fig. 6.4 A jet of fluid striking a plane surface. The surface is normal to the jet axis and the fluid eventually leaves parallel to the surface. It is assumed, and can be confirmed by observation that the jet does not rebound or splash back. The control volume is shown dotted.

As the pressure on the control surfaces which are in air or pass through the jet are atmospheric, i.e. zero gauge, and the pressure forces on the plate are considered as an external force, there are only two forces which have to be considered. The momentum equation then becomes

$$F_x + W = \rho u_2 a_2 (u_3 - u_2)$$

where W is the weight of the water in the control volume, u_2 the jet exit velocity, and u_3 the x direction velocity with which the water leaves the control volume, i.e. zero. Hence

$$F_x + W = -\rho u_2{}^2 a_2 = -\rho Q u_2 = -34\,\mathrm{kg\,s^{-1}} \times 7.33\,\mathrm{m\,s^{-1}} = -249.2\,\mathrm{N}.$$

To obtain F_x an estimate must be made of the weight of water in the control volume. If the plate lies 1 m below the nozzle exit and the mass of water running over the plate is the same as that in the column of water in the jet then

$$W = \rho g \times \text{volume of water} = 2\rho g \times \text{volume of jet}$$
$$= 2 \times 10^3 \, \text{kg m}^{-3} \times 9.81 \, \text{m s}^{-2} \times 1 \, \text{m} \times 4.64 \times 10^{-3} \, \text{m}^2$$
$$= 91 \, \text{N}.$$
$$F_x = -249.2 \, \text{N} - W = -249.2 \, \text{N} - 91 \, \text{N} = -350 \, \text{N},$$

i.e. the force on the control volume is opposing the motion and the force on the plate will be in the direction of the flow.

As the flow is symmetrical about the axis of the jet the forces in the y and z directions, z being normal to the plane of the paper, will be zero.

If the plate moves away from the nozzle the solution is best found by keeping the observer and control volume stationary relative to the plate and calculating the momentum flow rate using the relative velocities, see Problem 1c.

6.4 The force exerted by a fluid flow changing direction

The two examples given of the use of the momentum equation were both one-dimensional, i.e. there was only force in the x-direction. Consider a set of turning vanes as often fitted to the end of a ventilation duct, Fig. 6.5, to eject the air upwards to avoid the inlet duct. What will be the force on one vane?

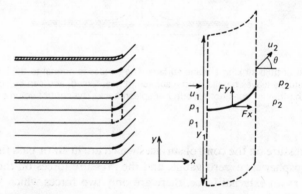

Fig. 6.5 The flow through turning vanes showing a convenient control volume for a force–momentum analysis of one vane.

Here again the choice of control volume is important. The downstream control surface must be close to the vanes for the jet soon mixes with the air outside and becomes like a wake with a non uniform flow distribution. The upstream control surface can lie in the plane of the leading edges of the vanes. It is most convenient to complete the control volume with two control surfaces which follow streamlines and thus have no momentum flow rate across them.

One difficulty is that information on the pressures and shear stresses on these control surfaces is not easy to find. However if these surfaces are chosen to lie exactly half way between the two adjacent turning vanes the pressure and shear forces on the two control surfaces will be equal and opposite and will give no resultant forces on the control volume.

Applying the mass continuity equation to the control volume

$$\rho_1 u_1 y_1 = \rho_2 u_2 y_1 \cos \theta,$$

where y_1 is the distance between turning vanes. Then if, as is usual in ventilating systems,

$$(p_1 - p_2) \ll p \text{ atmospheric}, \quad \rho_1 \simeq \rho_2$$

and hence $u_1 = u_2 \cos \theta$.

The force–momentum equations for the x and y directions become

$$p_1 y_1 + F_x - p_2 y_1 = 0 \quad \text{or} \quad F_x = (p_2 - p_1) y_1$$
$$F_y = \rho_1 u_1 y_1 u_2 \sin \theta = \rho u_1^2 y_1 \tan \theta$$

and the resultant force on the control volume can be found from the force components. This force will be the reaction provided to support the turning vane.

If, in the previous case of the jet striking a surface, the surface upon which the jet impinges is inclined at an angle θ as shown in Fig. 6.6, the fluid is again diverted along the surface, but now the flow is not symmetrical and the problem involves the change of momentum flow rate in two directions. More fluid flows up the surface than flows down. Some momentum is therefore retained by the fluid in the original jet direction after it has struck the surface. Additional experimental evidence would therefore be required to give the speeds and quantities of the fluid flowing up and down the surface, if the previous method is used.

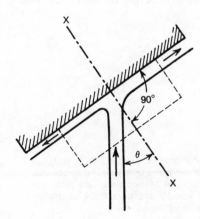

Fig. 6.6 A jet of fluid striking a plane surface which is inclined to the jet axis, showing a suitable control volume to calculate the force on the plate.

An alternative method, which will require an additional assumption, is to retain the position of the control volume relative to the plate from the previous analysis. Now consider the *component* of the oncoming momentum at right angles to the plate, on the line XX, that is $\rho a u^2 \cos \theta$. After the fluid has been diverted, none of this momentum remains as all the fluid is now moving at right angles to XX. A force $F = \rho a u^2 \cos \theta$ must therefore be exerted along the line XX.

Consider now the component of the oncoming momentum parallel to the surface, i.e. at right angles to XX, namely, $\rho a u^2 \sin \theta$. If this component is to suffer a change, then a force would be necessary parallel to the surface. Such a force must therefore act as a shear force on the fluid moving along the plate. But if it is assumed that the fluid is ideal, no shear forces are possible, so that the component $\rho a u^2 \sin \theta$ is not changed at all. There will therefore be no force exerted in a direction at right angles to XX.

The only force caused by the impact of a jet of *ideal* fluid is therefore the one normal to the surface concerned, in the direction XX, of magnitude $\rho a u^2 \cos \theta$. This force may of course be resolved into components in the original direction of the jet, $\rho a u^2 \cos^2 \theta$, and at right angles to it $\rho a u^2 \cos \theta \sin \theta$. If the fluid is viscous, then there may be a significant change of the momentum in the direction parallel to the surface and this must be added vectorially to the force normal to the surface. Still more experimental evidence is required to assess such a tangential force.

The forces due to the flow of a fluid round a bend in a pipe is another case where consideration of the change of momentum is necessary. Consider a pipe of uniform cross-sectional area a, laid horizontally with a 90° bend as shown, and with fluid of density ρ passing through at a velocity u, Fig. 6.7. If the gauge pressure intensity of the fluid, p, is constant throughout, then by Case i(*b*) of Chapter 2, the purely hydrostatic force on the bend in the x-direction is $F_{x_1} = pa$ acting outwards. Similarly the hydrostatic force in the y-direction is $F_{y_1} = pa$ also acting outwards, as shown.

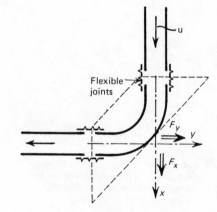

Fig. 6.7 Fluid being deflected through 90° in a bend of a pipeline. The forces due to the changes of momentum are resolved into the x and y directions to be used with the control volume shown.

Now consider the hydrodynamic forces acting on the bend due to the changes in the velocity of the fluid. The fluid entering the bend brings with it a momentum flow of $\rho Q u$ in the x-direction, where Q is the quantity of fluid arriving per second. After the bend, none of this x-momentum remains, because the fluid no longer has a velocity in that direction. A force $\rho Q u$ must therefore have been exerted by the bend on the fluid to cause the change of momentum flow rate. The reaction of this force, F_{x_2}, is the force of the fluid on the bend, where $F_{x_2} = \rho Q u$, acting in the direction of the x-arrow shown.

Similarly, the fluid originally has no y-momentum, but leaves the bend with a flow of $\rho Q u$ in that direction. A force is needed for this change of momentum flow rate and the reaction is the force of the fluid on the bend, F_{y_2}, where $F_{y_2} = \rho Q u$ acting in the direction of the y-arrow. The total force on the bend is thus composed of the four forces above, in the x-direction $F_{x_1} + F_{x_2} = pa + \rho Q u$; and in the y-direction $F_{y_1} + F_{y_2} = pa + \rho Q u$. Using the parallelogram of forces, the total force on the bend is $(pa + \rho Q u)\sqrt{2}$ acting in a direction at $45°$ to both x and y. To prevent movement of the bend by this force, a restraining force must be applied. On many pipes, the joints are designed to be strong enough to supply this force. On large pipes, such as are used for conveying water at high velocity and pressure to hydro-electric installations, it is uneconomical to make the joints so strong; a large concrete anchorage is therefore provided to secure the bend in place.

6.5 The force due to a flow along a surface

So far the force–momentum equation has been applied to flows where the velocity has been considered uniform across a cross section. Where solid boundaries are present there are shear forces and non-uniform velocity distributions occur. The simplest of these cases concerns the formation of a *boundary layer*, first mentioned in Chapter 4, on a flat plate parallel to an unrestricted flow such as a large river. Inside the thin boundary layer the shear forces reduce the velocity of flow, while outside the boundary layer the velocity of the flow is unaffected by the shear forces (Fig. 6.8).

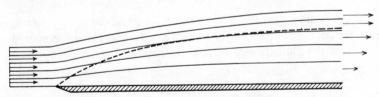

Fig. 6.8 The streamlines of a flow in a boundary layer over a flat surface. The dimensions normal to the surface are greatly exaggerated and in reality the curvature of the streamlines is negligible. Observe how the distance between the streamlines increases as the velocities decrease in the boundary layer. The control surface encloses the flow in the boundary layer which has been retarded by shear forces but excludes the unretarded flow outside the boundary layer.

The shear forces acting on any surface in a flow may be of considerable consequence and hence a method of estimating them must be found. As many solutions of the boundary layer problem are based on the force–momentum equation a simple example will be given here to be followed by an introduction to more detailed forms of analysis in Chapter 12.

Consider a control surface drawn normal to the flat plate from its surface at Z to cut the edge of the boundary layer at Q (Fig. 6.9). Draw another control surface along the streamline through Q upstream until a point P is reached where the streamline cuts the normal from the leading edge of the plate A. Close the control volume with the control surfaces PA, AZ. The application of the force–momentum equation to this control volume is relatively simple. As the surface is flat and parallel to an unrestricted flow the resultant pressure force on it will be zero in the flow direction. As a horizontal flow is being considered there can be no component of the weight force in the flow direction and finally the shear stresses are zero on all the control surfaces except AZ where the shear forces can be combined to form the external force F_x, which is the only force on the control volume.

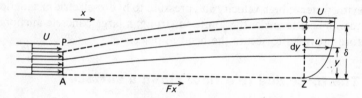

Fig. 6.9 A control volume for analysing the boundary layer on a flat plate by the force–momentum equation.

Considering unit width of the surface at right angles to the plane of the diagram, the force–momentum equation becomes

$$F_x = \int_Z^Q \rho_2 u \, (u \cos \theta_2 - U \cos \theta_1) \, dy \times 1 \text{ m}.$$

As the x axis is drawn parallel to the plate, $\theta_1 = 0$ and $\cos \theta_1 = 1$. Similarly as the thickness of the boundary layer on such a plate is very small compared with the length of the plate, (Table 12.1), θ_2 is small and hence $\cos \theta_2 \simeq 1$ so

$$F = \int_0^\delta \rho u (u - U) \, dy \times 1\text{m}, \tag{6.7}$$

where u is the velocity at a distance y from the plate at control surface ZQ and δ is the thickness of the boundary layer there. As there are no changes of pressure in the flow the density will be the same at all points and there is no need to have a subscript in ρ.

When evaluating the integral, as $U > u$, its value must be negative, i.e. F_x will be negative and the force on the control volume will oppose the flow. The shear force on the plate, which is the reaction to F_x, will therefore be in the flow direction.

An alternative derivation of the integral can be made assuming that the pressure and density are uniform and that the flow is parallel to the plate. Consider a unit width of the surface in a direction at right angles to the plane of the diagram (Fig. 6.9), that is, across the flow. A small increment of height dy in the diagram on the control surface QZ defines a streamtube of cross-sectional area dy high and unit length long across the flow. Through this streamtube, the mass of fluid passing in unit time is ρudy. Now every unit mass of fluid which passes through originally had a velocity U, so that its x-momentum change has been $(U - u)$, and therefore the x-momentum change per second of the fluid passing through the streamtube is ρud$y(U - u)$. All the fluid within the boundary layer has been retarded to some extent, so that the total change of momentum throughout the whole layer is

$$\int_{y=0}^{y=\delta} \rho u(U - u)\,\mathrm{d}y$$

and this equals the force F per unit width of the surface, which has caused the retardation, i.e. the drag force of the fuid on the surface between A and Z.

The integral cannot be evaluated without knowing how u varies with y, a matter of experiment or more advanced theory (Chapter 12). But under conditions of uniform pressure with a turbulent flow in the boundary layer an approximation which gives a fairly accurate estimate is $\dfrac{u}{U} = \left(\dfrac{y}{\delta}\right)^{1/5}$. Using this expression the integral can be made dimensionless by considering y/δ as the variable. Writing d$y = \delta\,.\,\mathrm{d}\left(\dfrac{y}{\delta}\right)$, substitution into the integral of equation (6.7) gives

$$F = \rho U^2 \delta \int_{y/\delta=0}^{y/\delta=1} \frac{u}{U}\left(1 - \frac{u}{U}\right)\mathrm{d}\left(\frac{y}{\delta}\right) = \frac{5}{42}\rho U^2 \delta \qquad (6.8)$$

The boundary layer thickness δ depends on the roughness of the underlying surface, on the fluid properties, and on the distance AZ. In some conditions, when the pressure is not constant along AZ, as occurs on the curved surfaces of an aircraft wing, or a spillway, more complicated velocity distributions may apply than the simple one of the foregoing example, so that quite different equations are obtained for F.

In the case where only a part of the boundary layer is being considered, the velocity distribution is non-uniform at both the entry and the exit of the control volume. It will then be necessary to evaluate two separate integrals to find the rate of change of momentum.

Example

A spillway discharges $12 \text{ m}^3 \text{ s}^{-1}$ of water per metre width on a plane surface sloping downward at 45°. At the top of the slope the depth of the flow is 2 m and there is a boundary layer 0.8 m thick whose velocity profile is approximately

$$\frac{u}{U} = \left(\frac{y}{\delta}\right)^{1/7}.$$

Calculate the velocity in the flow outside the boundary layer.

At a distance of 1.65 m further downstream the depth of flow is 1.5 m and the velocity outside the boundary layer is 8.56 m s^{-1}. What is the mean shear stress on the slab between the two control sections?

Solution

To find the velocity of flow integrate the velocity distribution and equate to the flow rate at the upstream control surface AD (Fig. 6.10). The flow rate is the sum of the uniform flow outside the boundary layer and the flow inside the boundary layer, the whole flow being considered incompressible.

$$Q = 12 \text{ m}^2 \text{ s}^{-1} = \int u \, dy = (d - \delta)U_1 + \int_0^{0.8 \text{ m}} U_1 \left(\frac{y}{\delta}\right)^{1/7} dy.$$

Making the integral dimensionless

$$12 \text{ m}^2 \text{ s}^{-1} = U_1 \left((d - \delta) + \delta \int_0^1 \left(\frac{y}{\delta}\right)^{1/7} d\left(\frac{y}{\delta}\right) \right)$$

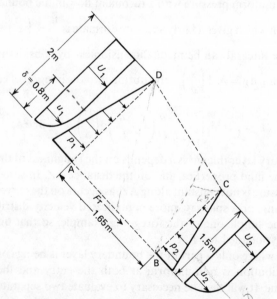

Fig. 6.10 A control volume for the flow down a spillway.

$$= U_1\left((d-\delta)+\delta\left[\frac{7}{8}\left(\frac{y}{\delta}\right)^{1/7}\right]_0^1\right) = u_1\left(d-\delta+\frac{7}{8}\delta\right)$$

$$= U_1\left(d-\frac{1}{8}\delta\right)$$

$$= U_1 \times 1.9\ \text{m}$$

Hence $U_1 = \dfrac{12\ \text{m}^2\ \text{s}^{-1}}{1.9\ \text{m}} = 6.32\ \text{m s}^{-1}.$

It will be necessary to calculate the thickness of the boundary layer, δ, at the downstream control surface BC. Using the expression above, but with the new values for the flow depth and the boundary layer thickness

$$12\ \text{m}^2\ \text{s}^{-1} = U_2\left(d-\frac{1}{8}\delta\right)$$

$$= 8.56\left(1.5\ \text{m}-\frac{1}{8}\delta\right)\text{m s}^{-1}.$$

Hence $\quad \delta = 8\left(1.5-\dfrac{12}{8.56}\right)\text{m} = 0.785\ \text{m}.$

If a control volume ABCD is drawn enclosing the water between the two specified control surfaces the force–momentum equation can be applied. As the flow both into and out of the control volume is non-uniform the momentum flow rates must be calculated separately, the change of momentum flow rate in the control volume found by subtraction, and equated to the resultant force on the control volume.

The momentum flow rate into control volume across **AD** can be calculated as the sum of the momentum flow rate of the uniform flow outside the boundary layer and the momentum flow rate inside the boundary layer

$$= \rho U_1^2(d-\delta)+\int_0^\delta \rho u_1^2\,dy$$

$$= \rho U_1^2(d-\delta)+\rho U_1^2\delta\int_0^1\left(\frac{u_1}{U_1}\right)^2 d\left(\frac{y}{\delta}\right)$$

$$= \rho U_1^2\left(d+\delta\int_0^1\left(\frac{y}{\delta}\right)^{2/7}d\left(\frac{y}{\delta}\right)-\delta\right)$$

$$= \rho U_1^2\left(d+\frac{7\delta}{9}-\delta\right) = \rho U_1^2\left(d-\frac{2\delta}{9}\right)$$

$$= 10^3\ \text{kg m}^{-3}\times 6.32^2\ \text{m}^2\ \text{s}^{-2}\left(2.0-\frac{2\times0.8}{9}\right)\text{m}$$

$$= 72.69\ \text{kN m}^{-1}.$$

Similarly the momentum outflow rate across BC

$$= \rho U_2^{\,2}\left(d - \frac{2\delta}{9}\right)$$

$$= 10^3 \text{ kg m}^{-3} \times 8.56^2 \text{ m}^2 \text{ s}^{-2}\left(1.5 - \frac{2 \times 0.785}{9}\right)\text{m}$$

$$= 97.13 \text{ kN m}^{-1}.$$

Hence change of momentum flow rate between the ends AD and BC of the control volume $= (97.13 - 72.69)\,\text{kN m}^{-1}$

$$= 24.44 \text{ kN m}^{-1}.$$

Resultant force on the control volume in the flow direction = resultant pressure force on AD, BC, + component of weight force down slope + shear force

$$= \int p_1 \,da_1 - \int p_2 \,da_2 + W \cos 45° + F_\tau = \text{change of momentum flow rate.}$$

The pressure force on a control surface normal to the flow $= \frac{1}{2}\rho g d^2 \cos 45°$ as the control surface is at 45° to the vertical.

$$\int p_1 \,da_1 = \frac{1}{2} \times 10^3 \text{ kg m}^{-3} \times 9.81 \text{ m s}^{-2} \times 4 \text{ m}^2 \times \frac{1}{\sqrt{2}} = 13.87 \text{ kN m}^{-1}$$

$$\int p_2 \,da_2 = 8.88 \text{ kN m}^{-1}$$

$$W \cos 45° = \rho g l \frac{(d + d_2)}{2}\cos 45°$$

$$= 10^3 \text{ kg m}^{-3} \times 9.81 \frac{\text{m}}{\text{s}^2} \times 1.65 \times \frac{3.5}{2} \times \frac{1}{\sqrt{2}} \text{ m}^2$$

$$= 20.03 \text{ kN m}^{-1}$$

Resultant force $= (13.87 - 8.88 + 20.03)\,\text{kN m}^{-1} + F_\tau$

$$= 25.0 \text{ kN m}^{-1} + F_\tau$$

Whence $F_\tau = 24.44 - 25.02 = -0.58 \text{ kN m}^{-1}$,

and mean shear stress $= \dfrac{F_\tau}{\text{area of surface in contact with flow}}$

$$= -\frac{0.58}{1.65} = -0.35\frac{\text{kN}}{\text{m}^2},$$

i.e. the shear stress opposes the motion.

6.6 The drag force on an object in a flow

So far the calculations for the force due to the loss of momentum flow rate in a boundary layer have only been for plane surfaces where the pressure forces on the plane have no component in the flow direction. The shear force is sometimes referred to as the *skin friction drag* action on the surface. If, however, the surface in contact with the flow is not plane then the pressure forces acting on it will have components in the flow direction and the resultant pressure force in the flow direction is called the *form drag*. If the surface is three-dimensional and is generating a lift force normal to the flow, as the aerofoil in Section 2.6, there is an extra *induced drag* due to the vortices in the flow shown in Fig. 11.14. The sum of the skin friction, form, and induced drag gives the *total drag* on the surface.

In the case of an *internal flow*, such as the flow through the nozzle described in Section 6.2, the total drag can be calculated directly by applying the force–momentum equation to the whole flow within the boundaries. For an *external flow* such as round a building, or vehicle the application of the force–momentum equation is not so obvious as the flow is effectively unbounded. The usual method of solution is to draw a control volume around the object which is bounded by streamlines which are sufficiently far away so that the pressure on them is effectively the undisturbed stream pressure, and the velocity the undisturbed stream velocity, e.g. Fig. 6.11. The upstream control surface AD is drawn normal to the flow where the flow is undisturbed, and the downstream surface BC is sufficiently far downstream for the flow in the wake to be steady. The velocity distribution Fig. 6.11(b) and the pressure distribution Fig. 6.11(c) in the wake BC are then all that need to be measured to apply the force–momentum equation to the control volume.

The only forces on the control volume are the external reaction F and the resultant pressure force. Working in gauge pressures the only control surface with a pressure force is that through the wake, so the resultant force in the flow direction for unit width is

$$F - \int_{-\delta}^{+\delta} (p - p_0) \, \mathrm{d}y$$

and this will equal the change of momentum flow rate in a similar expression to that derived for the boundary layer i.e.,

$$F - \int_{-\delta}^{+\delta} (p - p_0) \, \mathrm{d}y = \int_{-\delta}^{+\delta} \rho u (u - U) \, \mathrm{d}y$$

or

$$F = \int_{-\delta}^{+\delta} (p - p_0) \, \mathrm{d}y + \int_{-\delta}^{+\delta} \rho u (u - U) \, \mathrm{d}y$$

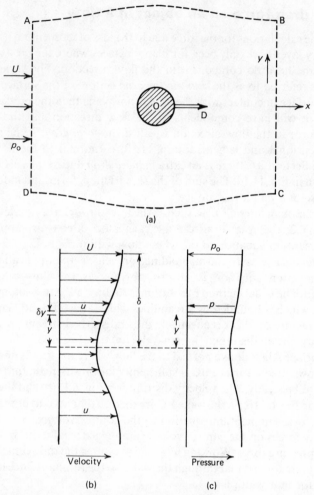

Fig. 6.11 The drag force on an object O in a fluid flow is accompanied by a reduction of velocity in the wake of the object. (a) Control volume. (b) Velocity u in the wake. The velocity is unchanged at the original value U at a distance δ from the line of symmetry. (c) Pressure in the plane wake on control surface BC.

As D, the drag force on the object, is equal and opposite to the reaction F, and $p_0 > p$, $U > u$ it is more usual to write the equation as

$$-F = D = -\int_{-\delta}^{+\delta} (p - p_0)\,\mathrm{d}y - \int_{-\delta}^{+\delta} \rho u(u - U)\,\mathrm{d}y$$

or

$$D = \int_{-\delta}^{+\delta} (p_0 - p)\,\mathrm{d}y + \int_{-\delta}^{+\delta} \rho u(U - u)\,\mathrm{d}y.$$

Example

A submarine cruising horizontally far below the water surface leaves a cylindrical wake in which the velocity may be taken to vary linearly from half the vessels speed of 10 m s^{-1} at its axis to zero at a radius of 10 m. The pressure in the wake may be assumed to be hydrostatic. If the density of sea water is 1025 kg m^{-3} calculate the drag of the submarine and the rate at which work is being done on it by the propulsion system.

Solution

The information is presented as seen by a stationary observer, and, hence appears as an unsteady flow problem, Fig. 6.12(a). If the flow is considered relative to the submarine then a control volume can be drawn to which the force–momentum equation can be applied, Fig. 6.12(b).

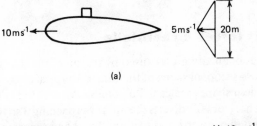

(a)

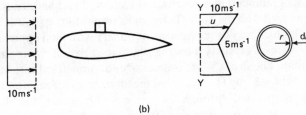

(b)

Fig. 6.12 The flow through a control volume used to calculate the drag of a submarine.

The change of momentum flow rate through the control volume is equal to the drag force on the submarine, D, i.e.

$$D = \int \rho u (U - u)\, \mathrm{d}a.$$

As the wake is axially symmetrical

$$\mathrm{d}a = 2\pi r\, \mathrm{d}r,$$

and for $0 \leqslant r \leqslant 10$ m, $u = 5\left(1 + \dfrac{r}{10 \text{ m}}\right)$ m s^{-1}.

Hence $D = \int_0^{10\text{ m}} 1025 \times 5\left(1 + \frac{r}{10\text{ m}}\right)\left[10 - 5\left(1 + \frac{r}{10\text{ m}}\right)\right]2\pi r\,dr\,\frac{\text{kg}}{\text{m}^3}\frac{\text{m}^2}{\text{s}^2}$

$$= 51\,250\pi \int_0^{10\text{ m}} \left(1 + \frac{r}{10\text{ m}}\right)\left(1 - \frac{r}{10\text{ m}}\right)r\,dr\,\frac{\text{N}}{\text{m}^2}$$

$$= 51.25\pi \int_0^{10\text{ m}} \left(r - \frac{r^3}{100\text{ m}^2}\right)dr\,\frac{\text{kN}}{\text{m}^2}$$

$$= 51.25\pi \left[\frac{100}{2} - \frac{10^4}{400}\right]\text{kN} = 4.025\text{ MN}.$$

Rate at which work is being done on submarine $= D \times U$

$$= 4.025\text{ MN} \times 10\,\frac{\text{m}}{\text{s}} = 40.25\text{ MW}.$$

6.7 Drag and lift forces on aerofoils

The lift on the wings of an aircraft is caused by reason of the changes of air velocity due to the cross-sectional shape of the wings. A typical cross section is shown in Fig. 6.13 (these shapes are called *aerofoils*). An aerofoil is merely a body which if moving horizontally diverts the fluid approaching it so that there is a change of momentum in a vertical direction; this change causes a vertical force to be exerted on the body. Take, for example, the aerofoil section shown in Fig. 6.13, which may be regarded as stationary with fluid approaching it at a velocity U. (This is precisely the same case as the aerofoil moving at velocity U into still fluid.) The shape, which is necessarily asymmetrical with respect to the direction of the velocity U, is such that the fluid leaves the aerofoil at velocity u, having been diverted through an angle θ. Originally the fluid has no momentum in the lift or y-direction (at right angles to that of U), but finally has a momentum of $\int \rho u^2 \sin\theta\,dy$ in that direction. This change of momentum requires a force L to accomplish it and this is the lift on the aerofoil, which appears as an excess of pressure on the lower surface over the pressure on the upper surface.

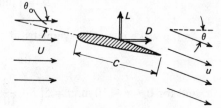

Fig. 6.13 An aerofoil is a device for producing a change of momentum in the oncoming fluid. The flow direction is changed by the angle θ, which is not necessarily the same for all streamlines.

Theoretically, then, the lift on an aerofoil can be found by an experiment measuring u and θ, and evaluating the integral $L = \int \rho u^2 \sin \theta \, \mathrm{d}y$. Actually it is difficult to determine θ sufficiently accurately, because the flow behind an aerofoil is always unsteady; the direction of flow undergoes great and sudden changes, so making it difficult to measure the mean value. The experimental determination of the lift is therefore usually done directly by fixing the aerofoil to a balance in a wind-tunnel, or by finding the pressure intensity at a number of places around the aerofoil and integrating the elementary pressure forces so found (see Chapter 2).

It is usually found that u is not greatly different in magnitude from U for a well-designed aerofoil, so that if θ' is the *mean* angle over which the fluid is deflected, L is proportional to $U^2 \sin \theta'$. It is also evident that θ' will be fixed by θ_0, the angle that the aerofoil itself makes with the direction of U (this angle is called the *angle of incidence*). An aircraft of a given weight must therefore fly at a larger θ_0 at low speeds than at high speeds, so that the product $U^2 \sin \theta'$ remains constant. At the lowest speed of all, when the aircraft is landing, θ_0 may have to be so large that a radical change of the flow pattern takes place round the aerofoil, and breakaway would occur if the shape of the aerofoil were not changed (see Chapter 4). If breakaway does occur the air is not then deflected on the average by such a large angle δ, so the lift suddenly decreases and the aircraft *stalls*. Flaps (see Fig. 6.14) and slats are often fitted to the aerofoil so as to increase θ' locally. In this way the lift may be preserved at low speed, and control of the aircraft retained.

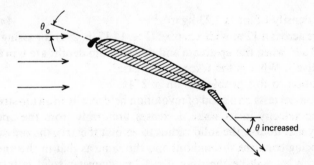

Fig. 6.14 A slat and flap on an aerofoil at a large angle of incidence diverts the air through a large angle and preserves the lift at low forward speeds.

The design and testing of aerofoils forms a highly specialized part of fluid mechanics practice. It is clearly advantageous to make aerofoils to give a high value of L/D, but other considerations often have to be taken into account such as the change of position of the lift force on the wing with incidence and behaviour near the stall.

Problems

1. (a) A nozzle discharges $0.08 \, \text{m}^3 \, \text{s}^{-1}$ of water at a velocity of $16 \, \text{m s}^{-1}$ against a large flat plate normal to the jet. What force is required to hold the plate still? *Ans.* 1.28 kN

   (b) If the same jet strikes a curved vane tangentially and all the water is deflected through $35°$, what force is required to hold the vane? *Ans.* 768 N at $72\frac{1}{2}°$ to jet dirn

   (c) What would be the force on the curved vane if it is moved away from the direction of the jet at $5 \, \text{m s}^{-1}$? *Ans.* 363 N

2. Part of an oil pipe-line consists of two straight, horizontal sections joined by a $30°$ bend. The pipe has an internal diameter of 0.6 m. The bend is anchored by a block of concrete in order to relieve the loads on the connections between the bend and the straight sections of pipe.

   (a) What horizontal force must the block exert when the pipe is full of oil, stationary, at a gauge pressure of $54 \, \text{kN m}^2$? *Ans.* 7.9 kN

   (b) What additional horizontal force must the block exert if the oil flows with a mean speed of $4.2 \, \text{m s}^{-1}$? How will the pressure distribution on the inside surface of the bend compare with that in (a)? *Ans.* 2.2 kN

   (c) State the direction of the forces in (a) and (b). Assume the density of the oil is $845 \, \text{kg m}^{-3}$.

3. The wake behind a two-dimensional body in an air-stream of $40 \, \text{m s}^{-1}$ is 10 cm wide. Assuming that the speed in it decreases linearly in each half of the wake to $20 \, \text{m s}^{-1}$ at the centreline, estimate the drag/metre width of the body.

   The density of air is $1.23 \, \text{kg m}^{-3}$ *Ans.* 32.8 N

4. A sluice across a 12 m wide channel (Fig. 14.17, p. 291) is passing a flow of $142 \, \text{m}^3 \, \text{s}^{-1}$ when the upstream and downstream depths are 6 m and 1.2 m respectively. What is the force on the gate?
   (A solution to this problem is on p. 274).

5. Wind-tunnel tests on a solid of revolution held axially in an air-stream show that the velocity in its wake decreases uniformly from the undisturbed velocity u at double the solid radius to zero at the axis, the pressure in the wake being constant throughout and the same as that in the undisturbed stream. What will be the drag if a 1.2 m diameter solid is travelling at $100 \, \text{m s}^{-1}$ through still air at 500 mm mercury pressure and $0°\text{C}$? *Ans.* 6470 N

   If the pressure in the wake decreases uniformly towards the axis, being $\frac{1}{8}\rho u^2$ below the undisturbed pressure there, what is now the drag? *Ans.* 8130 N

7
Pressure intensity and velocity changes in moving fluids

7.1 Flow in a single streamtube

It is shown in Chapter 6 that a force is exerted when there is a change in the flow of momentum of a moving fluid. This force, calculated by means of the force–momentum equation is the total of several component forces all acting on the control volume where the momentum change takes place. As well as this total force, the engineer often requires more detailed knowledge of the distribution of both pressure intensity and velocity. A relationship between velocity and pressure intensity is thus required, and it can be obtained for certain flow conditions.

Consider a control volume which is a part of a streamtube in a steady non-uniform flow of a fluid. The velocity of flow is uniform over any one cross section of the streamtube, though further along the streamtube, and in adjacent streamtubes, the velocity may be different. In a length δs of the streamtube, the cross-sectional area changes from a to $a + \delta a$; the velocity changes from u to $u + \delta u$; the density from ρ to $\rho + \delta\rho$; the pressure intensity changes from p to $p + \delta p$; and the height of the centroid of the end planes changes from z above an arbitrary datum level to $z + \delta z$ (see Fig. 7.1). There is a mean shear stress τ on the surface in the direction of flow and an external force δF. Notice how all the variables are assumed to increase in the direction of motion. There can be no flow into or out of the streamtube except through the ends, so that the changes in cross-sectional area are connected with changes in u and ρ. Hence for steady flow the mass continuity equation is

$$(\rho + \delta\rho)(a + \delta a)(u + \delta u) = \rho a u,$$

which, taking first order infinitesimals, becomes

$$\rho a\,\mathrm{d}u + \rho u\,\mathrm{d}a + a u\,\mathrm{d}\rho = 0.$$

The total force in the direction of motion has five components. Firstly, the varying pressure intensities on the sides of the tube cause a force F_1 acting in the direction of motion. If it is assumed that δs is small, then F_1 is the product of the *mean* pressure intensity on the sides and the projected area in the desired direction, δa (see Case 2 of Chapter 2).

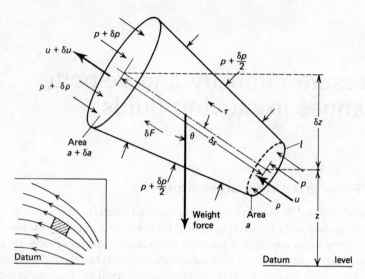

Fig. 7.1 Small portion of a streamtube with flow from right to left. Pressure intensities, velocity and areas are marked, and the positive direction of increase of all variables is the direction of motion. The inset figure shows a general view of the streamlines surrounding the element of streamtube in the larger diagram.

That is
$$F_1 = \delta a \left(p + \frac{\delta p}{2} \right) = p\delta a,$$

ignoring the second order term.

Secondly, the pressure intensities p and $p + \delta p$ exert forces on the plane ends of the streamtube in opposite directions, so that the force in the *direction of motion* is F_2 where

$$F_2 = -(p + \delta p)(a + \delta a) + pa$$
$$= -p\delta a - a\delta p.$$

Thirdly, the element of streamtube has a weight force F_3 in the direction of motion, because its mass is $\rho (a + \delta a/2) \, \delta s$; the total weight force acts vertically downwards so that the component required, F_3, is

$$F_3 = -\rho \left(a + \frac{\delta a}{2} \right) \delta s \, g \cos \theta,$$

where θ is the angle between the element's axis and the vertical.

But
$$\cos \theta = \frac{\delta z}{\delta s}$$

so that
$$F_3 = -\rho \left(a + \frac{\delta a}{2} \right) \delta s \, g \frac{\delta z}{\delta s}$$
$$= -\rho a g \delta z,$$

ignoring the second order term. The fourth, shear, force depends on the area of the surface of the streamtube projected in a direction normal to the flow. If l is the perimeter of the streamtube then

$$F_4 = \tau l \delta s,$$

while the final, external, force is

$$F_5 = \delta F.$$

As the change of momentum flow rate for the control volume is the product of the mass flow rate and the change of velocity, applying the force–momentum equation to the element of the streamtube gives

$$\rho a u \delta u = F_1 + F_2 + F_3 + F_4 + F_5$$
$$= p \delta a + (-p \delta a - a \delta p) + (-\rho a g \delta z) + \tau l \delta s + \delta F$$

Replacing the finite increments by infinitesimals and ignoring second order terms

$$\rho a u \, du = -a \, dp - \rho a g \, dz + \tau l \, ds + dF$$

For convenience, divide throughout by $\rho a g$

$$\frac{u}{g} \, du + \frac{dp}{\rho g} + dz - \frac{\tau l \, ds}{\rho a g} - \frac{dF}{\rho a g} = 0.$$

Integrating all terms on both sides this becomes

$$\frac{u^2}{2g} + \int \frac{dp}{\rho g} + z - \int \frac{\tau l}{\rho a g} \, ds - \int \frac{dF}{\rho a g} = \text{Constant} \qquad (7.1)$$

For many engineering purposes a liquid may always, and a gas may sometimes, be considered to be incompressible; that is, ρ does not vary with p. In this case, a simplification can be made, as the second term can now be integrated so that the equation reads

$$\frac{u^2}{2g} + \frac{p}{\rho g} + z - \int \frac{\tau l}{\rho a g} \, ds - \int \frac{dF}{\rho a g} = \text{Constant}$$

This equation appears complex, but in one special case can be greatly simplified. If the streamtube considered lies outside all boundary layers and wakes of the flow the shear stress, τ, will be zero, and if no external force is applied the equation can be written

$$\frac{u^2}{2g} + \frac{p}{\rho g} + z = \text{Constant} = H \qquad (7.2)$$

This equation is sometimes called *Bernoulli's equation* after its first enunciator, and applies only to incompressible fluids in steady flow. For compressible fluids under similar flow conditions (7.1) becomes

$$\frac{u^2}{2g} + \int \frac{dp}{\rho g} + z = \text{Constant}$$

and is known as *Euler's equation.*

It will be seen that each term of the Bernoulli equation has the dimensions of a length, so that it can be regarded as a quantity of energy in a volume of fluid of unit weight. H is often called the *Total Head* of the fluid in the particular streamtube concerned. Since it will be seen that the distance s along the streamline does not appear above, H does not vary with s, and Bernoulli's equation can be thus stated—that along any one streamtube of incompressible fluid the Total Head remains constant if the only forces are pressure and weight forces.

The final two terms in equation 7.1 are more difficult to evaluate than the terms which comprise the total head. The shear and external forces are non-conservative, i.e. they involve either the addition or removal of energy from the fluid or the conversion of mechanical to internal energy or 'heat' irreversibly. In contrast a motion described accurately by Bernoulli's equation is conservative (or 'reversible' in thermodynamic terminology) and balances the changes of kinetic energy, represented by $u^2/2g$, and potential energy, Z, with the *flow work*, $P/\rho g$, performed by the pressure forces.

The shear and external force terms cannot be generalized and the effects of the shear forces and the work done on or by the flow by means of external forces has to be evaluated for each flow situation. Particular examples of these effects will be considered in the cases of the flow in pipes (Chapter 13), open channels (Chapter 14), and hydraulic machinery (Chapter 15). In the absence of any particular information it is usual to represent the combined shear force and external force effects in equation 7.1 by, E, thus allowing the derivation of the *incompressible steady flow energy equation*

$$\frac{u^2}{2g} + \frac{p}{\rho g} + Z + E = \text{Constant}. \tag{7.3}$$

Writing the energy equation, as it will be referred to from now on, for two control surfaces at A and B in the same streamtube gives

$$\frac{u_A{}^2}{2g} + \frac{p_A{}^2}{\rho g} + Z_A = \frac{u_B{}^2}{2g} + \frac{p_B}{\rho g} + Z_B + E$$

or
$$H_A = H_B + E.$$

It should also be understood that strictly the value of H applies only to one streamtube. Under special circumstances H may indeed be the same for several adjacent streamtubes but in general it is not, when the total energy changes from place to place across the direction of motion, i.e. the total head is non uniformly distributed.

7.2 Limitations of Bernoulli's equation

If it is desired to apply Bernoulli's equation to find velocity changes from pressure changes, or vice versa, it is necessary to limit the applications to cases of steady flow of incompressible fluid wherein there is no change of total head along any streamtube. If these limitations are observed, then between two places A (upstream) and B (downstream), both on one streamtube,

Total head at A = Total head at B

or
$$\frac{u_A{}^2}{2g} + \frac{p_A}{\rho g} + z_A = \frac{u_B{}^2}{2g} + \frac{p_B}{\rho g} + z_B$$

or
$$\frac{u_A{}^2 - u_B{}^2}{2g} + \frac{p_A - p_B}{\rho g} + z_A - z_B = 0 \tag{7.4}$$

In this way it will be seen that a change of pressure p is accompanied by changes of velocity or height or both.

In engineering practice, the limitation which proves to be the most onerous is that which prohibits shear forces, for real fluids are all viscous to some extent and produce shear forces in the direction of motion. These forces cause energy to be used in overcoming them, this energy being degraded into a form of energy which is not included in the Bernoulli equation. This is low-grade thermal energy, which cannot be reconverted at a later stage into any of the three forms of energy included in H (see footnote on p. 45). The total head H is therefore diminished as energy is degraded, and the steady flow energy equation (7.3) must be used to analyze the flow with the loss of total head represented by E. In general, E must be found experimentally or by more advanced theoretical reasoning but if E is small compared with the other terms in the equation then the Bernoulli equation may still be an acceptable approximation, particularly if an empirical correction is used to make allowance for the small discrepancy (Chapter 8).

If the fluid concerned is a gas, then some of the thermal energy is taken into account when it changes the properties of the fluid by the Gas Laws. It is usually assumed that no heat is allowed to escape through the boundaries of the flow, and an energy balance equation is derived for compressible fluids which corresponds to the Bernoulli equation for incompressible flow.

Shear forces occur in both laminar and turbulent flows. In the former, wherever a velocity gradient du/dy is formed, a shear stress $\tau = \mu \frac{du}{dy}$ appears with it. In the latter, shear forces and their associated degradation of energy occur by reason of the irregular motions in the flow. There are, at any one instant, places where the instantaneous velocity is different from the mean velocity. On a river in flood, for example, it is easy to see from a bridge overhead that there are patches of water which move for a short time relative to the surrounding water. Eventually the motions of these patches die out and are replaced by those of a quite different arrangement of patches. There is a tendency too for these patches to have rotary motion within them, so that, in general, turbulent motion is said to have *eddies* in it. Eddies have comparatively large velocity gradients du/dY at their boundaries (Y is a direction at right angles to the velocity u at a certain point). These gradients, by the definition of viscosity in Chapter 1, produce local shear forces $\tau = \mu \dfrac{du}{dY}$ in the fluid and cause a consequent degradation of energy into heat, which is usually much greater than the degradation if laminar flow was occurring at the same velocity. Sometimes the degradation of energy E is sufficiently small that it can be neglected, even though the flow is turbulent, in which case the simple Bernoulli equation (7.2) may be used. One of these occasions is when the cross-sectional area of the flow is decreasing; for reasons that will be discussed in Chapter 12 on the Boundary Layer, it is common experience to find that if the streamlines representing a flow are converging in the direction of the motion, then the turbulence is decreased, so that the energy degradation is small. Consequently, in a convergence, Bernoulli's equation predicts pressure and velocity changes accurately enough for most engineering purposes. If, however, a flow is diverging in the direction of motion, then the turbulence is increased, energy degradation is large, and the Bernoulli equation gives quite misleading results.

Example.

Calculate the total head of the flow at the entry and exit of the nozzle described on page 86 if the inlet of the nozzle is 0.17 m above its exit. Comment on the applicability of the Bernoulli equation to such a flow.

Solution

It is most convenient to take the level of the exit to be the datum for measurements of height and to work in gauge pressures. Hence at the outlet the total head is only due to the velocity

$$H_2 = \frac{u_2{}^2}{2g} + \frac{p_2}{\rho g} + Z_2 = \frac{7.33^2}{2 \times 9.81} \frac{\text{m}^2\,\text{s}^{-2}}{\text{m}\,\text{s}^{-2}} = 2.74 \text{ m.}$$

At the inlet the total head is

$$H_1 = \frac{u_1{}^2}{2g} + \frac{p_1}{\rho g} + z$$

$$= \frac{1.84^2}{2 \times 9.81} \, \text{m} + \frac{2.55 \times 10^4}{10^3 \times 9.81} \frac{\text{N m}^{-2}}{\text{kg m}^{-3} \, \text{m s}^{-2}} + 0.17 \, \text{m}$$

$$= 0.17 + 2.60 + 0.17 = 2.94 \, \text{m}.$$

Comparing the inlet and outlet total head it can be seen that there is a decrease of

$$\frac{2.94 - 2.74}{2.94} \times 100 = 6.8 \%$$

of total head between the two control surfaces which indicates that the Bernoulli equation cannot be applied exactly. The reasons are probably two. Firstly some work will be done against the shear forces. Secondly the total head has been calculated as if the whole flow were a single streamtube, i.e. the flow is uniformly distributed at each control surface. This will make little difference at the inlet where the velocity head is small, but it will cause an underestimate of the total head at the exit. See Chapter 4 page 51.

7.3 Power requirements

Consider a flow where all the fluid undergoes a change ΔH of its total head H as it passes from one cross section to another further downstream. Such a change would be negative (decreasing H) when going downstream if there is a degradation of energy into heat and work is done by the flow as in a turbine; or it would be positive (increasing H) going downstream if work is being done on the fluid by a pump. The change $E = \Delta H$ is an energy change per unit weight of fluid, so that if the discharge of the flow is Q (volume per unit time), then the mass flow is ρQ, and the weight of fluid passing a given point in unit time is $g\rho Q$. Since each unit of weight of the fluid changes its head by ΔH, then the rate at which energy is being degraded or given to the fluid is $g\rho Q \Delta H$ per unit time. Such a rate of change of energy is the *power* degraded or supplied.

$$\text{Power} = \rho g Q \Delta H \tag{7.5}$$

Power will be required by a flow in a number of circumstances. If the potential energy z of the fluid is increased (that is, the flow is uphill) while the pressure and velocity remain constant, then power must be supplied to the fluid to achieve this increase of energy. In addition, there will be a degradation of energy to heat, caused by the friction at the solid boundaries of the flow. Power will therefore also be required to balance this drain of energy from the fluid.

Example

The nozzle described in the previous example has its water supply pumped from a sump in which the water surface is 6 m below the nozzle exit. What is the minimum power which would have to be supplied to the flow by the pump? If the water eventually flowed back into the sump after leaving the nozzle, by how much would its temperature have risen?

The specific heat of water is $4.2 \, \text{kJ kg}^{-1} \text{K}^{-1}$.

Solution

The total head of the flow as it leaves the nozzle measured relative to the sump surface $= 6.0 + 2.74 = 8.74 \, \text{m}$. This is the minimum head which the pump must supply.

$$\text{Minimum power} = \rho g Q \Delta H$$
$$= 10^3 \, \text{kg m}^{-3} \times 9.81 \, \text{m s}^{-2} \times 0.034 \, \text{m}^3 \, \text{s}^{-1} \times 8.74 \, \text{m}$$
$$= 2915 \, \text{kg m}^2 \, \text{s}^{-3.}$$
$$= 2915 \, \text{N m s}^{-1}$$
$$= 2915 \, \text{W} = 2.915 \, \text{kW}.$$

When the water has returned to the sump the power supplied by the pump will all have gone into raising the temperature of the water, i.e. increasing its internal energy, hence

$$\text{Power} = \rho Q C_P \Delta T$$
$$= 10^3 \, \text{kg m}^3 \times 0.034 \, \text{m}^3 \, \text{s}^{-1} \times 4.2 \, \text{kJ kg}^{-1} \text{K}^{-1} \Delta T$$
$$= 142.8 \, \text{kJ s}^{-1} \text{K}^{-1} \Delta T = 142.8 \, \text{kW K}^{-1} \Delta T$$
$$= 2.915 \, \text{kW}$$

and
$$T = \frac{2.915}{142.8} \, \text{K} = 0.02 \, \text{K}$$

Notice how very small the rise in temperature is. It takes the same energy to heat water by 1 K as it does to raise it some 428 m.

Problems

1. Find the hydraulic forces acting on a 90° reducing bend joining two pipes (30 cm and 15 cm bores) when the pressure is $150 \, \text{kN m}^{-2}$ in the 30 cm pipe and when
   (a) there is no flow, and
   (b) when the flow is $0.3 \, \text{m}^3 \, \text{s}^{-1}$ of water.
   Note In (b) Bernoulli must be used for pressure intensities.
   Ans. (a) 2700 N: 10 800 N. (b) 11 870 N: 7170 N.
2. A horizontal water-pipe reduces from 50 cm diameter to 30 cm diameter. The pressure at the downstream end of the reducer is $160 \, \text{kN m}^{-2}$ and the flow is $0.3 \, \text{m}^3 \, \text{s}^{-1}$. What is the pressure at the upstream end, if the energy degraded in the reducer is $1.0 \, \text{J N}^{-1}$.   *Ans.* $162 \, \text{kN m}^{-2}$.
3. A stream of fluid of density $800 \, \text{kg m}^{-3}$ has energy taken from it between the points A and B at the rate of 75 kW. At A, the velocity is $1.5 \, \text{m s}^{-1}$ and

the cross-sectional area is $1\,\text{m}^2$; at B, 1 m higher, the area is $0.4\,\text{m}^2$. What is the absolute pressure at B when the pressure at A is (i) atmospheric, (ii) $150\,\text{kN}\,\text{m}^{-2}$ gauge? What pressure at A gives cavitation pressure at B?

$Ans.$ (i) $24.4\,\text{kN}\,\text{m}^{-2}$ absolute; (ii) $174.4\,\text{kN}\,\text{m}^{-2}$ absolute; $75.6\,\text{kN}\,\text{m}^{-2}$ absolute.

4. Water is to be pumped, at the rate of $0.03\,\text{m}^3\,\text{s}^{-1}$, from a river through a 10 cm diameter pipe which discharges to atmosphere at a height of 7 m above the river. The combined efficiency of the pump and the pipe is 25 per cent. What power will be required to drive the pump?

Note Remember velocity energy. *Ans.* 9.1 kW.

5. If a circular cylinder moves sideways through an ideal fluid, show from the equation to the streamline pattern (Chapter 5) that the relative velocity at the surface of the cylinder is $2U \sin \theta$, where U is the relative velocity of the cylinder axis with respect to the still fluid. Determine the pressure distribution around the surface, and the force tending to separate the two halves of the cylinder if cut on the diameter in the plane of the motion.

$$Ans.\ \frac{p - p_0}{\frac{1}{2}\rho U^2} = 1 - 4 \sin^2 \theta; \quad F = p_0 d - \frac{5d}{6}\rho U^2.$$

6. A source, $0.5\,\text{m}^3\,\text{s}^{-1}$, is placed 20 cm directly upstream of a sink, $0.25\,\text{m}^3\,\text{s}^{-1}$, in a uniform stream of $2\,\text{m}\,\text{s}^{-1}$ and 30 cm deep. Plot the streamlines and determine how far the stagnation point is distant upstream from the source. Find the velocity and the difference in pressure from the undisturbed stream pressure at two points on the streamline passing through the stagnation point, one point being 45 degrees forward of the source and the other level with it.

7. Show that in a hydrostatic fluid of constant density $(p/\rho g) + z$ is a constant everywhere. How is this modified when the fluid is moving?

8

Applications of Bernoulli's equation

8.1 Flow in a converging pipe

Consider a steady incompressible flow in a pipe which decreases from a cross-sectional area a_1 to a_2 in a distance not much greater than about $3\sqrt{a_1}$ (Fig. 8.1). (This restriction is necessary to avoid consideration of long, gently tapering pipes wherein the shear forces become of importance, rendering inaccurate the estimate of the pressure changes which will be derived.) The pressure of the fluid at these cross sections can be found by connecting suitable manometers to *tapping holes* in the pipe walls. The fluid velocity, which is assumed uniform and parallel over each cross section, increases from u_1 to u_2 where $u_1 a_1 = u_2 a_2 = Q$, the discharge of fluid through the pipe. It is assumed that the pressure distribution at either section varies with depth in the same way as occurs in a static fluid, so that $(p/\rho g) + z$ is constant all over any one cross section. Since the velocity is uniform also, then the total head

$H = \dfrac{p}{\rho g} + z + \dfrac{u^2}{2g}$ is the same for every streamtube passing each section;

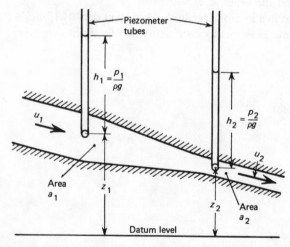

Fig. 8.1 Flow in a converging pipe. The piezometer tubes show the pressure at two places.

114

furthermore, since a convergence of the flow decreases the turbulence, and reduces the energy degradation to a negligible amount, the total head H is constant along every streamline, so that Bernoulli's equation may be directly applied between the two sections. That is

$$\frac{u_1{}^2}{2g} + \frac{p_1}{\rho g} + z_1 = \frac{u_2{}^2}{2g} + \frac{p_2}{\rho g} + z_2$$

or

$$\frac{u_2{}^2 - u_1{}^2}{2g} = \left(\frac{p_1 - p_2}{\rho g}\right) + (z_1 - z_2).$$

But

$$u_1 = u_2 \frac{a_2}{a_1},$$

so that

$$\frac{u_2{}^2}{2g}\left(1 - \left(\frac{a_2}{a_1}\right)^2\right) = \left(\frac{p_1 - p_2}{\rho g}\right) + (z_1 - z_2)$$

or

$$u_2 = \left\{1 - \left(\frac{a_2}{a_1}\right)^2\right\}^{-\frac{1}{2}} \sqrt{\left(2g\left\{\frac{(p_1 - p_2)}{\rho g} + (z_1 - z_2)\right\}\right)}$$

and

$$Q = a_2 u_2 = a_2 \left\{1 - \left(\frac{a_2}{a_1}\right)^2\right\}^{-\frac{1}{2}} \sqrt{\left(2g\left\{\frac{(p_1 - p_2)}{\rho g} + (z_1 - z_2)\right\}\right)}$$

Actually, of course, there is a small energy degradation due to the shear forces even when the convergence is well designed. Further, it has been assumed that uniform flow exists at both cross sections, although this is nearly true at the downstream section, it is not so at the upstream one, as there is a boundary layer in the pipe, with the fluid close to the walls travelling more slowly than that at the centre. The total energy at the upstream section is therefore not the same for all streamlines: it is lower for ones near the wall than for ones at the centre. However, it is convenient to retain $\dfrac{u_1{}^2}{2g}$ in the Bernoulli equation, where u_1 is now the *mean* velocity of flow $\bar{u}$, even though the actual mean velocity energy is a few per cent higher than $\bar{u}^2/2g$ (see example in Chapter 4). The error so incurred, together with the error due to the ignored energy degradation, is compensated by introducing a numerical coefficient C_d, called the *coefficient of discharge*, so that if Q is the actual discharge

$$Q = C_d a_2 \left\{1 - \left(\frac{a_2}{a_1}\right)^2\right\}^{-\frac{1}{2}} \sqrt{\left(2g\left\{\frac{(p_1 - p_2)}{\rho g} + (z_1 - z_2)\right\}\right)} \qquad (8.1)$$

Experiments conducted with well-shaped convergences in pipes, having long straight sections upstream, and air or water as fluids, show that C_d is about 0.98; in other words, the measured discharge through a convergence is about 98 per cent of that predicted by the Bernoulli equation which has been developed for flow in a single streamtube without shear forces. This constant and high proportion is fortunate, for it allows the engineer to use contractions as meters in pipelines to measure the quantity of fluid passing. It is only necessary to know the pressure difference between two cross sections of the convergence; this is done by drilling small holes (tapping points) in the wall of the pipe and connecting them to a suitable differential manometer. A knowledge of the reading of the manometer, $(p_1 - p_2)/\rho g$, together with the constants $(z_1 - z_2)$, a_1 and a_2, enables the discharge Q to be computed.

Contractions in pipelines from one diameter pipe to another, Fig. 8.2 (*a*), are, however, rare because it is always desirable to keep the mean speed of the fluid fairly low to avoid undue degradation of energy to heat by the shear forces in the flow (see Chapter 13). If a convergence is to be used for measuring the discharge, then it is desirable to return the fluid to its original, lower speed as quickly as possible downstream of the convergence. Such an arrangement, Fig. 8.2 (*b*), is called a *Venturi meter*, of which the smallest cross section is called the *throat*.

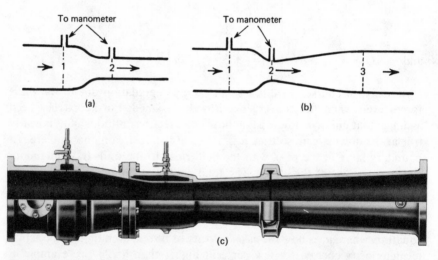

Fig. 8.2 (a) A contraction in a pipe used as a flow-measuring apparatus. (b) The pipe downstream of a contraction is expanded again to form a *Venturi meter.* (c) A sectioned Venturi meter: Notice that the pressure at both the upstream pipe position and at the throat is taken at several tapping points around the periphery which are connected to common annular passages, and thence to the manometer (Photo by G. Kent Ltd.)

In the diverging cone downstream of the throat there are conditions where it is difficult to prevent separation and the production of eddies, and there is consequently an energy degradation. In this diverging outlet cone, the energy equation can be applied, that is

$$\frac{u_1{}^2}{2g} + \frac{p_1}{g\rho} + z_1 = \frac{u_3{}^2}{2g} + \frac{p_3}{g\rho} + z_3 + E.$$

since $\qquad u_1 = u_3, \qquad \dfrac{p_1 - p_3}{g\rho} = E + (z_3 - z_1).$

Thus there will always be a drop in pressure between the ends of the meter, and the amount of degraded energy represented by this drop in a large Venturi meter may be of economic importance. Care is therefore taken to make E a minimum. Good design of the divergence can make E about 10 per cent of $u_2{}^2/2g$ and this is done with a cone of semi-vertex angle of about 6°. A longer cone of smaller angle gives higher values of E; this is because the degradation of energy by the ordinary fluid friction forces in the longer cone now becomes larger than the saving of energy due to the more gentle divergence.

A Venturi meter may be made for any size pipe, though those in large water mains are expensive pieces of equipment. It is not necessary for either the pipe or the throat of the meter to be circular, as only the areas a_1 and a_2 are important. One way of making Venturi meters in large pipes is to insert a longitudinal diaphragm as in Fig. 8.3, thus giving a somewhat D-shaped throat section.

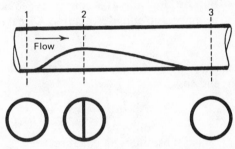

Fig. 8.3 Venturi meter made by welding a plate into the bore of a pipe so that the throat section is D-shaped.

Example

A small Venturi meter has entrance and exit areas of $20\ \text{cm}^2$ and a throat area of $3.6\ \text{cm}^2$. The meter is tested with water when fitted with two mercury U-tube manometers, the first being connected between the entrance of the meter and the throat, the second between the entrance and the exit. In a typical test the flow rate was found to be $4.8\ \text{litre s}^{-1}$ while the manometers showed differences of mercury–water surface levels of 74.8 cm between the entrance and the throat, and 17.0 cm between the entrance and exit.

Using the results of the test estimate the coefficient of discharge of the meter at this flow rate and the rate of loss of energy between the throat and the exit of the meter. Compare this with the rate of loss of energy between the inlet and the throat of the meter.

Solution

Although it is possible to apply equation 8.1 directly the calculation of the discharge coefficient can be made simply from first principles. Assume the flow is steady, inviscid, imcompressible and the velocities uniformly distributed at each control surface and use the Bernoulli equation to derive a calculated flow Q_c. Applying Bernoulli's equation between the entrance, 1, and throat, 2, assuming the meter to be horizontal

$$\frac{p_1}{\rho g} + \frac{u_1'^2}{2g} = \frac{p_2}{\rho g} + \frac{u_2'^2}{2g},$$

whence
$$u_2'^2 - u_1'^2 = \frac{2(p_1 - p_2)}{\rho}, \tag{8.2}$$

where ρ is the density of water. But $(p_1 - p_2)$ is measured by the mercury–water manometer, so $p_1 - p_2 = (\rho_m - \rho)gh$, (see page 30) where ρ_m is the density of mercury, and h the difference in level of the interfaces.

$$\frac{p_1 - p_2}{\rho} = \frac{(\rho_m - \rho)}{\rho} gh = 12.6 \times 9.81 \text{ m s}^{-2} \times 0.748 \text{ m} = 92.4 \text{ m}^2 \text{ s}^{-2}.$$

But by applying the continuity equation for the steady flow of an incompressible fluid, (equation 4.9 page 52)

$$a_1 u_1' = a_2 u_2'$$
$$20\, u_1' = 3.6\, u_2',$$

and so

$$u_2'^2 - u_1'^2 = u_2'^2 \left(1 - \left(\frac{u_1'}{u_2'}\right)^2\right) = u_2'^2 \left(1 - \left(\frac{3.6}{20}\right)^2\right)$$

Hence substituting in equation 8.2

$$u_2'^2 \left(1 - \left(\frac{3.6}{20}\right)^2\right) = 2 \times 92.4 \text{ m}^2 \text{ s}^{-2}$$

and so
$$u_2' = 13.8 \text{ m s}^{-1}.$$

So calculated flow, $Q_c = a_2 u_2' = 13.8 \times 3.6 \times 10^{-4} = 4.975 \times 10^{-3} \text{ m}^3 \text{ s}^{-1}$. Actual flow $Q_a = 4.8$ litre s$^{-1} \times 10^{-3} = 4.8 \times 10^{-3} \text{ m}^3 \text{ s}^{-1}$.

Hence
$$C_d = \frac{Q_a}{Q_c} = \frac{4.8}{4.975} = 0.965.$$

To find rate of loss of energy after the throat consider the flow between the throat, 2, and the exit, 3, and apply the energy equation using the actual flow velocities.

$$\frac{u_2^2}{2} + \frac{p_2}{\rho} = \frac{u_3^2}{2} + \frac{p_3}{\rho} + gE.$$

Whence $gE = \dfrac{u_2^2 - u_3^2}{2} + \dfrac{p_2 - p_3}{\rho}$

Now $\qquad u_2 = \dfrac{4.8 \times 10^{-3} \text{ m}^3 \text{ s}^{-1}}{3.6 \times 10^{-4} \text{ m}^2} = 13.33 \text{ m s}^{-1}$,

and $\qquad u_3 = \dfrac{4.8 \times 10^{-3}}{20 \times 10^{-4}} = 2.4 \text{ m s}^{-1}$.

also $\qquad \dfrac{p_2 - p_3}{\rho} = 12.6\,(-0.748 + 0.170) \text{ m} \times 9.81 \text{ m s}^{-2} = -71.44 \text{ m}^2 \text{ s}^{-2}$.

Then $\qquad gE = \dfrac{(13.33^2 - 2.4^2)}{2} - 71.44 = 14.56 \text{ m}^2 \text{ s}^{-2}$.

Rate of loss of energy $= \rho g Q E$
$$= 4.8 \text{ kg s}^{-1} \times 14.56 \text{ m}^2 \text{ s}^{-2}$$
$$= 69.9 \text{ W}.$$

In the contraction:

$$gE = \dfrac{u_1{}^2 - u_2{}^2}{2} + \dfrac{p_1 - p_2}{\rho}$$

$$= -86.0 + 92.4 = 6.4 \text{ m}^2 \text{ s}^{-2}$$

Rate of loss of energy $= 30.7$ W i.e. loss of energy in the contraction is less than half that in the expansion.

8.2 Orifices

In the preceding section it appears that the final diverging portion (sometimes called a *diffuser*) is not an essential part of a Venturi meter; it is only used to minimize the energy degradation in the meter as a whole. It certainly increases the capital cost of the meter. In places where it is not necessary to minimize E a perfectly satisfactory meter is a convergence without any diffuser, as shown in Fig. 8.4. Such an arrangement is called a *streamlined orifice*, for which precisely the same flow equation, equation 8.1, is used as for the Venturi

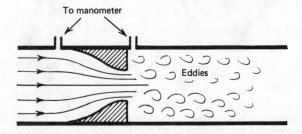

Fig. 8.4 A possible flow-measuring apparatus of the pipe contraction type—sometimes called a *streamlined orifice*. The eddies downstream cause a large degradation of energy. Note the similarity of the flow to the flow through a nozzle.

meter. Orifices can be designed to fit between the flanges at the ends of two lengths of pipe.

But even streamlined orifices are sometimes unnecessarily expensive to install and use. A simpler type of orifice is the so-called *sharp-edged orifice*, which is simply a hole drilled in a flat plate, the hole being bevelled to a sharp edge (Fig. 8.5). The bevelled side of the plate faces downstream. In the figure, streamlines of the flow are shown which approach and pass through the orifice. It will be seen that the jet of high speed fluid continues to contract for some distance downstream of the plane of the orifice, becoming parallel sided at a

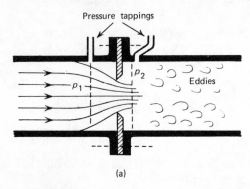

(a)

(b)

Fig. 8.5 (a) A cheaper form of measuring apparatus—the sharp-edged orifice. Note how the flow continues to contract downstream from the plane of the orifice to the minimum cross-sectional area, known as the vena contract. (b) A sharp-edged orifice fitted in a pipe showing the tapping points in the walls of the pipe upstream and downstream of the orifice.

distance of about the diameter of the orifice. Here the pressure is uniform all over the cross section, so that the pressure at the walls is that of the whole of the fluid stream; in the curved, contracting section the pressure at the flow boundaries may not be the pressure throughout the stream (see Chapter 11). Consequently, the Venturi meter formula may be used for such a contraction, taking section 1 as upstream and section 2 at the place'where the jet first becomes parallel again. Downstream of section 2 the jet breaks down into a highly turbulent area wherein the flow is returned to its original area, and in doing so suffers a large degradation of energy into heat. The formula requires a measurement of the jet area a_2, which is inconvenient and difficult for the engineer, who however can measure the size of the orifice itself to a high degree of accuracy. He therefore substitutes a_0, the orifice area, for a_2, preserving the accuracy of the formula by multiplying throughout by another coefficient C_c so that

$$Q = C_d C_c a_0 \left\{ 1 - \left(\frac{a_0}{a_1}\right)^2 \right\}^{-\frac{1}{2}} \sqrt{\left(2g\left\{\frac{(p_1 - p_2)}{g\rho} + (z_1 - z_2)\right\}\right)}.$$

To a first order of approximation $C_c = \dfrac{a_2}{a_0}$, if the effect of the term $(a_0/a_1)^2$ is ignored in the denominator, so that C_c is often called the *coefficient of contraction*, for it now expresses the amount by which the jet is smaller than the orifice from which it emerges.

For a sharp-edged orifice with a 45° bevel on it, and with smooth, well-finished surfaces everywhere, C_c is usually about 0.63. Thus the product $C_d.C_c$ is about 0.62, and is usually called the *coefficient of discharge* of the orifice. The coefficient C_d, which was the *coefficient of discharge* of the Venturi meter, is now called the *coefficient of velocity*, as it indicates the ratio between the actual velocity u_2 and that predicted by Bernoulli's equation.

8.3 Orifices in large tanks or reservoirs

The Bernoulli equation for flow through a contraction is much simplified if the upstream area of flow a_1 is very large compared with a_2. An example is when an orifice, streamlined or sharp-edged, is in the bottom or side of a large tank (Fig. 8.6). Cross section 1 is now the whole cross section of the tank, and it is convenient to consider it as at the free water surface, a height z_1 above datum. Here the pressure is atmospheric and the gauge pressure p_1 is zero. The pressure of the out-flowing jet is also atmospheric at the place where it becomes parallel sided, if from a sharp-edged orifice, or at the orifice outlet if streamlined. Hence, the Bernoulli equation is now

$$\frac{u_1^2}{2g} + z_1 = \frac{u_2^2}{2g} + z_2.$$

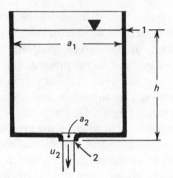

Fig. 8.6 A rounded orifice in the bottom of a large tank. Cross-sectional area a_1 is very large compared with a_2. The water level in the tank is shown by the triangle mark.

But since $a_1 \gg a_2$, $u_1 \ll u_2$, so that $u_1^2/2g$ can be neglected.

Thus
$$u_2 = \sqrt{2g(z_1 - z_2)}$$

and
$$Q = a_2 \sqrt{2g(z_1 - z_2)}.$$

As before, a coefficient of discharge is applied to care for the neglected degradation of energy to heat, so that $Q = C_d a_2 \sqrt{2g(z_1 - z_2)}$. If the orifice is sharp-edged, a coefficient C_c is applied, as before, if a_2 is taken as the area of the orifice instead of the area of the jet.

The equation may be still further simplified by writing $z_1 - z_2 = h$, where h is the vertical distance between the free water surface and the outlet or parallel portion of the jet, so that

$$Q = C_d a_2 \sqrt{(2gh)} \tag{8.3}$$

8.4 Pressure distribution and lift on aerofoils

In Chapter 2 the method was shown of finding the lift force on an aerofoil by integration of the pressure distribution on it. If a model or full sized aerofoil is tested in a wind tunnel, this pressure distribution can be found experimentally by connecting tapping points all around it to suitable manometers. But in the design stage of an aerofoil, it may be necessary to predict the pressure distribution and so the lift before any such model has been tested. Provided the velocity all around the aerofoil is known, and this can be done by calculating the streamline pattern by the methods outlined in Chapter 5, then the pressure can be found as follows.

Imagine the aerofoil being stationary and air being forced past it, as in a wind tunnel (Fig. 8.7). Well upstream of the aerofoil the speed and pressure are uniformly U and p_0 (a great deal of effort is expended in wind tunnel design in order to ensure this uniformity). The streamlines, originally uniformly spaced upstream, are distorted by the aerofoil, usually being brought closer together on the upper surface and further apart on the lower. The spacings become nearly uniform again well downstream of the trailing edge. Thus the air is

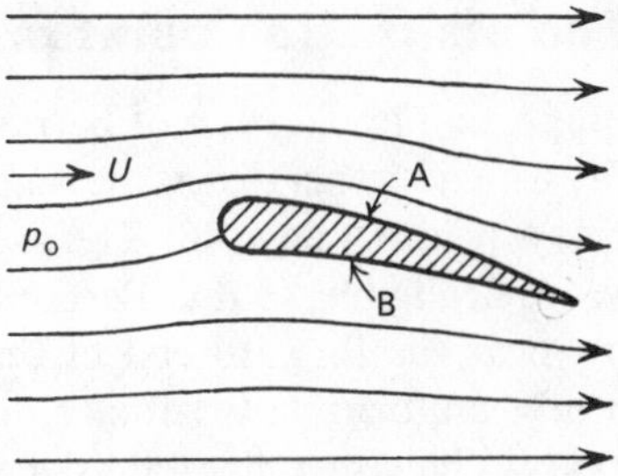

Fig. 8.7 Aerofoil under test in a wind tunnel. Approximate streamline pattern shown. Note that the streamlines tend to be closer together above the upper surface than in the undisturbed stream, and vice versa below the lower surface.

accelerated above and decelerated below the aerofoil. Velocities, such as those at A and B, u_A and u_B respectively, can be measured from the spacing of the streamlines there. Now consider Bernoulli's equation for streamlines between the upstream position and A or B. There are, in well designed aerofoils, only small shear forces acting on the flow over the upper or lower surfaces. Consequently the total head along these streamlines is constant and the pressure p_A or p_B can be found by

$$\frac{U^2}{2g}+\frac{p_0}{g\rho}=\frac{u_A{}^2}{2g}+\frac{p_A}{g\rho}=\frac{u_B{}^2}{2g}+\frac{p_B}{g\rho},$$

ignoring the changes of z which are usually small compared with the changes of the other terms. In this way, the excess or deficiency of pressure $(p_A-p_0)/g\rho$ and $(p_B-p_0)/g\rho$ can be computed for every point on the aerofoil surface. Comparison of the pressure distribution thus found with the experimental values found from wind tunnel tests usually shows fair or good agreement near the leading edge, with increasing errors towards the trailing edge as the shear forces become more important. If the main flow breaks away from the aerofoil, leaving an eddy adjacent to the surface, then large errors may be expected, for the theoretical streamline pattern will not show breakaway at all. As will be explained later, breakaway only occurs on the upper side, that is, the side of the aerofoil where the velocity is greater than that of the oncoming stream and where the pressure is therefore lower than that upstream.

An alternative method of using a theoretical streamline pattern around an aerofoil is to find the thickness of the boundary layers around the aerofoil by much more advanced theory and experiment. The streamlines just outside the boundary layer suffer no degradation of energy at all, and Bernoulli's equation can therefore be applied precisely to find the pressure at points on them. It is then assumed that these pressures are those produced at the surface of the aerofoil. Again, some error is inevitable near the trailing edge where the boundary layers are thick. A combination of a streamline plotting technique and of Bernoulli's equation can thus give a first approximation to the lift force on an aerofoil.

8.5 The stagnation point pressure

An important application of Bernoulli's equation is in finding the pressure intensity which is generated at the stagnation point of a solid body subjected to a fluid flow (Fig. 4.4 page 46). The shear forces, and therefore degradation of energy, are negligible, so that Bernoulli's equation holds good for all parts of the flow near the forward end of the body. In Fig. 8.8, the streamlines are shown of a uniform flow, laminar or turbulent, approaching a symmetrical solid body. The streamlines are deflected, being crowded together near the shoulders of the body at AA. In particular, the fluid velocity near the stagnation point B is decreased, for it will be seen that the streamlines become wider apart there. The velocity near B is therefore much lower than the velocity U in the undisturbed flow upstream at C. At B itself the stagnation streamline meets the solid surface, which is also a streamline and just at the stagnation point the flow velocity is practically zero.

Applying Bernoulli's equation to the streamline at B, the relation between p_0, the undisturbed stream pressure, and p_B, the stagnation point pressure, is given by

$$\frac{U^2}{2g} + \frac{p_0}{g\rho} = \frac{u_B{}^2}{2g} + \frac{p_B}{g\rho},$$

but since $u_B = 0$,
$$p_B - p_0 = \rho\,\frac{U^2}{2} \tag{8.4}$$

or, in words, 'the pressure at a stagnation point is higher than the pressure in the undisturbed stream by the product of the fluid density and the undisturbed kinetic energy per unit mass of fluid.'

The property of a solid body in having a stagnation point on or near its leading edge gives a convenient way of finding the undisturbed velocity of a stream, by introducing a solid body. A hole is drilled in the body at the stagnation point and is connected by tubing to a manometer, which therefore measures p_B. The undisturbed pressure p_0 can be measured in two ways: if the flow is an internal flow then another tapping point in the wall of the pipe or duct containing the flow somewhat upstream of the solid body can be connected to the other limb of the manometer, which therefore now shows $p_0 - p_B$; or another tapping can be placed on the solid body in a position such as D (Fig. 8.8), where experiment has shown the pressure to be such that $p_D = p_0$. Both arrangements are used extensively in experimental work, the solid body being made as a small cylinder put on a stem which permits a traverse across the stream (see Fig. 8.9). Arrangement (*a*) is sometimes called a *Pitot* tube, named after the inventor, or a *total pressure* tube, as the stagnation pressure $\rho U^2/2 + p_0$, can be thought of as the total pressure of the fluid can apply at that particular level. Arrangement (*b*) is called a *Pitot-static* tube, implying that the *static* pressure p_0 is measured by the same instrument that

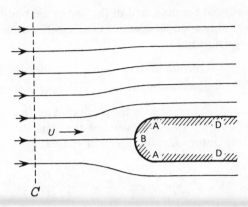

Fig. 8.8 Streamlines of flow around a symmetrical round-nosed object in a stream. Note that the streamlines some distance away are hardly affected but that those near B have wider intervals than those at C.

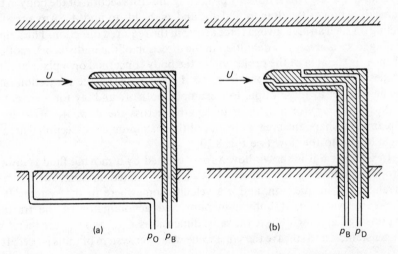

Fig. 8.9 Two methods of finding the velocity at a point in a pipe, using the stagnation pressure on a blunt nosed object. (a) is a *total pressure* tube with a wall tapping point; (b) is a *Pitot-static* tube.

measures the total pressure, the difference in the two pressures being the *dynamic pressure* $\rho U^2/2$.

The stagnation pressure is also used in another important way. As has already been explained, the forces on a solid body (such as the lift and drag) can be found by a direct weighing technique, or by the method of finding the rate of change of momentum of the fluid caused by the drag. But a mere statement of the force is without meaning unless it is accompanied by statements of the

126   *Applications of Bernoulli's equation*

velocity and density of the fluid, and of the size of the body. It is therefore a convenience to be able to quote instead a *Coefficient of Drag,* C_{drag} (or C_{lift}, as the case may be) defined as

$$C_{drag} = \frac{\text{measured drag}}{\text{reference drag force}} \text{ or } C_{lift} = \frac{\text{measured lift}}{\text{reference lift}}$$

The reference forces are defined as the product of an area appropriate to the body concerned and the excess pressure at the stagnation point, $(p_B - p_0)$. Thus, if $D =$ drag force, $L =$ lift force, $A =$ area, $U =$ undisturbed flow velocity, then

$$C_{drag} = \frac{D}{A_1 \rho U^2/2}; \quad C_{lift} = \frac{L}{A_2 \rho U^2/2}.$$

To ensure that the reference forces are truly comparable with the measured ones, it is necessary that they shall act in the same direction as their respective measured forces. This is done by projecting the cross section of the body in the same direction as the drag to find A_1, and as the lift to find A_2. Any pressure acting on such an area gives a force acting in the required direction. Thus, since a drag force is always by definition in the direction of the undisturbed motion, then A_1 is the area of the projection of the body concerned on a plane at right angles to the direction of motion; and A_2 is the area of the projection on a plane parallel to the motion. For example, both A_1 and A_2 for a sphere of radius r are πr^2; for a cylinder of length l across the flow, $A_1 = 2lr$; for a streamlined shape the area A_1 is that of the maximum cross-sectional area at right angles to the flow (see Fig. 8.10).*

In Chapter 6 it is shown how a force exerted by a moving fluid is always proportionate to $\rho A u^2$ (where A is an area related in some way to the arrangement in question, and u a velocity somewhere in the system). It is therefore evident that both numerator and denominator of the fraction expressing C_{drag} or C_{lift} have the same dimensions. C_{drag} and C_{lift} are therefore dimensionless and so have the same value whatever system of units is used. It is

---

* The definition of C_{drag} above is for the general case of a 'bluff' body which is not specifically designed to create a lift force, though it may incidentally do so. In the special case of an aerofoil which is primarily designed to give lift, it is usual to define

$$C_{drag} = \frac{D}{A_2 \rho U^2/2},$$

that is, the relevant area for drag is taken as the one which is truly appropriate to the *lift* force, the plan area of the aerofoil. This somewhat illogical definition is defensible only because the aeronautical engineer is mainly concerned with the variation of C_{lift} and C_{drag} as the angle of incidence θ changes. Since the projected area A_1 in the direction of motion will change with θ, then A_1 will not be a constant for a given aerofoil. It is much more convenient to have a constant area for expressing C_{drag}, and the *chord* area A_2 (i.e. that appropriate to C_{lift}) is the one usually taken.

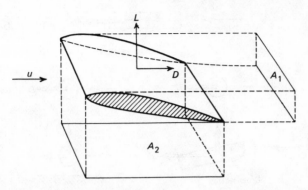

Fig. 8.10 Perspective view of a solid body placed asymmetrically in a fluid flow and therefore producing a lift force *L* and a drag force *D*. The projected areas A_1, A_2 appropriate to the drag and lift coefficients are shown.

not to be expected that either C_{drag} or C_{lift} is a constant value even for one shape; both coefficients vary widely under different conditions. As a general rule, well streamlined shapes, leaving little wake behind them, have low values of C_{drag}, perhaps of the order of 0.1, but bluff bodies giving rise to large eddies behind them may have a C_{drag} as high as 1.5 or so. For a given shape of body, C_{drag} may vary for different speeds, or viscosities of the fluid, and size of the body, and there is considerable evidence (to be discussed in a later chapter) that the combination of variables $uA^{1/2}/v$ is important in deciding the value of C_{drag} for any shape. The way in which C_{drag} varies with $uA^{1/2}/v$ is not the same for all shapes.

The coefficient of lift is essentially concerned with bodies which are not symmetrical about a line in the direction of the undisturbed flow. Thus, unless there is some asymmetry in the flow itself, such as may exist when there is a solid boundary near the body, the symmetrical shapes such as the sphere and cylinder shown in Fig. 8.11(*a*) and (*b*) do not produce any lift at all (and consequently $C_{\text{lift}} = 0$). An asymmetrical body such as an aerofoil, Fig. 8.10, or a symmetrical body held asymmetrically to the flow, Fig. 8.11(*c*), does produce a lift force, depending on the degree of asymmetry of the flow. It is found in fact that the coefficient of lift of an aerofoil increases with the angle of incidence (i.e. the angle between the axis of the aerofoil and the direction of the oncoming stream) until a position occurs when breakaway takes place on the upper surface.

Example

Assuming that the maximum pressure recorded on the aerofoil in the example on page 24 is at the stagnation point calculate the velocity of flow in which it was tested if the density of the air was 1.23 kg m^{-3}.

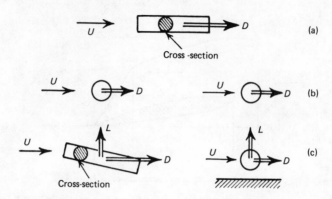

Cross-section

Cross-section

Fig. 8.11 (a) A cylinder end-on to the flow gives drag force D only, no lift.
(b) Cylinders and spheres held symmetrically to the flow give drag forces only.
(c) Cylinders held asymmetrically to the flow, or having a fixed boundary near by, give both drag and lift forces.

What was the maximum velocity over the surface of the wing and its lift coefficient?

Solution

The stagnation pressure supported a column of 25 mm of water in a manometer above the undisturbed pressure whence

$$p_B - p_0 = \rho_w gh = 10^3 \text{ kg m}^{-3} \times 9.81 \text{ m s}^{-2} \times 0.025 \text{ m}$$
$$= \tfrac{1}{2}\rho_a U^2 = \tfrac{1}{2} \times 1.23 \text{ kg m}^{-3} \times U^2$$
$$U^2 = \frac{2 \times 9.81 \times 25}{1.23} = 399 \text{ m}^2 \text{ s}^{-2}$$

and $U = 19.97 \text{ m s}^{-1}$, which is the velocity of flow in which the model was tested. At the point of maximum velocity the pressure is a minimum i.e. -75 mm of water.

$$\text{Hence } \hat{U}^2 = \frac{2 \times 9.81 \times 75}{1.23} = 1196 \text{ m}^2 \text{ s}^{-2}$$
$$\hat{U} = 34.6 \text{ m s}^{-1},$$

i.e. nearly twice the velocity of the undisturbed flow. Using the normal definition of lift coefficient, with a lift force of 48.6 N m^{-1},

$$C_L = \frac{L}{\tfrac{1}{2}\rho U^2 \times \text{wing plan area}}$$
$$= \frac{2 \times 48.6 \text{ N m}^{-1}}{1.23 \times 19.97^2 \times 0.25 \text{ kg m}^{-3} \text{ m}^2 \text{ s}^{-2} \text{ m}}$$
$$= 0.79.$$

This is a fairly typical lift coefficient for an aerofoil.

Appendix: Other methods of flow measurement

Pitot tubes, Venturi meters and orifice plates are not the only methods available for continuously measuring velocities or discharges of fluids. There are many others, some of which use dynamic forces and pressures (and so involve Bernoulli's equation), and some which do not.

There are arrangements which distort a flow in a more complicated way than the simple ones shown in this chapter, and which therefore give a pressure change proportional to the square of the velocity. Pressures across a diameter of a pipe bend (see Chapter 11), or near any change of pipe section, or indeed due to the friction in a length of pipe, may all be used. In open channels, wall distortions may be used to create a difference of water level: these are discussed in Chapter 14.

Among devices which use the flow pressures directly are meters which change their internal geometry as the flow increases, thus giving a direct indication. Pistons or discs may move through apertures, or aerofoil section blades change incidence against a spring system, and such devices are well known in industrial practice. Another group of devices is concerned with the kinematics of the flow; these do not measure forces or pressures directly. They include rotary meters of several sorts, propellers, paddle wheels (see Fig. 4.6), anemometers and similar instruments designed to give the minimum resisting force to the flow. All such instruments require calibration before use.

A laboratory device is the hot-wire (and hot-film) method, where a wire is heated by an electric current but the heat is removed by the passing fluid. The wire temperature depends on the velocity, though not by a linear law, and this in turn controls the electrical resistance, which is measured. Laser–Doppler anemometers can also be used to measure velocities which vary rapidly with time. An electromagnetic device is now frequently used in industrial 'work, and it has the advantage that there are no obstructions in the pipe. A magnetic field is produced across the pipe, and the fluid, acting as a moving conductor, creates an electromotive force between electrodes at right-angles to the field. The disadvantage lies in the complexities of the electronics.

Much more detailed accounts of these methods are to be found in the bibliography.

Problems

1. A large pipe, carrying water at a pressure of 30 kN m^{-2} gauge, leads to an orifice near the top of an airtight tank in which initially there is perfect vacuum. The orifice diameter is 5 cm and $C_d = 0.65$. Find the time required for 6000 litres to flow into the tank. Vapour pressure of water is 1.2 m absolute.   *Ans.* 429 s.

2. A 20 cm water pipe has in it a Venturi meter of throat diameter 12.5 cm, which is connected to a mercury manometer showing a difference of 87.8 cm. Find the velocity in the throat and the discharge. If the upstream pressure is 690 kN m^{-2}, what power would be given up by the water if it was allowed to discharge to atmospheric pressure.
Ans. 16 m s^{-1}: 0.197 m^3 s^{-1}: 136 kW.

3. An aerofoil is so shaped that the velocities along the upper and lower surfaces are respectively 25 per cent greater than, and 25 per cent smaller than, the velocity of the oncoming stream. What is the lift force on such a

wing, 15 m long and 3 m chord, at 320 km h^{-1}? What is the lift coefficient? How nearly can such conditions be achieved in practice? Air density $\frac{1}{800}$ of water density.

Ans. 222 kN $C_{\text{lift}} = 1.0$.

4. A 30 cm diameter axial flow fan supplies 2 m^3 s^{-1} of air drawn from the atmosphere to a 60 cm diameter pipe by means of a well-designed diffuser fitted between the fan and the pipe. A manometer connected across the fan indicates a pressure rise equivalent to 5 cm of water.

   Assuming both the friction losses and the tangential component of velocity downstream of the fan to be negligible, calculate

   (a) the gauge pressure at the entry to the 60 cm diameter pipe;
   (b) the longitudinal force transmitted by the fan to its driving motor;
   (c) the longitudinal force exerted at the flange between the 60 cm pipe and the diffuser.

   Assume the density of air constant and equal to 1.25 kg m^{-2}.

   Ans. 4.69 cm water: 34.7 N: 83.2 N.

5. At a place in a pipeline, the bore changes suddenly from one cross-sectional area to another larger one downstream. If the pressure intensity on the annular area between the two bores is found experimentally to be the same as the pressure in the smaller bore, show that the change of pressure between places upstream and downstream of the enlargement is

$$p_1 - p_2 = \rho u_2 (u_1 - u_2)$$

and the energy degradation is $E = \dfrac{(u_1 - u_2)^2}{2g}$,

where u_1 and u_2 are the velocities in the smaller and larger bores respectively.

Hint Use momentum theorem for pressure forces: at the expansion eddies form, so Bernoulli's equation must have an unknown degradation E.

6. Find the time to empty a tank 6 m square and 2 m deep through a 20 cm diameter faired orifice which is 1 m below the tank bottom.

   Ans. 6 m, 21 s.

9

A summary of calculations concerning forces on and pressures and velocities in moving fluids

There are two ways of calculating the interaction of velocities, forces and pressures in moving fluids. The first is the force momentum equation (Chapter 6): and the second is the Bernoulli equation, or energy equation (Chapter 7). Both these ways are derived from Newton's Second Law of Motion and so deal with the same variable, namely the forces on, and the consequent motion of an element of fluid, but the information required and the results obtained are not the same for both these two ways. The momentum theorem is applied to the conditions at the boundaries of a control volume fixed relative to solid surfaces, and the changes of momentum arriving and leaving the volume are assessed to find the total forces applied to the fluid. In the Bernoulli equation or energy equation the changes of energy along a streamtube in steady motion are used to relate the velocity and pressure changes within the flow. The pressure intensity at the solid surfaces bounding the flow may then be integrated to find the forces normal to them. The table shows the similarities and differences of the two methods, and the following examples show how one or the other or occasionally both methods may be used according to the circumstances.

Example (a) Flow through an orifice (Chapter 8, p. 120)
Due to the convergence of the flow, the energy conditions are known—the total energy is virtually constant throughout, so the Bernoulli equation can be applied to find the flow rate. The force momentum equation cannot be used because the total force in the direction of motion is not known. Thus, at the boundaries of the control volume ABCD in Fig. 9.1, the pressure intensity all over the side CD is known, so the pressure force on this side is also known. But the pressure intensity distribution on the side AB is not known, for it falls towards atmospheric pressure at the orifice as the fluid accelerates. Consequently the pressure force on AB is not known unless experiments are specifically made to find the pressure intensity there or the streamline pattern is found by some method.

Example (b) Flow through a boundary layer (Chapter 6, p. 93)

131

132 Forces on and pressures and velocities in moving fluids

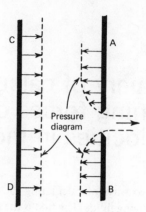

Fig. 9.1 Flow through a sharp-edged orifice in the side of a tank in which the total energy of the fluid is kept constant. The increase of velocity near the orifice decreases the pressure intensity there so that the total pressure force on the jet by the tank walls is unknown. The 'control' volume is the rectangle ABCD.

Since a boundary layer is caused by a tangential shear force on the fluid, there is a degradation of energy into heat within the fluid. As more and more fluid is slowed down by this force while the pressure remains constant, the total head, $z + p/\rho g + u^2/2g$ steadily falls within the layer, the energy degrading to heat. The energy equation cannot therefore be used unless this energy degradation is known from independent evidence. The momentum theorem can, however, be used to find the sum of the forces on the fluid (pressure plus shear forces) if the changes of velocity in the layer are known.

Example (c) Flow in a horizontal nozzle
It is possible occasionally to apply both the force momentum and the Bernoulli equations.

Consider the nozzle shown in Fig. 6.3(a) carrying a steady incompressible flow but rotated to make the flow horizontal. The weight forces are no longer in the flow direction and from equation 6.6 on page 88 the force on the control volume in the flow direction is given by

$$F = \rho Q(u_2 - u_1) - p_1 a_1.$$

Alternatively as the shear stresses are small, the control volume shown in Fig. 6.3(c) shows that the sum of the external forces, F, and the resultant of the pressure forces in the flow direction is zero, i.e.

$$F + \int_{a_1}^{a_2} p\,\mathrm{d}a = 0.$$

Using the previous sign convention, the area projected in the flow direction, da, is assumed to increase in the flow direction, i.e. the ring in Fig. 9.2 has an area $-\mathrm{d}a$.

Then
$$F = -\int_{a_1}^{a_2} p(-\mathrm{d}a),$$

Table 9.1 Capabilities of the force momentum equation and the energy equation

	Force momentum	Energy
Applicable	To any fluid flow, providing an acceleration term is considered when flow is non-steady	To steady flows in which the energy changes are zero or are known independently
Information required	Velocity distribution in the stream at one end of a control volume *and either* total forces (pressure and shear) on boundaries of *or* velocity distribution at other end of control volume control volume	Velocity and pressure at one point on streamline, with independent knowledge of energy changes *and either* pressure variation along *or* velocity variation along streamline streamline
Solution gives	Average final velocity of *or* Total force stream	Velocity variation along *or* Pressure variation along streamline streamline
Solution will not give	Actual velocity distribution within control *or* Distribution of pressure volume or at the and shear forces boundaries	Tangential forces due to friction
Best application	When energy changes are unknown and only an overall knowledge of the flow is required, e.g. total resultant forces, mean velocities	When energy changes are known and detailed information on the flow is required, e.g. velocity and pressure distributions

Notes
1. If the flow is compressible, separate information is required to find the density changes.
2. If both energy changes and total forces are known, both methods are applicable and give the same results to any given set of data. *Alternatively*, the two equations can be used simultaneously to determine another unknown variable which would otherwise be found purely by experiment.

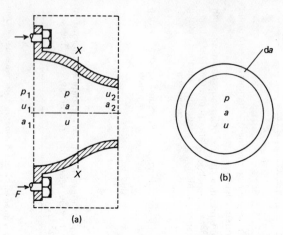

Fig. 9.2 (a) The control volume for the flow through a cylindrical nozzle. (b) Section at control surface XX.

and if the variation of pressure in the nozzle can be determined, F may be found and compared with the value derived from the force momentum equation. As the flow is incompressible, is assumed to be uniformly distributed, and the shear forces are small Bernoulli's equation may be applied between the control surfaces at inlet and $X - X$ giving

$$p_1 + \tfrac{1}{2}\rho u_1^2 = p + \tfrac{1}{2}\rho u^2$$

and if the continuity equation is used

$$Q = a_1 u_1 = a u$$

so that the unknown pressure p is found as

$$p = p_1 + \frac{\rho Q^2}{2}\left(\frac{1}{a_1^2} - \frac{1}{a^2}\right).$$

The integral can now be evaluated, a being the only variable.

$$
\begin{aligned}
F &= \int_{a_1}^{a_2}\left(p_1 + \frac{\rho Q^2}{2}\left(\frac{1}{a_1^2} - \frac{1}{a^2}\right)\right)\mathrm{d}a \\
&= \int_{a_1}^{a_2} p_1\,\mathrm{d}a + \frac{\rho Q^2}{2}\int_{a_1}^{a_2}\left(\frac{1}{a_1^2} - \frac{1}{a^2}\right)\mathrm{d}a \\
&= p_1(a_2 - a_1) + \frac{\rho Q^2}{2}\left(\frac{a_2}{a_1^2} - \frac{a_1}{a_1^2} + \frac{1}{a_2} - \frac{1}{a_1}\right) \\
&= \left(p_1 + \frac{\rho Q^2}{2a_1^2}\right)a_2 - p_1 a_1 + \frac{\rho Q^2}{2}\left(\frac{1}{a_2} - \frac{2}{a_1}\right).
\end{aligned}
$$

Applying Bernoulli's equation between the inlet and exit of the nozzle

$$p_1 + \frac{\rho Q^2}{2a_1^2} = p_1 + \frac{\rho u_1^2}{2} = \frac{\rho u_1^2}{2} = \frac{\rho Q^2}{2a_2^2}. \tag{9.1}$$

Substituting

$$F = \frac{\rho Q^2}{2a_2} - p_1 a_1 + \frac{\rho Q^2}{2a_2} - \frac{\rho Q^2}{a_1} = \rho \frac{Q^2}{a_2} - \rho \frac{Q^2}{a_1} - p_1 a_1$$

$$= \rho Q(u_2 - u_1) - p_1 a_1.$$

This expression *is exactly the same as was derived from the force–momentum equation.* In this particular case where both the force–momentum and Bernoulli equation can be applied they will produce identical solutions to the problem.

For a nozzle whose inlet and outlet areas are known and for a given fluid there are only three unknown values, F, Q, and p_1. As two equations are available, one of these unknowns can be eliminated leaving a single equation connecting the remaining two unknowns. Using equation 9.1 Q may be eliminated from the force–momentum equation leaving the Force determined directly from the inlet pressure p_1.

$$F = -\frac{p_1 a_1}{(a_1 + a_2)}(a_1 - a_2)$$

Note that if the nozzle had been vertical, as in the example on page 87 the weight force would have to be included in the force–momentum equation and the difference of elevation between the inlet and outlet of the nozzle would have to be included in Bernoulli's equation. After much algebraic manipulation the expression becomes

$$F = \frac{a_1}{a_1 + a_2}(2\rho g y_1 a_2 - p_1(a_1 - a_2) - W).$$

Substituting the values from the example on page 87

$$F = \frac{1.85}{1.85 + 0.464}\left(2 \times 10^3 \times 9.81 \times 0.17 \times \frac{4.64 \text{ kg m}}{10^3 \text{ m}^3 \text{ s}^2}\text{m}^3 - 2.55 \times 10^4 \times \frac{1.386 \text{N}}{10^2}\right) - 16.3 \text{ N} = -286.5 \text{ N}$$

which compares with the value based on the measured flow rate of

$$F_x = -301.4 \text{ N}$$

giving an error of

$$\delta F = \frac{301.4 - 286.5}{301.4} \times 100 = 5\%$$

This error might well be acceptable for design purposes, the whole design then being performed on the basis of calculations. However, if the actual force had to be known accurately the construction of a model and the experimental determination of the force might be necessary. The implications of model testing will be considered in the next chapter.

Example (d) Flow in a Borda's mouthpiece

In some flows the application of both the force momentum and Bernoulli equations enable a property to be calculated without recourse to experiment.

A Borda's mouthpiece is a short length of pipe of cross section a_0 projecting into a tank of fluid, wherein the pressure is equivalent to a head h. A jet of fluid of cross-sectional area a_1 separates from the pipe edge and is ejected through it. Since the flow is converging, Bernoulli's equation clearly applies, as there is little, if any, degradation of energy to heat. Also, the whole force system on the fluid is fully known: for example, on the boundaries of the control volume ABCD in Fig. 9.3, the pressure exerted on the fluid by the tank wall CD is the same as that by AB. The projecting pipe has removed all the motion to within the tank, so that there are no fluid velocities against the surface AB, changing the pressure there, contrary to the case of an ordinary orifice considered in example (a). Thus the excess of pressure force on CD above that on AB is only that due to the head h on the area a_0, namely $\rho g h a_0$, and this is the force producing the flow, the fluid being accelerated from a standstill to a velocity u. The change of momentum flow rate is then $\rho Q u = \rho a_1 u^2$. Equating the force to the change of momentum flow rate, $\rho a_1 u^2 = \rho g h a_0$. The unknown variable a_1 can only be determined

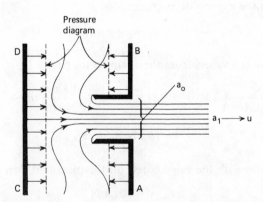

Fig. 9.3 Flow through a Borda's mouthpiece. Streamlines are shown, and the velocity near the wall AB is low, so the pressure intensity is sensibly the same as that on CD. The pressure force is thus known and the momentum theorem may be used to find a_1. The control volume is the rectangle ABCD.

by using Bernoulli's equation, which gives the velocity in terms of the head h or $u = \sqrt{(2gh)}$, assuming that the area of the tank is large compared with a_0. Substitution of this equation for u into the momentum relation gives

$$\rho a_1 2gh = \rho g h a_0$$

or

$$a_1 = \tfrac{1}{2} a_0.$$

Thus

$$Q = a_1 u = 0.5 a_0 \sqrt{(2gh)}.$$

The coefficient of discharge of this special type of orifice is therefore 0.50, a value found quite without experiment, solely because both the Bernoulli equation and the force–momentum equation can be applied: they have been used as simultaneous equations to determine a_1, which could only be found in the case of the sharp-edged orifice (p. 131) by experiment.

Example (e) A jet striking a surface

The simple case of the jet of fluid striking an extensive plane surface normal to the jet axis (Fig. 9.4) shows clearly the limitations and capabilities of both the force momentum and Bernoulli equations. The force momentum equation gives the total force exerted by the jet, providing the surface is sufficiently large to turn all the jet through 90°. It will not however give any information about the distribution of pressure over the surface, nor anything about the internal geometry of the flow pattern; for instance, the radius of curvature of the streamlines of the jet near the surface is neither known nor given. Only the geometry of the flow at the ends of the control volume need be known.

The Bernoulli equation gives the pressure distribution along the plate, and so also gives the total force (by integration), but only if the velocity is known at all these points. In turn the velocity can only be known from a flow pattern. It is clear that there will be a stagnation point on the axis of the jet and adjacent to the surface, but the velocities and curvatures of the

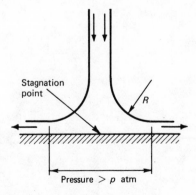

Fig. 9.4 A jet of water striking a flat normal surface must satisfy the force–momentum equation, but the maximum pressure intensity (at the stagnation point) must satisfy Bernoulli's equation. Between these two requirements, R is fixed.

Stagnation point

R

Pressure $> p$ atm

streamlines elsewhere are unknown until a full internal flow pattern can be drawn. If such a pattern is known the pressure distribution can be inferred and by integration the force is again evaluated.

With one dependent variable, the force, computed by two independent methods, it is clear that another dependent variable can be solved. In essentials this can be the flow pattern, of which the radius of curvature of the free surface of the jet is one feature (see Fig. 9.4). This radius is thus fixed by the equivalence of the momentum flow of the incoming stream with the integrated pressure distribution along the plane. The properties of flows in which the streamlines are curved are discussed in Chapter 11.

10
The analysis of experiments in fluid mechanics

10.1 The idea of dimension

It is apparent at an early stage in the study of fluid mechanics that purely theoretical analyses rarely produce exact solutions to the problems set by real fluids flowing under specified conditions. Experiment is nearly always needed to correlate the theory with the actual measurements of the variables concerned. Coefficients must be introduced to bridge the gap thus disclosed, and to account for the errors caused by incorrect or simplifying assumptions in the theory. Such assumptions are necessary because the present stage of development of mathematics is inadequate to deal with the problems of real fluids. In many arrangements of the boundaries of flows in which engineers are interested theory is inadequate so that recourse must be made to experiment to verify and extend analysis by theory.

A problem with such experiments in fluid mechanics is the large number of variables, all of which are interrelated, and all of which may affect the measurements being studied. Experiments cannot always be arranged so that only one independent variable is changed at a time, with the other variables temporarily kept constant. It is only this method that allows the effect of each variable to be studied separately. It is much more often found that circumstances force experiments to be arranged so that a greater number of variables must change at the same time, so that the effect of each variable cannot be separately studied. It is highly desirable, however, to present the results of these experiments so that the effect of each can be seen. A way of finding the correct way of presenting the information, and of arranging the variables, is by using the Method of Dimensions. This method uses the axiom that the dimensions of all the terms of any physically correct equation must be the same. For example, if one term of an equation is represented by the dimensions of a velocity, then *all* other terms in the equation must have the dimensions of velocity. To ensure that this is so, it is necessary to break down the dimensions of mechanical properties into the three primary dimensions, those of Mass M, Length L, and Time T. Thus a velocity has the dimensions (length/time) = LT^{-1}: an acceleration is LT^{-2}: force, being the product of mass and acceleration, is MLT^{-2}: density, being a mass per unit volume, is

ML^{-3}: velocity gradient, being a velocity per unit length, is T^{-1}: the coefficient of viscosity, being a force per unit area per unit velocity gradient, is $ML^{-1}T^{-1}$: and so on. In incompressible fluid flows all dimensions can be expressed in terms of M, L and T.

10.2 Dimensionless quantities

Now consider a phenomenon where one variable (the *dependent* variable) might be controlled by several others. These others are called *independent* variables because any one of them can be changed, and while affecting the dependent, does not affect any of the other independents. An example is the force F produced by a jet of fluid striking a large flat surface normal to the jet axis (the direct analysis of this phenomenon is given in Chapter 6). Clearly F might change when the velocity U of the jet changes; when the viscosity μ and the density ρ of the fluid changes; or when the size of the jet changes. The last variable may be expressed as the diameter d of a circular jet of the same cross-sectional area as the actual jet. This assumes that the shape of the jet does not affect F; a preliminary experiment could be carried out to show that this assumption is indeed true. Notice that although the variable μ was not considered in the direct analysis of Chapter 6, it is included here as one variable that might affect F. Experiment will eventually decide if it is indeed relevant or not.

Now the dependence of F on the other variables can be conveniently expressed as

$$F = \phi(u, d, \rho, \mu)$$

the symbol ϕ meaning solely 'a function of' and not implying anything about the form of the function. It does not imply that F varies linearly with the product $ud\rho\mu$, for instance. But any function can be expressed as a series comprised of a number of terms each being made up of the product of the variables brought to suitable powers. That is,

$$F = u^{a_1}d^{\,b_1}\rho^{c_1}\mu^{d_1} + u^{a_2}d^{\,b_2}\rho^{c_2}\mu^{d_2} + \dots$$

where $a_1, a_2, a_3, \dots : b_1, b_2, b_3, \dots : c_1, c_2, c_3 \dots$ are indices. Dividing both sides of the equation by the first term on the right-hand side

$$F/u^{a_1}d^{b_1}\rho^{c_1}\mu^{d_1} = 1 + u^{a_2-a_1}d^{\,b_2-b_1}\rho^{c_2-c_1}\mu^{d_2-d_1} + \dots$$

Since the first term on the right-hand side is a number, it is dimensionless, so that by the axiom already mentioned, *all* terms are dimensionless. In particular, the term on the left-hand side can be written

$$[F/u^a d^{\,b}\rho^c\mu^d] = 0$$

the sign $[\ \ ]$ meaning 'the dimensions of'.

Thus $$[F] \equiv [u^a d^b \rho^c \mu^d]$$

and by inserting the dimensions of each variable, an equation in M, L and T is found,

$$MLT^{-2} \equiv (LT^{-1})^a (L)^b (ML^{-3})^c (ML^{-1}T^{-1})^d$$
$$= L^a T^{-a} L^b M^c L^{-3c} M^d L^{-d} T^{-d}$$

The indices for each of the three primary dimensions M, L, and T may now be collected to form three equations

$$\text{for } M, \quad 1 = c + d$$
$$\text{for } L, \quad 1 = a + b - 3c - d$$
$$\text{for } T, \quad -2 = -a - d$$

These three simultaneous equations have four unknown quantities in them, only three of which can be determined in terms of the fourth. There are not enough equations to solve for all four unknowns, and it is purely a matter of experience (and intuition) to know for which three unknowns the equations are to be solved. In this case it is convenient (and a well-known form of the solution is found) if a, b, and c are solved in terms of d: and so by the usual methods of solution of simultaneous equations

$$a = 2 - d : b = 2 - d : c = 1 - d,$$

so that

$$F = u^{2-d} d^{2-d} \rho^{1-d} \mu^d + \text{other terms all involving } u, d, \rho \text{ and } \mu.$$

The 'other' terms on the right-hand side all involve u, d, ρ and μ; since the same dimensional argument can be put to each term, they will all come out into the form

$$u^{2-d'} d^{2-d'} \rho^{1-d'} \mu^{d'}$$

the index d' being a different value for each term. It will be seen that the product $u^2 d^2 \rho$ is common to all terms and that the difference between the terms lies in the value of $(u^{-1}\rho^{-1}d^{-1}\mu)^d$. The whole expression for F may therefore be rewritten

$$\frac{F}{u^2 d^2 \rho} = \left(\frac{\mu}{u\rho d}\right)^{d_1} + \left(\frac{\mu}{u\rho d}\right)^{d_2} + \left(\frac{\mu}{u\rho d}\right)^{d_3} + \dots$$

or, replacing the series by the symbol ϕ again,

$$\frac{F}{u^2 d^2 \rho} = \phi \frac{ud}{v} \quad \text{where } v = \mu/\rho$$

or

$$F = u^2 d^2 \rho \phi \left(\frac{ud}{v}\right) \tag{10.1}$$

It will therefore be seen that the dimensionless quantity $F/u^2 d^2 \rho$ is a function of (ud/v), and this conclusion is based solely upon the axiom that an equation for F in terms of the four variables u, d, ρ and μ must be dimensionally homogeneous.

10.3 Dimensional analysis

Here, then, is a clue to one way of presenting the results of experiments on this particular phenomenon. Having changed u, d, ρ and μ in a number of tests, and in each one having measured F, a graph may be drawn up of $F/u^2 d^2 \rho$ plotted against (ud/v). All the data will fall on one curve (experimental error excluded), which is the graph of the function ϕ. In this way the effect of every variable is easily seen on one diagram only. If, of course, the data all fall on a straight line parallel to the axis of (ud/v) in a certain range of values of (ud/v), the quantity $F/u^2 d^2 \rho$ does not depend at all on (ud/v). This is what occurs in the simple case of the jet striking a flat surface ($F = u^2 a \rho \sin \theta \cos \theta$: see Chapter 6), when F does not depend on v except when ud/v is small (treacle, thick oil, at low speeds and small jet diameters), for which conditions the assumptions of Chapter 6 do not hold (no shear forces, fluid leaves surface tangentially).

The method of analysis is shown above as applied to the phenomenon of a jet striking a flat surface. Precisely the same analysis results if the phenomenon to be investigated is that of a force caused by any fluid motion past a solid surface, for example, the drag of a cylinder immersed in an extensive stream, or the drag due to the boundary layer caused by a stream passing tangentially along a plate. The only difference will be in the method of defining the length measurement corresponding to the jet diameter d. In the former case the cylinder diameter d fixes the scale of the fluid phenomenon; in the latter case the distance x from the leading edge of the plate fixes the thickness of the boundary layer and so the drag force. The dimensional analysis will give $F/u^2 d^2 \rho = \phi_2(ud/v)$ and $F/u^2 x^2 \rho = \phi_3(ux/v)$ respectively. The variables in both these expressions are essentially the same as that for the first case taken, but the functions ϕ_2 or ϕ_3 will be quite different from ϕ. *Thus the method of dimensional analysis is valid only for comparing the same phenomenon at different scales and speeds*, but not valid at all for comparing two different phenomena. A change in the geometry of the boundaries changes the phenomenon to be considered, and the function determined experimentally for one set of boundaries is of no use in predicting the function for another set. The effect on the flow around a circular cylinder of changing the dimensionless variable ud/v is shown in Fig. 10.1.

In (a), aluminium dust is fed into the airstream downwind of the cylinder through the tube shown. The wake is illuminated from above by a plane beam of light, the outline of the cylinder in this plane being shown by the dotted circle. Allowance has been made for perspective effects. Within the wake,

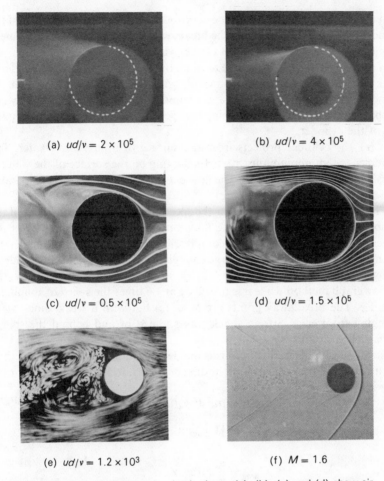

(a) $ud/v = 2 \times 10^5$

(b) $ud/v = 4 \times 10^5$

(c) $ud/v = 0.5 \times 10^5$

(d) $ud/v = 1.5 \times 10^5$

(e) $ud/v = 1.2 \times 10^3$

(f) $M = 1.6$

Fig. 10.1 Examples of flow around a simple shape: (a), (b), (c) and (d) show air flowing past a circular cylinder in a wind tunnel; (e) shows the flow past a circular cylinder standing in a water stream; (f) shows a sphere travelling at supersonic speed through still air. The flow relative to the cylinder or sphere is from right to left in every case. Photograph (f) by U.S. Army Ballistic Research Laboratories, Aberdeen Proving Ground, Maryland.

where the velocities are low, each particle of dust circulates for a long period before it is picked up by the fast flow at the edge of the wake and blown away downstream. Thus with a moderately long exposure the breakaway points and the wake can be seen.

In (*b*), the same cylinder is used as in (*a*) but the speed of the air has been doubled. Observe that the breakaway points have moved further back and the wake is narrower than in (*a*).

In (c), filaments of smoke have been introduced into the air upstream of the cylinder. In addition to showing the breakaway points and the wake, filament lines outside the wake and ahead of the cylinder can now be seen.

In (d), the air speed is the same as in (c) but the diameter of the cylinder is three times as big. Compare (c) and (d) with (a) and (b). Observe that the rearward movement of the breakaway points and the narrowing of the wake can be produced either by increasing the speed of the air or by increasing the size of the cylinder.

In (e), the cylinder projects from the surface of a stream of water. The movement of aluminium dust particles floating on the surface of the water is revealed by a time photograph. The flow pattern for water in (e) is comparable with the flow for air in (a) and (c).

Whatever the fluid and whatever the speed or the size of the cylinder, if the number ud/v is kept constant the flow pattern will not change unless the velocity becomes so high that compressibility effects become important. Changes in air density are then large enough to produce optical effects due to changes in refractive index. In (f) a sphere is photographed by the shadowgraph method while it is travelling at 1.6 times the speed of sound, i.e. $M = 1.6$. The breakaway points are as in (b) but with a weak shock wave springing from each. Strong shock waves ahead of and behind the sphere appear as black and white lines.

The dimensional analysis confirms the derivation in Chapter 8 of the drag and lift coefficients. For a sphere the dimensionally derived expression for the drag is

$$F = \rho u^2 d^2 \phi_4(ud/v).$$

But

$$D = \tfrac{1}{2} C_{\text{drag}} \rho u^2 \frac{\pi}{4} d^2$$

hence

$$C_{\text{drag}} = \frac{8}{\pi} \phi_4(ud/v),$$

i.e. the drag coefficient will be a function of the dimensionless variable (ud/v). This is found by experiment to be the case (Fig 10.2).

10.4 Omission of variables

It may be argued that the result of the above method depends on the variables chosen to start with, and upon the choice of the three variables whose indices are solved by the simultaneous equations. So it does. If a variable has been omitted at the beginning which is in fact a relevant one for the phenomenon, then the final experimental plot of ϕ will show a scatter of the points which is not due to experimental error. A search must then be made for the omitted variable, a new dimensional analysis made and a revised method of

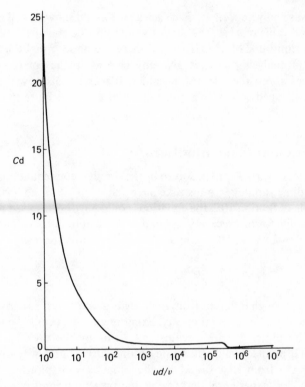

Fig. 10.2 The drag coefficient (C_α) of a sphere as a function of the dimensionless group ud/v (the Reynolds number).

plotting found. If, on the other hand, a variable is included that is really not important to the problem, then the analysis will show finally two groups of variables from which the function ϕ is produced. Both groups would therefore appear to control the phenomenon. In the preceding example, if the surface tension T (force per unit length) had been included as a possible variable, a dimensionless group $u^2 d\rho/T$ would have appeared in the analysis thus,

$$\frac{F}{u^2 d^2 \rho} = \phi\left(\frac{ud}{v}, \frac{u^2 d\rho}{T}\right)$$

A set of experiments could now be carried out to determine if T really affected the phenomenon. For this purpose (ud/v) must be kept constant while $u^2 d\rho/T$ is varied. A plot of $F/u^2 d^2 \rho$ against $u^2 d\rho/T$ would then show a straight line parallel to the axis of $u^2 d\rho/T$ showing that in fact T was irrelevant to the problem. Alternatively, if a large number of experiments were carried out and all the variables, u, d, ρ, μ and T changed, then a plot of $F/u^2 d^2 \rho$ against (ud/v) would show all the points on one graph, no matter how T changed. If, of course, T was really important, then such a plot would show several graph lines, one for each value of $u^2 d\rho/T$. The plot would thus be one of $F/u^2 d^2 \rho$ against ud/v, with 'contours' of $u^2 d\rho/T$. It will therefore be seen that the method of dimensional analysis does not give any clue as to the correctness of the assumptions about the relevant variables. It merely tells how the variables should be grouped so that experiments will decide which are the important ones.

10.5 Dimensionless numbers

The choice of variables to be solved by the simultaneous equations must also affect the dimensionless groups (like ud/v and $u^2 d\rho/T$) in the final result. In the example of Section 10.2, the equations could have been solved for b, c and d in terms of a, for instance, instead of a, b and c in terms of d. That is $b = a$, $c = a - 1$, $d = 2 - a$, so that $F = u^a d^a \rho^{a-1} \mu^{2-a}$ + other terms.

or
$$F = \frac{\mu^2}{\rho} \phi_1 \left(\frac{ud\rho}{\mu} \right). \tag{10.2}$$

This answer, though different from the previous one, *is just as correct*. It only presents the variables in another way. Experimental data could be just as well expressed by plotting $F\rho/\mu^2$ against $ud\rho/\mu$ as by the previous way. The experimental points would still all lie on one curve, but it would be a quite different curve from that found by the other way of plotting. It is only convention, and experience, that makes the presentation of data by certain dimensionless groups more common than by others. The common forms of dimensionless groups used in fluid mechanics are

$\dfrac{ud}{v}$ (often called the Reynolds number: d is a length measurement, not necessarily a diameter).

$\dfrac{u}{\sqrt{dg}}$ (often called the Froude number: g is the acceleration due to gravity, see section 10.7.)

$\dfrac{u}{a}$ (often called the Mach number: a is the speed of sound, see section 10.6.)

$\dfrac{T}{u^2 d\rho}$ (often called the Weber number: T is the surface tension)

Sometimes unusual groups are needed for special purposes. For example, consider the terminal velocity u of a sphere of diameter d falling through a fluid of density ρ and viscosity μ. If the density of the material of the sphere is ρ', then the downward weight force acting on the sphere is $\dfrac{\pi}{6} d^3 (\rho' - \rho)g$ (for by Archimedes' principle there is an upthrust equal to the weight of fluid displaced). Now at the steady terminal velocity the downward force exactly equals F, the hydrodynamic drag force caused by the motion. But the dependence of F can be investigated by dimensional analysis, which for this case is exactly the same as for the jet hitting a plate, equation 10.1.

That is
$$F = \frac{\pi}{6} d^3 (\rho' - \rho) g = u^2 d^2 \rho \phi \left(\frac{ud}{v} \right)$$

or
$$\frac{dg}{u^2} \left(\frac{\rho' - \rho}{\rho} \right) = \phi_1 \left(\frac{ud}{v} \right).$$

The experimental data for spheres falling through a fluid could therefore be correctly presented by plotting $\dfrac{dg}{u^2} \left(\dfrac{\rho' - \rho}{\rho} \right)$ against $\left(\dfrac{ud}{v} \right)$. One curve would result for all experiments, but it would be an inconvenient one to use later if it was desired to find, in advance of experiment, what speed would result from given values of d, ρ, ρ' and μ: the unknown variable u features in the quantities on both axes and a trial and error method would be necessary to determine a value of u.

It is much more convenient to use the unusual dimensional analysis (equation 10.2) already given above as a second way of finding F,

that is
$$F = \frac{\mu^2}{\rho} \phi \left(\frac{ud}{v} \right)$$

But
$$F \propto d^3 (\rho' - \rho) g$$

so that
$$\frac{d^3 g}{\mu^2} \rho(\rho' - \rho) = \phi_2 \left(\frac{ud}{v} \right),$$

and the corresponding method of plotting is $\dfrac{d^3 g}{\mu^2} \rho(\rho' - \rho)$ against ud/v. In this method u only occurs in the quantities on one axis so that by putting the given values of d, μ, ρ and ρ' into one variable, a value can be instantly read from the curve of ud/v, and u found directly.

10.6 Use of dimensional analysis in experiments

Having performed a dimensional analysis of the possible variables for a phenomenon, the result can be used in the organization of experiments which will illustrate every aspect of the phenomenon. Suppose it has been found by the analysis that four variables A, B, C, D are interconnected by several dimensionless groups N_1, N_2, N_3, etc., each of which may contain some of the variables A, B, C, D and also other variables such as μ, T, K, etc.

That is
$$A/B^x C^y D^z = \phi(N_1)(N_2)(N_3).$$

From the examples given it will be seen that N_1, N_2, N_3 might be Reynolds, Mach or Weber numbers or any other dimensionless group. The experimental data may therefore be presented as shown in Fig. 10.3, as a set of graphs of $A/B^x C^y D^z$ against N_1, with 'contour curves' of N_2, each graph being for a certain value of N_3. In this way the effect of every group is presented separately, though of course there will be many graphs if the number of relevant dimensionless groups N is large. If one group, N_2 say, is in fact irrelevant, then all the curves of each graph will coincide, leaving only N_1 and N_3 as relevant groups. Notice that the effect of grouping the variables is that three fewer graphs are necessary to present the data than would be necessary if dimensionless grouping were not used at all. For example, in the case that has already been described of a force F depending on u, d, ρ and μ, *four* graphs would be necessary to show a complete set of experimental data in the direct way. That is, F against u, with d, ρ and u remaining constant: F against d, with u, ρ and u constant: F against ρ: and F against μ. Contrast these four graphs with the one dimensionless plot of $F/\rho d^2 u^2$ against $ud\rho/\mu$ which shows all that the four graphs show.

It is a long and tedious job to investigate fully a phenomenon in this way, if there are many variables concerned, and it is often impossible to select fluids which have the requisite properties so that tests can be made by changing one group, N_1, without changing another, N_2. Technical advances slowly make it more possible to make full analyses of problems of fluid mechanics; for example, the advent of detergents has permitted control to be exerted over the surface tension, without greatly changing density or viscosity. The effect of the Weber number can thus now be more fully investigated and presented by the graphical method of Fig. 10.3. Engineers, however, can rarely wait long enough for full investigations to be made of a phenomenon, so that their data is from experiments that only cover a small range of the relevant variables. These experiments must frequently be carried out with the boundaries of the flow at a reduced scale from that which will eventually be used, so as to decrease the cost of any modifications which may be necessary. For example, it would be too expensive and risky to carry out experiments on full-size aeroplanes: reduced-scale models are held stationary exposed to an air flow in wind tunnels. But it is now necessary to relate the measurements on a model to the corresponding ones on the full-size aeroplane (*prototype*). Having measured the force F_m on a model of size l_m, exposed to a flow of speed u_m in a fluid of density ρ_m and viscosity μ_m, what will be the force F_p on the prototype at speed u_p?

The corresponding forces F_m and F_p can be very easily found if the numerical values of the dimensionless groups $N_1, N_2, N_3 \ldots$ are the same for both model and prototype. In this way $\phi(N_1, N_2, N_3 \ldots)$ for the model must be the same as $\phi(N_1, N_2, N_3 \ldots)$ for the prototype, and the other variables are then simply related, with no further experimental data. For example, take the case of an aeroplane to be tested to find the drag force F. Clearly the velocity u through still air, the size l (exemplified for instance by the wing span), the fluid

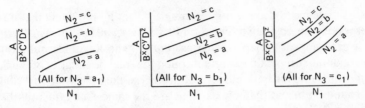

Fig. 10.3 The way in which 4 variables, *A*, *B*, *C* and *D* can be expressed as functions of 3 non-dimensional groups N_1, N_2 and N_3 each composed of other variables. The phenomenon to which these curves refer is entirely imaginary. Dimensional Analysis has given the indices *x*, *y*, *z*: *a*, *b*, *c* are numerical values of N_2: a_1, b_1, c_1 are numerical values of N_3.

density ρ and viscosity μ are of importance in finding *F*. The surface tension *T* is not of importance (because there is no liquid-gas interface involved), and it is assumed that the speed is low compared with the velocity of sound so that compressibility effects are negligible, and any effect of the change of Mach number can be ignored.

A dimensional analysis then gives

$$F/u^2 l^2 \rho = \phi(ul\rho/\mu).$$

Now relate both the model condition (suffix m) and prototype condition (suffix p). That is,

$$F_m/u^2{}_m l^2{}_m \rho_m = \phi(u_m l_m \rho_m/\mu_m)$$

and

$$F_p/u^2{}_p l^2{}_p \rho_p = \phi(u_p l_p \rho_p/\mu_p).$$

The model must be of course exactly the same geometrical shape as the prototype so that the fluid boundaries are the same; if this is so, then the curve of the function ϕ is the same for both prototype and model. The numerical value of ϕ can however only be the same in the two cases if

$$u_m l_m \rho_m/\mu_m = u_p l_p \rho_p/\mu_p$$

or

$$\frac{u_m}{u_p} = \frac{l_p}{l_m} \cdot \frac{\rho_p}{\rho_m} \cdot \frac{\mu_m}{\mu_p} \qquad (10.3)$$

If u_m and u_p are so connected, then

$$F_m/u^2{}_m l^2{}_m \rho_m = F_p/u^2{}_p l^2{}_p \rho_p$$

or

$$F_p = F_m \left(\frac{u_p}{u_m}\right)^2 \left(\frac{l_p}{l_m}\right)^2 \left(\frac{\rho_p}{\rho_m}\right) \qquad (10.4)$$

Substituting for u_m from above,

$$F_p = F_m \left(\frac{l_m}{l_p} \cdot \frac{\rho_m}{\rho_p} \cdot \frac{\mu_p}{\mu_m}\right)^2 \left(\frac{l_p}{l_m}\right)^2 \frac{\rho_p}{\rho_m}$$

$$= F_m \frac{\rho_m}{\rho_p} \left(\frac{\mu_p}{\mu_m}\right)^2$$

Thus a simple method exists of predicting F_p from F_m providing that u_m is the corresponding speed given by equation 10.3. Now it will be shown in Chapter 12 that a grouping of variables such as $(ul\rho/\mu)$—the Reynolds number—is a measure of the sort of eddies present round the solid boundaries of the flow. If $ul\rho/\mu$ is the same for both model and prototype then the same sort of eddies exist round both, and the flow patterns are the same. Consequently the drag coefficient $2F/u^2 l^2 \rho$ is the same for both model and prototype. Running the model at the corresponding speed of equation 10.3 is a guarantee that the breakaway and the eddies in the wake occur at the same position in the model as in the prototype (though of course if there are slight differences of shape or roughness then differences of flow are inevitable—a dimensional analysis is always for two precisely similarly shaped bodies).

Example

What will be the Reynolds number of the flow around the aerofoil in the example on page 128, if the viscosity of the air was $1.8 \times 10^{-5} \, \text{N s m}^{-2}$?

At what speed would a 1 metre chord aerofoil of the same section and at the same angle of incidence operating at an altitude where the air density is $0.4 \, \text{kg m}^{-3}$ and the viscosity $1.5 \times 10^{-5} \, \text{N s m}^-$ have exactly the same lift coefficient?

Solution

The Reynolds number is $\rho u l / \mu$ and an appropriate length, l, has to be chosen for its calculation. As only the chord is known

$$Re = 1.23 \frac{\text{kg}}{\text{m}^3} \times 19.97 \frac{\text{m}}{\text{s}} \times 0.25 \, \text{m} \times \frac{10^5 \, \text{m}^2}{1.8 \, \text{N s}} \times \frac{\text{N s}^2}{\text{kg m}}$$

$$= 3.4 \times 10^5.$$

Note that the Reynolds number is dimensionless and that numerically its value is large.

For the larger aerofoil to have the same lift coefficient at the same incidence as the smaller, the Reynolds numbers of the two flows must be the same. For the larger aerofoil

$$u = \frac{\mu Re}{\rho l} = \frac{1.5}{10^5} \times \frac{3.4 \times 10^5}{0.4 \times 1} = 12.75 \, \text{m s}^{-1}$$

i.e. the velocity would have to be very low for an aircraft.

For a small-scale model $l_p/l_m > 1.0$, and if it is tested in the same fluid as the full-size aeroplane, $\rho_p/\rho_m = 1.0$ and $\mu_m/\mu_p = 1.0$. To achieve the corresponding speed it follows that $u_m/u_p > 1.0$: the model must have a higher relative wind speed than its prototype if the flow pattern and drag coefficient are to be exactly the same. For example, a 1/5 scale model should be tested at 1000 km h^{-1} in atmospheric air if the prototype is to fly at only 200 km h^{-1}. Such high speeds cause additional phenomena to occur, due to the compressibility of the air, which have not been expected in the dimensional analysis because the Mach number was considered small and so not included (Fig. 10.1(f)). The flow conditions around the model are therefore dissimilar to those round the prototype, and the model drag gives a misleading idea of the prototype's drag, if equation 10.4 is now used. With modern high-speed aircraft the effect is magnified, a 1000 km h^{-1} prototype needing an experiment to be carried out at 5000 km h^{-1} on a 1/5 size model. One way of getting similarity of eddies without the drawback of such high speeds is to increase the density of the fluid in which the model is tested, by using a heavy gas in an enclosed room within which is the wind tunnel, or by compressing the air to many atmospheres pressure (Fig. 10.4). These arrangements are undoubtedly more complicated than the usual atmospheric wind tunnel, but they enable the Reynolds number for the model to be more nearly equal to that of the prototype. If such special tunnels are not available, then it is customary to test the model at a speed far below the corresponding speed and assume that $\phi(ul\rho/\mu)$ does not change when extrapolating from the model to the prototype. Fortunately it is found by experience that this procedure does not give large errors for well-streamlined aeroplane shapes.

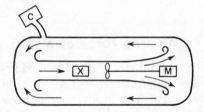

Fig. 10.4 Diagram of a compressed-air wind tunnel. M is the motor driving the fan that circulates the air compressed by the compressor C. X is the model under test.

If the prototype aircraft is to fly at high speeds then it is found that the speed of sound (Chapter 1) becomes important and that the flow pattern is greatly affected by compression waves. Inserting the variable a into the dimensional analysis a revised equation is found

$$\frac{F}{u^2 l^2 \rho} = \phi\left(\frac{ul\rho}{\mu}\right),\left(\frac{u}{a}\right)$$

where $a = \sqrt{(K/\rho)} = \sqrt{(\gamma RT)}$, is the velocity of sound waves in the undisturbed air. To obtain true similarity of the flow conditions it would be necessary to have both

$$u_m l_m \rho_m / \mu_m = u_p l_p \rho_p / \mu_p \quad \text{and also} \quad u_m / a_m = u_p / a_p$$

That is $\quad \dfrac{u_m}{u_p} = \dfrac{l_p}{l_m} \cdot \dfrac{\rho_p}{\rho_m} \cdot \dfrac{\mu_m}{\mu_p}$ and $\quad \dfrac{u_m}{u_p} = \sqrt{\left(\dfrac{K_m}{K_p} \cdot \dfrac{\rho_p}{\rho_m} \right)}$

or $\qquad \dfrac{l_p}{l_m} \cdot \dfrac{\rho_p}{\rho_m} \cdot \dfrac{\mu_m}{\mu_p} = \sqrt{\left(\dfrac{K_m}{K_p} \cdot \dfrac{\rho_p}{\rho_m} \right)}.$

It is impossible to obtain this relationship with air or any other usual gases or liquids, for which the possible range of ρ, μ and K are very limited. In the previous paragraph it is shown how difficult it is to get the flow patterns exactly similar by equalizing the two Reynolds numbers; it is, however, easy to get the compressibility phenomena similar because it is only necessary to make

$$\dfrac{u_m}{u_p} = \dfrac{a_m}{a_p} = \sqrt{\left(\dfrac{K_m}{K_p} \cdot \dfrac{\rho_p}{\rho_m} \right)}$$

That is, if the model is tested in the same fluid as the prototype, $K_p = K_m$ and $\rho_m = \rho_p$ so that $u_m/u_p = 1.0$. The model speed is therefore precisely the same as the prototype no matter what the scale of the model may be, if only the compressibility effects are to be made similar. The drag forces due to viscous effects are not now in scale, but again it is found in practice that $\phi(ul\rho/\mu)$ is sufficiently constant over certain ranges of speed that $\dfrac{F_m}{F_p} = \left(\dfrac{u_m}{u_p} \right)^2 \left(\dfrac{l_m}{l_p} \right)^2 \left(\dfrac{\rho_m}{\rho_p} \right)$ as before (equation 10.4).

10.7 The drag force of a ship

An important experimental method has been developed to find from model tests the drag force experienced by a ship. It is a matter of great economic importance to know in advance the power and fuel required to drive a ship at a given speed, and model tests are the only certain way of doing so.

Comapared to the case just described of the drag of a deeply-immersed object (aeroplane or submarine), the new factor to be introduced is the production of surface waves. It is well known that a moving ship produces both a bow-wave and a stern-wave while a pattern of waves (the *wake*) streams out behind. This wave pattern is not by any means coincident with the mass of eddies (*eddy wake*) produced and left behind by the shear of the water on the ship's hull. As the waves travel away from the ship they become lower, and their energy is degraded into thermal energy by the viscosity of the water: this energy must be continuously supplied by the ship to preserve the whole wave system which does not change. The supply of this wave energy causes a drag force which is additional to the frictional drag of the ship's hull as it is being forced through the water. Now the effect of a series of waves is to raise some water above the mean sea-level and to depress other parts of the surface. Such changes of level, which are an increase of potential energy, require a quantity of energy to produce them proportional to the product gh, where h is the height of the waves. The term g must therefore be included in the dimensional analysis because the gravitational acceleration could be changed (if we had the means to do so) without changing any other of the variables u, l, ρ and μ. The term h is not included in the analysis because the height of the waves cannot be changed independently of the other variables: h is in fact a dependent variable fixed by the independent variables in the same sort of way as they fix F.

Thus the dimensional analysis is performed on the variables F, u, l, g, ρ, μ, where l is a length characterizing the size of the ship, the overall length for example. One answer is

$$F = u^2 l^2 \rho \phi(lg/u^2), \ (\mu/\rho u l)$$
$$F = u^2 l^2 \rho \phi_1(ul/v), \ (u/(gl)^{1/2}),$$

ϕ_1 merely meaning 'another function of'. The drag of a surface ship is therefore a function of both the Reynolds number ul/v and also of the Froude number $u/(lg)^{1/2}$. The first expresses, as before, the effect of viscosity on the drag force: the second expresses the effect of surface-wave formation on the drag. In most ship designs the contributions of shear and wave formation to the total drag are of the same order. It is not justifiable to neglect either, but in very high speed ships such as hydroplanes and speed boats, the shear component is relatively small.

The significance of the Reynolds number is shown in Chapter 12 to be that its numerical value indicates the type of eddy pattern caused by shear stresses. In a similar sort of way the numerical value of the Froude number indicates the type of wave pattern caused by the passage of a ship through the water. The speed C of a simple set of waves depends on the wave-length L, the distance between successive wave crests. In deep water $C = (gL/2\pi)^{1/2}$, as will be shown in Chapter 14. Now the ship causes a wave pattern which travels at the same speed u as the ship so that the wave-length of the waves in the pattern is fixed by u, a low speed giving short wave-lengths, and vice versa.* The number of waves in the pattern along the ship's length depends therefore on the speed. If there is only one wave in the ship's length, $L = l$, and if this wave still exactly follows the simple deep water speed law

$$u = C = (gl/2\pi)^{1/2} \quad \text{or} \quad u/(gl)^{1/2} = (1/2\pi)^{1/2} = 0.4.$$

Shorter waves, with more of them in the length l, give lower values of $u/(gl)^{1/2}$. Thus the numerical value of the Froude number $u/(gl)^{1/2}$ indicates the number of waves in the pattern, and indeed the type of pattern itself. The speed law for a forced pattern of waves such as this is slightly different from the simple deep-water law, but C is still proportional to $L^{1/2}$. It is found by experience that

*The speed of the waves can be shown by suddenly stopping the ship, when the bow wave continues at the original ship's speed.

great changes come over the pattern at about $u/(gl)^{1/2} = 1.0$, and at higher
values of the Froude number, the ship rides upon one wave-crest, 'planing'
over it as a high speed motor-boat does. The part of the total drag force due to
the wave generation undergoes a great modification because of these changes,
there being a distinct decrease at about $u/(gl)^{1/2} = 1.0$. However, the part due
to frictional drag is always increasing nearly proportionately to u^2 so that the
combined total drag may not show this decrease, see Fig. 10.5. Ocean-going
ships are always designed for $u/(gl)^{1/2} < 1.0$; (300 m ship at 30 knots gives
$u/(gl)^{1/2} = 0.284$: its model, which might be about 5 m in length and weigh
700 kg would travel at $1.96\,\mathrm{m\,s^{-1}}$, not an excessive speed).

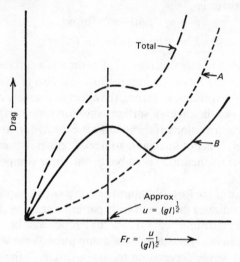

Fig. 10.5 A typical drag curve for a ship which produces waves on the surface of the
sea. *A* is the frictional drag: *B* is the wave drag which is greatly variable with the Froude
number $Fr = u/(gl)^{\frac{1}{2}}$ The total drag often shows a point of inflexion near $Fr = 1$.

Because the pattern of the waves affects the drag it is essential to tow models
of ships at the correct (or corresponding) speed to give the same pattern as does
the full-size prototype. See Figs. 10.6 and 10.7. To do this

$$u_m/(gl_m)^{1/2} = u_p/(gl_p)^{1/2} \quad \text{or} \quad u_m/u_p = (l_m/l_p)^{1/2}.$$

But it is also desirable to tow at such a speed that the flow patterns are similar
and give the same pattern of turbulence in the boundary layers on the ship's
surface. That is

$$u_m l_m/\nu_m = u_p l_p/\nu_p$$

or
$$\frac{u_m}{u_p} = \left(\frac{l_p}{l_m}\right)\frac{\nu_m}{\nu_p}$$

Fig. 10.6 A 26 000-'ton' ship, 180 m long, travelling at its designed speed. The wave pattern at the bow and stern caused by its passage through the water can be seen. A ton' refers neither to the weight force nor to the mass of the ship. It is the internationally agreed name for an enclosed volume of 100 ft^3 = 2.83 m^3.

Fig. 10.7 A 5 m model of the ship of Fig. 10.6 travelling at the corresponding speed. Notice how the wave pattern caused by the model is similar to that of the full-size ship, though at a reduced scale. (Photo by J. Brown & Co., Clydebank.)

It will therefore be seen that the requirement of u_m/u_p for waves to be similar is not the same for the flow patterns to be similar. Equating the right-hand sides of the equations for u_m/u_p, the values of v_m/v_p will be found which ensures that both the eddies and waves are similar, that is

$$\left(\frac{l_m}{l_p}\right)^{1/2} = \left(\frac{l_p}{l_m}\right)\frac{v_m}{v_p} \quad \text{or} \quad \frac{v_m}{v_p} = \left(\frac{l_m}{l_p}\right)^{3/2}.$$

Thus if true similarity is to be achieved for both eddies and waves, the fluid in which a small-scale model is to be tested must be less viscous than the water in which the prototype will float. For a $1:20$ scale model (and this would be an unusually large one for a merchant ship),

$$v_m/v_p = (1/20)^{3/2} = 1/89, \quad \text{when} \quad u_m/u_p = (1/20)^{1/2} = 1/4.47.$$

There are, in fact, no cheap, safe fluids of so small a viscosity that complete similarity can be obtained in this manner.

The engineer is therefore forced to compromise, in a somewhat similar manner to that used in the case of high-speed aircraft. He tests the model by towing it in water at the correct Froude number for the scale used, and measures the total drag F_m. The wave pattern is therefore correctly formed because $u_m/u_p = \sqrt{(l_m/l_p)}$. The friction drag F_{fm} of the model is estimated by assuming that it is the same as would occur if a flat surface of the same area is towed end on at the same speed u_m. The usual friction equation (Chapter 12) of the form $F_{fm} = C_f \rho u^2_m A/2$ is used where C_f is found from boundary layer theory and experiment. The wave drag of the model F_{Wm} is then found as

$$F_{Wm} = F_m - F_{fm}.$$

The wave drag of the prototype ship, F_{Wp}, can now be scaled up from F_{Wm} by the scaling law.

$$F_{Wm}/F_{Wp} = u^2{}_m l^2{}_m \rho_m / u_p{}^2 l_p{}^2 \rho_p$$

Because usually u_p and l_p are much larger than u_m and l_m, $F_{Wm} < F_{Wp}$. Having found F_{Wp}, the friction drag F_{fp} is calculated for the ship by the same friction equation, though C_f for the prototype is almost certainly different from that for the model as the surface of a prototype ship is much rougher than that of the model. Finally the two components F_{fp} and F_{Wp} are added to form the total drag F_p of the ship.

It is usually admitted that the division of the total drag of aeroplanes and ships into two additive and therefore independent portions is quite arbitrary, and that probably the shock-wave or surface-wave resistance is affected by the frictional resistance, and vice versa. Much more advanced research is needed to elucidate these problems for any particular aeroplane or ship form. However, the errors of the above simplified method are thought to be sufficiently small, and are at any rate less than the errors involved in attempting to measure the total drag of prototype ships and aeroplanes and thus to prove the validity of these admittedly crude methods.

10.8 Flow over a weir

The engineer often desires to know the discharge of fluid over a weir (a wall across a stream) and the way in which it depends on the height h (or *head*) of the upstream surface above the top of the weir. In one special case, that of the broad-crested weir, a direct analytical solution can be worked out, depending on Bernoulli's equation (see Chapter 14). Another sort of weir is a sharp-crested wall, an elevation of which is shown in Fig. 10.8. A derivation of the law connecting Q with h can again be made using Bernoulli's equation. This derivation is unsatisfactory, because inappropriate assumptions must now be made in order to solve Bernoulli's equation to obtain the discharge. Experimentally determined coefficients must be introduced to account for the errors introduced by these assumptions, and these coefficients are so large that it seems quite unreal to have employed such a precise statement of hydrodynamics as Bernoulli's equation for a starting-point. It is far more satisfying to treat the problem from the outset as an experimental one, and to use dimensional analysis to see how the variables should be grouped.

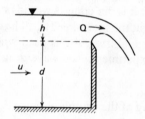

Fig. 10.8 Definition sketch for flow over a weir.

Consider then a uniform sharp-crested weir of height d across a stream of width b. The discharge per unit width, Q/b is the same for all points along the weir and will depend upon the head h of the undisturbed stream above weir crest, upon viscosity μ, density ρ and surface tension T. In addition, the gravitational acceleration g must also be included because gravitational forces are responsible for the falling of the jet over the weir, causing a deformation of the surface. This is exactly the same reason as was used for including g in the dimensional analysis for ship drag forces. Thus Q/b depends on h, d, u, g, ρ, T, μ and dimensional analysis gives one possible grouping of the variables as

$$Q/b = h^{3/2} g^{1/2} \phi(\mu/\rho h \sqrt{gh}), (T/\rho gh^2), (d/h), (u/\sqrt{gh})$$

or $$Q = bg^{1/2} h^{3/2} \phi(Re), (We), (d/h), (Fr)$$

where Re is the form of the Reynolds number applicable to a weir, We is the surface tension (Weber) number, Fr the Froude number, and (d/h) ensures geometrical similarity. As $Q/b = u(h + d)$ the Froude number criterion will be satisfied automatically if the model and prototype flows are geometrically similar. In effect the dimensionless ratio $Q/bg^{1/2}h^{3/2}$ is a form of Froude number.

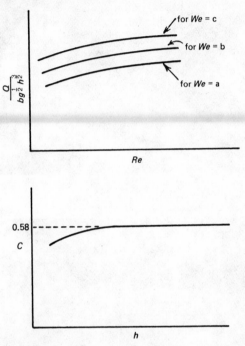

Fig. 10.9 The result of dimensional analysis and experiment on the discharge Q over a sharp-edged weir. (Top) Schematic diagram of how the results might be presented if the surface tension number We is an important variable. (Bottom) Since experiment shows We does not affect $C = Q/bh^{3/2}$, and since a function of $Re = (v/h^{3/2}\sqrt{(2g)})$ can be simplified to a function of h, if v and $\sqrt{(2g)}$ are constants, the simplified presentation of C against h is the one usually employed.

Experiments should therefore be made changing all the variables and the results could be plotted as curves of $Q/bg^{1/2}h^{3/2}$ against the Reynolds number $Re = \mu/\rho h\sqrt{(gh)}$, each curve for a different $We = T/\rho gh^2$. This method of plotting is illustrated in Fig. 10.9. Experiments such as these have been carried out many times, and it has been found that all curves are coincident whatever the value of We, except when h is less than about 1.3 cm. For most purposes, then, the flow over a sharp-edged weir is unaffected by the surface tension of the fluid. However, the surface tension can be of importance in other ways even in prototype structures, Fig. 10.10. Further, over a wide range of Re, $Q/bg^{1/2}h^{3/2}$ is nearly constant at a value of 0.58 for this shape of weir crest.

Fig. 10.10 The flow over Bendora Dam, Australia (a) A 1:40 scale model in the Snowy River Commission's hydraulic laboratory. (b) The bottom of the jet of water at Bendora with a flood corresponding to that of the model study. (Photos by Snowy Mountain Hydro-Electric Authority.)

Other shapes of weir crest have also been tested, but the values of $Q/bg^{1/2}h^{3/2}$ are different from that found for the sharp edge. As will be described in Chapter 14, a value of 0.54 is found for a broad-crested weir providing h is sufficiently small that the flow becomes parallel to the crest. For many designs of curved or *ogee* weirs which are often used in engineering work, $Q/bg^{1/2}h^{3/2}$ may be as high as about 0.75 though it is not constant at this value: such weirs are therefore not suitable as flow-measuring devices unless they have been previously calibrated (see Fig. 10.11).

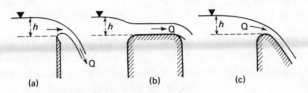

Fig. 10.11 The effect of the shape of the weir on the value of the coefficient $C = Q/bg^{\frac{1}{2}}h^{\frac{3}{2}}$. (a) A sharp-crested weir gives $C = 0.58$, if h is greater than about 1.3 cm. (b) A broad-crested weir gives $C = 0.54$ if it is long enough to create parallel flow over its crest (see Chapter 14). (c) An ogee weir may give C up to 0.75, but this value usually changes considerably with h.

The effects of surface tension are not always similar in model and prototype if the flow velocities are such that air is entrained at the free surface as in Fig. 10.10. While in such cases a model is able to give a good reproduction of the discharge coefficient, it is unable to reproduce the air entrainment characteristics of the prototype, i.e. the 'white water'. Entrained air may increase the volume of the flow sufficiently to require higher side walls on long sloping spillways. As model studies are of limited assistance the engineer has to base the diagram on empirical information from other prototype structures. The surface tension also has an effect on waves on the water surface as shown in Fig. 14.7.

10.9 Conclusion

The method of dimensional analysis of the variables of a physical phenomenon enables the effect of many variables to be studied together. It is not necessary to make any assumptions about the relevance of any particular variable as only experiment will show if such a variable should have been included or not. The method merely suggests the grouping of variables in order to show their effect, but gives no hint of whether any variable is in fact of importance or not. Thus dimensional analysis is peculiarly suited to problems in fluid mechanics which is pre-eminently an experimental science.

Problems

1. Define the Coefficient of Drag of a solid object in a fluid stream. With what variables does it change? What weight must a 13 cm diameter sphere have, if it is to fall in air at N.T.P. at the same speed as a 1 cm sphere falling in water, and weighing 0.01 N in air? (v for air $= 13$ v for water)

Ans. 0.9×10^{-4} N.

2. Particles of stone, density 2650 kg m^{-3}, fall freely at the following speeds in water at 20°C, $\mu = 0.010$ poise.

Particle diameter: mm	0.2	0.5	1.0	2.0	5.0	10
Speed: cm s^{-1}	1.8	2.5	10.2	17	30	42

Plot a curve of speeds for similarly shaped particles of 'Perspex', density 1200 kg m^{-3}, falling in water at 10°C, $\mu = 0.013$ poise. Why should Perspex sand be considered for the bed of civil engineering models of rivers?

Ans. First pair of data gives $d = 0.48$ mm and
$u = 1.0$ cm s^{-1} for Perspex.

3. When tested with water, a metering nozzle 2.54 cm in diameter gave

Pressure Difference:	1.08	3.24	5.38	6.47	m of water
Flow:	2.32	4.01	5.20	5.66	$\times 10^{-3}$ m^3 s^{-1}.

A similar nozzle but 7.6 cm in diameter is to be used to measure air discharging to the atmosphere. Tabulate the air discharges (m^{3}s^{-1}), and pressure differences in cm of water gauge for the range of similarity of the water tests.

Water density $\rho = 1000$ kg m^{-3}, $\mu = 0.0114$ poise

Air $\rho = 1.25$ kg m^{-3}, $v = 0.146$ stokes

Ans. 2.5 cm W.G. gives 0.0906 m^3 s^{-1}, etc.

4. What is the ratio of the windage torques of an electrical alternator running at constant speed in an atmosphere of (*a*) air, (*b*) hydrogen, if the flow in the clearance passages is fully turbulent? Derive an expression for the torque T by dimensional analysis on the variables diameter D, speed N, clearance C, viscosity μ, and density ρ.

Density of hydrogen is 1/14 density of air, and viscosity of hydrogen is 0.51 the viscosity of air.

Ans. $T = N^2 D^5 \rho \phi \left(\dfrac{C}{D} \right) \left(\dfrac{\mu}{\rho N D^2} \right)$ is one grouping: 1/14.

5. A model is to be made of the tail race of a hydro-electric power house. The largest pump available in the laboratory has a maximum flow of 0.07 m^3 s^{-1} and the maximum flood discharge in the actual tail race is 320 m^3 s^{-1}. What

is the largest undistorted scale that can be used for the model? What is the
ratio of the Reynolds numbers for the flow in full-size and model tail race.

Ans. 1/28 : 149 to 1.

6. A one-fourteenth scale model of an aircraft is to be tested in a wind-tunnel,
under conditions dynamically similar to those encountered by the full-scale
prototype flying at a true speed of 350 m s^{-1} at an altitude of 12 km (where
the temperature and pressure are 216.5 K and 18.25 kN m^{-2}). In the wind-
tunnel the temperature is always 288 K, but the pressure can be varied.
What tunnel pressure will be required? Assume the viscosity of air varies
with temperature $T^{3/4}$ approximately.

Show that the forces on the model will be 10% of the corresponding
forces on the prototype. Is the experiment likely to be practicable?

Ans. 365 kN m^{-2}.

Hint. Consider both Mach and Reynolds numbers.

11

Curvature of streamlines

11.1 Pressures in curved flows

The simple applications of Bernoulli's equation in Chapter 8 were made on the assumption that the total energy H of a flow is the same for every streamline, and the same along every streamline. Such a situation is nearly true in a number of cases, but in others there is a change of H across the flow. One important case is that of straightline flow with shear stresses applied at one of its boundaries so that there is a boundary layer wherein the fluid is slowed down: the pressure is constant all over planes normal to the flow so that the total energy $H = u^2/2g + p/\rho g$ decreases near the boundary. Properties of this sort of flow will be discussed in Chapter 12.

Another important case is when the streamlines of the flow are curved. To make the flow follow a curved path it is necessary to have a centripetal force, i.e. a radial pressure gradient. Consider a flow which changes direction around a curve of constant radius without changing speed. If a streamline at radius, r, is drawn then a streamtube can be defined by drawing a second streamline at a radius $r + \delta r$, and the speed of flow, u, will remain unchanged. Drawing two radial control surfaces at an angle 2θ to one another defines a control volume as part of the streamtube, Fig. 11.1. The force–momentum equation can be applied to the control volume, and if the flow is curved in the horizontal plane, the only horizontal forces will be due to the pressures in the fluid.

The radial pressure force on the control volume can be found by taking each control surface in turn and multiplying the pressure force on it by the area projected radially, inward forces being considered positive.

$$\text{Radial pressure force on AB} = (p + \delta p) \times \text{chord AB} \times \text{depth}$$
$$= (p + \delta p) \times 2(r + \delta r)\sin\theta \times d,$$
and the force on DC $\quad = -p \times 2r\sin\theta \times d,$

while for the two control surfaces AD, BC the total radial force

$$= -2\left(p + \frac{\delta p}{2}\right)\delta r \sin\theta \times d.$$

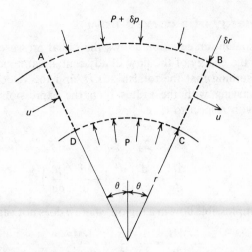

Fig. 11.1 The flow in a curved streamtube. The control volume is formed by streamlines and radial control surfaces.

Then if F_c is the centripetal force on the flow through the control volume ABCD,

$$F_c = 2(p + \delta p)(r + \delta r)d \sin \theta - 2prd \sin \theta - 2 \left(p + \frac{\delta p}{2} \right) d\delta r \sin \theta$$

$$= 2d \sin \theta \left[(p + \delta p)(r + \delta r) - pr - \left(p + \frac{\delta p}{2} \right) \delta r \right]$$

$$= (2r\delta p + \delta r \delta p)d \sin \theta.$$

If second order terms are neglected the force–momentum equation becomes

$$2rd\delta p \sin \theta = \text{change of radial momentum flow rate}$$
$$= \rho u d\delta r (u \sin \theta - (-u \sin \theta))$$
$$= 2\rho u^2 \, d\delta r \sin \theta,$$

or $\qquad\qquad r\delta p = \rho u^2 \, \delta r.$

In the limit $\delta \to 0$, $\quad \dfrac{\delta p}{\delta r} = \dfrac{dp}{dr} = \dfrac{\rho u^2}{r},$

i.e. the radial pressure gradient is the product of the fluid density and the centripetal acceleration. Note the similarity to the hydrostatic equation,

$$\frac{dp}{dz} = -\rho g$$

11.2 The energy of a curved flow

The force–momentum equation allows the radial pressure gradient to be calculated, but the energy of the flow in adjacent streamtubes has yet to be considered. Assuming that the total head, H, and the velocity, u, are both continuously varying with the radius then the expression for H can be differentiated with respect to r, i.e.

$$H = \frac{u^2}{2g} + \frac{p}{\rho g}$$

so

$$\frac{dH}{dr} = \frac{d}{dr}\left(\frac{u^2}{2g}\right) + \frac{d}{dr}\left(\frac{p}{\rho g}\right).$$

Assuming incompressible flow in a region where g does not vary with position gives:

$$\frac{dH}{dr} = \frac{1}{2g}\frac{du^2}{dr} + \frac{1}{\rho g}\frac{dp}{dr}$$

$$= \frac{1}{2g}2u\frac{du}{dr} + \frac{1}{\rho g}\frac{\rho u^2}{r},$$

i.e.

$$\frac{dH}{dr} = \frac{u}{g}\left(\frac{du}{dr} + \frac{u}{r}\right).$$

This relationship can be expressed in terms of angular velocities by considering the flow along two streamlines at radii r and $r + \delta r$ from a common centre of curvature O, Fig. 11.2. If the velocities of flow along the streamlines at

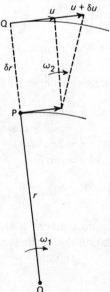

Fig. 11.2 The angular velocities in a flow along two curved streamlines.

P and Q are u and $u + \delta u$ respectively, the angular velocity of P relative to O, ω_1, and Q relative to P, ω_2, are defined as

$$\omega_1 = \frac{u}{r} \quad \text{and} \quad \omega_2 = \frac{\delta u}{\delta r} = \frac{du}{dr},$$

as $\delta r \to 0$. Hence if the total angular rotational velocity of PQ is Ω,

$$\Omega = \omega_1 + \omega_2$$

and

$$\frac{dH}{dr} = \frac{u\Omega}{g} = \frac{u}{g}(\omega_1 + \omega_2) = \frac{u}{g}\left(\frac{du}{dr} + \frac{u}{r}\right).$$

These relationships between the gradient of total head in a flow and the velocity distribution, or angular velocity have important consequences in fluid flows which will be demonstrated by considering a number of particular cases.

11.3 Flows with small curvature, i.e. $r \to \infty$

As the radius of curvature of the flow increases, i.e. the flow approximates more closely to a linear flow, the second term in the equation, u/r, becomes small compared with du/dr and:

$$\frac{dH}{dr} \to \frac{u}{g}\frac{du}{dr} \quad \text{i.e.} \quad \omega_1 = 0 \quad \text{but} \quad \omega_2 \neq 0.$$

Typical of this type of flow is the boundary layer flow past a gently curved surface where the changes in head across the flow depend on the velocity gradient caused by the shear stresses in the flow. If, of course, there is no gradient of velocity $\omega_2 = 0$, then the flow will be uniform and the total head will be the same for all streamtubes.

11.4 The curvature of flows which were initially uniform

If the uniform stream of the preceding paragraph were passed through a curved passage no work would be done on the flow if shear forces are small, and the head in each streamtube would remain the same. Hence $dH/dr = 0$ but r is now finite,

i.e.

$$\frac{dH}{dr} = 0 = \frac{u}{g}\left(\frac{du}{dr} + \frac{u}{r}\right),$$

and so

$$\frac{du}{dr} = -\frac{u}{r}.$$

Separating the variables and integrating gives

$$\frac{du}{u} = -\frac{dr}{r},$$

whence $$\log_e u = -\log_e r + \log_e C,$$

where C is the constant of integration,

or $$u = \frac{C}{r}.$$

This means that if a uniform flow follows a curved path and no work is done on it to change the energy in the stream tubes, then the velocity decreases as the radius increases, and vice versa as shown in Fig. 11.3. This is the *free* or *potential vortex* motion described in Chapter 5 where the plotting of streamlines is discussed in detail.

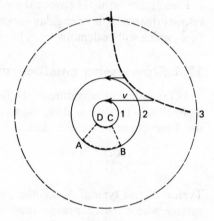

Fig. 11.3 Velocities and streamlines in a free vortex. The control volumes are for calculating the circulation (Section 11.7).

There are two limits to the existence of the free vortex, firstly as the radius of curvature, r, decreases the velocity will increase, eventually becoming infinite at $r = 0$. Secondly as the free vortex motion assumes uniform energy distribution throughout the flow it cannot exist where energy is added to or extracted from the flow, for example by shear forces, (see Section 11.8).

Example

A 50 m canal of rectangular cross-section has a flow at a depth of 5 m. The velocity of flow at a depth of 3 m is equal to the mean flow velocity over the whole cross-section of the canal at 0.5 m s^{-1}. Above this depth the water is flowing slightly faster and nearer the bed the flow velocity has been reduced due to the shear forces.

The flow enters a gentle bend in the canal, the radius of the inside bank being 1000 m and the outside bank 1050 m. Assuming that the depth of flow remains constant estimate the velocity distribution in the bend at a depth of 3 m if the shear forces can be neglected.

Using the velocity distribution calculated find the difference in pressure head between the inside and outside of the bend and check that the assumption of uniform depth was reasonable. How will this head difference affect the flow above and below the depth of 3 m?

Solution

For a gentle bend such as this the flow may approximate to a free vortex if the boundary layers on the banks are neglected. If u_1 is the velocity at the inner bank of radius r_1 and u_2 and r_2 at the outer bank then

$$u_1 r_1 = u_2 r_2 = ur = C$$

Equating the flow through a strip of depth δd at a depth of 3 m before and in the bend

$$Q = ub\delta d = 0.5 \times 50\, \delta d \text{ m}^2 \text{ s}^{-1}$$

$$= \int_{r_1}^{r_2} u\,\delta d\,dr$$

or $$25 \text{ m}^2 \text{ s}^{-1} = \int_{r_1}^{r_2} u\,dr = \int_{r_1}^{r_2} \frac{C}{r}\,dr$$

$$= \left[C\log_e r \right]_{r_1}^{r_2} = C\log_e \frac{r_2}{r_1}$$

As $$r_1 = 1000 \text{ m}, r_2 = 1050 \text{ m},$$

$$\log_e \frac{r_2}{r_1} = \log_e 1.05 = 0.0488,$$

and $$C = \frac{25}{0.0488} = 512.4 \text{ m}^2 \text{ s}^{-1}$$

Calculating the values of the velocity at 10 m intervals (see Table 11.1), and plotting in Fig. 11.4 shows that the velocities will be slightly higher on the inside of the bend.

Table 11.1 Variation of velocity across a bend in a canal.

Radius/(m)	1000	1010	1020	1030	1040	1050
Velocity $= \dfrac{C}{r} /(\text{m s}^{-1})$	0.5124	0.5073	0.5024	0.4975	0.4927	0.4880

Knowing the velocity distribution it is possible to find the difference of pressure head, h, by integrating the expression for the radial pressure gradient divided by ρg.

$$\frac{1}{\rho g}\frac{dp}{dr} = \frac{d}{dr}\left(\frac{p}{\rho g} \right) = \frac{dh}{dr} = \frac{1}{\rho g}\frac{\rho u^2}{r} = \frac{u^2}{rg}$$

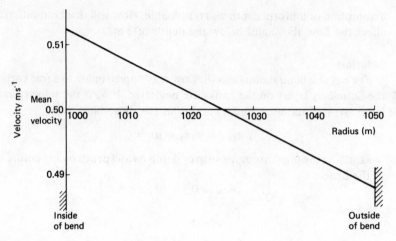

Fig. 11.4 The velocity distribution in a free vortex in a canal bend. Note the false zero of velocity.

Hence

$$\int_{r_1}^{r_2} \frac{dh}{dr} dr = h_2 - h_1 = \int_{r_1}^{r_2} \frac{u^2}{rg} dr = \int_{r_1}^{r_2} \frac{C^2}{r^3 g} dr = -\frac{C^2}{2g}\left[\frac{1}{r^2}\right]_{r_1}^{r_2}$$

so

$$h_2 - h_1 = \frac{C^2}{2g}\left(\frac{1}{r_1^{\,2}} - \frac{1}{r_2^{\,2}}\right) = \frac{u_1^{\,2} - u_2^{\,2}}{2g}$$

$$= 0.0012 \,\text{m} = 1.2 \,\text{mm}$$

A difference of depth of 1.2 mm in 5 metres is hardly significant.

The faster flowing water nearer the surface would need a larger pressure head difference to make it follow the curvature of the bend while the slower water near the bend would require a smaller pressure head difference. This would mean a departure from the hydrostatic pressure distribution in the vertical plane and the possibility of vertical velocities, (see Section 11.8).

11.5 A flow with a uniform angular velocity

This case is again one of circular flow, but the velocity increases proportionally with the radius so that all the fluid has the same angular velocity ω. The velocity distribution and streamlines are shown in Fig. 11.5.

Thus $u = \omega r$ or $u/r = \omega$ and $du/dr = \omega$.

The head equation therefore becomes

$$dH/dr = u/g(\omega + \omega) = 2\omega u/g.$$

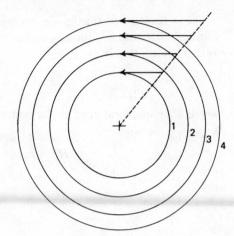

Fig. 11.5 Velocities and streamlines in a forced vortex.

Between two radii r_1 and r_2 the total head of the fluid thus increases by

$$\int_{H_1}^{H_2} dH = \int_{r_1}^{r_2} \frac{2\omega u}{g} \, dr = \frac{2\omega^2}{g} \int_{r_1}^{r_2} r \, dr$$

or

$$H_2 - H_1 = \frac{\omega^2}{g} (r_2{}^2 - r_1{}^2).$$

This sort of circular flow is called a *forced* or *flywheel* vortex (see also Chapter 5). It can be generated by the fluid being whirled around in a container so that every part has the same angular velocity; or it exists when a paddle rotates in a large mass of fluid, though outside the paddle the conditions are more nearly those of a free vortex (Fig. 11.6). In the forced vortex the total energy of each streamline is different, so that although Bernoulli's equation could be applied along each streamline, it cannot be applied to points on different streamlines.

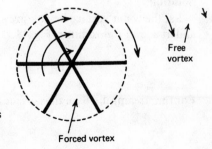

Fig. 11.6 A paddle rotating in a fluid creates a forced vortex within its diameter and a free vortex outside it. In both vortices there is a rise of pressure in an outward radial direction.

The increase of pressure ($p - p_0$) between the centre and a certain radius r of a forced vortex can be found by considering the increase of H from the centre to that radius, that is

$$H - H_0 = \frac{\omega^2 r^2}{g}$$

where H_0 is the total head at the centre.

But since
$$u = 0 \quad \text{at} \quad r = 0,$$
$$H_0 = p_0/\rho g$$

and
$$H = \frac{p}{\rho g} + \frac{u^2}{2g} = \frac{p}{\rho g} + \frac{\omega^2 r^2}{2g},$$

thus
$$\frac{p}{\rho g} + \frac{\omega^2 r^2}{2g} - \frac{p_0}{\rho g} = \frac{\omega^2 r^2}{g} \quad \text{or} \quad \frac{p - p_0}{\rho g} = \frac{\omega^2 r^2}{2g}.$$

It will therefore be seen that large pressure differences can be set up in forced vortices. In fact, they are excellent devices for increasing the total energy of a fluid. Flow work is done on the fluid which increases its pressure, and this flow work plus the equal increase in kinetic energy of the fluid represents the total work done on the fluid. Fluid at low pressure can be fed into the centre of a suitable vortex where it is whirled around and ejected at the periphery with a much-enhanced pressure. This is the principle of a centrifugal pump which is described in Chapter 15. In the other direction, high-pressure fluid can be made to do work in a forced vortex, being ejected at low pressure at the centre, having given up its energy; this is the principle of water turbines.

Example

A washing machine has a drum which has a diameter of 0.5 m, and rotates about a vertical axis at 100 revolutions per minute. What would be the shape of the water surface in the drum and the total head of the flow on its circumference relative to the water at its axis?

Solution

Let the water surface be at a pressure $p_0 = 0$ and level $y = 0$ on the axis, and at a level y at radius r, Fig. 11.7. For the stream tube A at $y = 0$, r, then

$$H_A = \frac{P_A}{\rho g} + \frac{\omega^2 r^2}{2g} = \frac{\omega^2 r^2}{g}.$$

For the streamtube B on the surface at radius r, elevation y, the total head is

$$H_B = y + \frac{\omega^2 r^2}{2g}.$$

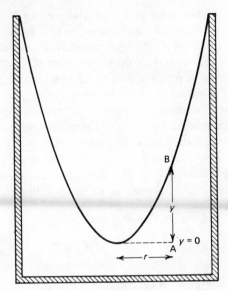

Fig. 11.7 The shape of a free liquid surface in a rotating drum.

As the total head in a forced vortex depends only on the radius $H_A = H_B$, i.e.

$$\frac{P_A}{\rho g} + \frac{\omega^2 r^2}{2g} = y + \frac{\omega^2 r^2}{2g},$$

or

$$y = \frac{P_A}{\rho g} = \frac{\omega^2 r^2}{2g} = Cr^2,$$

where C is a constant with respect to both y and r. The cross section of the water surface is a parabola and the surface a paraboloid of revolution, Fig. 11.7. At the circumference of the drum the radius is R and the level Y where:

$$Y = \frac{\omega^2 R^2}{2g} = \left(\frac{100}{60} \times 2\pi\right)^2 \times \frac{0.25^2}{2 \times 9.81} \frac{s^{-2} m^2}{m\, s^{-2}}$$

$$= 0.35\,m,$$

and,

$$H_R = Y + \frac{\omega^2 R^2}{2g} = 2Y = 0.7\,m,$$

that is, the total head in a forced vortex at any radius is twice the kinetic head at that radius.

11.6 A comparison of forced and free vortices

The free vortex may be thought of as the natural response of a fluid to unsymmetrical motion. It occurs almost anywhere there is non-uniform acceleration such as in a bath where, not too close to the plug-hole, the flow approximates to the combination of a hydrodynamic sink (see Section 5.7(c)) and a free vortex. The free vortex is an example of the conservation of angular momentum; as the radius of rotation of the flow decreases so its velocity must increase to preserve its angular momentum. In contrast the angular momentum of a flow in a forced vortex increases with radius as both its momentum and the moment about the axis of rotation increase. This means that forces and hence accelerations have to be imposed on the forced vortex flow which are not necessary in the case of a free vortex.

If a particle with a tangential velocity u has a low outward radial velocity, then when moving outward a distance δr in a time δt it would naturally decrease its velocity to u_b to conform to a free vortex motion Fig. 11.3, such that

$$ur = u_b(r + \delta r),$$

or

$$u_b = \frac{ur}{r + \delta r} = u\left(1 + \frac{\delta r}{r}\right)^{-1} = u\left(1 - \frac{\delta r}{r} + 0(\delta^2)\right).$$

But if a forced vortex motion is impressed on the flow then the particle would have to increase its velocity from u_b to u_f the forced vortex velocity, Fig. 11.5, where

$$u = \omega r, \text{ and } u_f = \omega(r + \delta r),$$

or

$$u_f = \frac{u}{r}(r + \delta r) = u\left(1 + \frac{\delta r}{r}\right).$$

The change of velocity from u_b to u_f in time δt will require a tangential acceleration, a, where,

$$a = \frac{u_f - u_b}{\delta t} = \frac{1}{\delta t}\left(u\left(1 + \frac{\delta r}{r}\right) - u\left(1 - \frac{\delta r}{r} + 0(\delta^2)\right)\right)$$

$$= \frac{u}{\delta t}\left(\frac{2\delta r}{r} + 0(\delta^2)\right).$$

Taking the limit as $\delta \to 0$ then the relation becomes

$$a = \frac{2u}{r}\frac{dr}{dt} = 2\frac{u}{r}u_r = 2\omega u_r,$$

where u_r is the radial velocity. This acceleration is known as the Coriolis acceleration and is of importance in the flow of large masses of air or water.

With a large mass of fluid on the surface of the earth any radial inflow or outflow tends to behave like a free vortex, but the earth's rotation imparts Coriolis accelerations to the fluid in the boundary layer in an attempt to form a forced vortex to match the solid body rotation of the earth. In the atmosphere the large cyclones and anti-cyclones are free vortices with radial inflows and outflows respectively. It is the Coriolis accelerations given to these flows which makes the surface winds flow parallel to the lines of constant pressure (isobars) rather than normal to them as in linear fluid flow.

11.7 The strength of a vortex motion

The fact that in a free vortex the total head remains constant has consequences beyond merely determining the velocity distribution. For example as H is constant the head equation becomes

$$\frac{dH}{dr} = \frac{u}{g}(\omega_1 + \omega_2) = 0,$$

and the only non-trivial solution to this equation is that $\omega_1 = -\omega_2$. Referring to Fig. 11.2, it can be seen that this results in the line PQ having zero angular velocity, i.e. its motion is *irrotational* and its *vorticity* is zero. These terms are often used to describe flows for which the velocity potential of Chapter 5 can be calculated and even lead to the apparently contradictory description of the free or *irrotational vortex*. It is this property of irrotationality which enables a free vortex flow to occur without energy dissipation due to shear forces within the flow itself even in a real fluid.

As a free vortex does not produce vorticity or rotational motion in a flow it is necessary to define another property which will quantify the strength of the vortex. This can be done by considering an element of a control surface, da across which there is a flow with a velocity V, whose components u, v, are normal and parallel to da respectively, Fig. 11.8. If the control surface is closed to form a control volume then the volumetric flow rate into or out of the control volume, Q, as explained in Chapter 4, is given by integrating the expression uda around the control volume in an anti-clockwise direction,

$$Q = \oint u\,da.$$

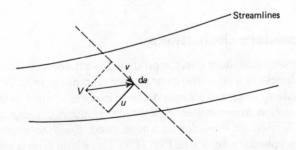

Fig. 11.8 The components of velocity normal and parallel to a control surface.

Considering the other component of velocity, v, along the control surface, it is possible to define a quantity Γ such that

$$\Gamma = \oint v\, \mathrm{d}a$$

where Γ is a measure of the flow around the perimeter of the control volume and is known as the *circulation*.

Example

Show that for a free vortex flow the circulation is the same for any control volume enclosing the vortex, but zero for any control volume not enclosing the vortex.

Solution

As the streamlines in a vortex flow are circles the simplest control volume around which to calculate the circulation is a streamline, (Fig. 11.8), along which the velocity is constant.

Hence $\qquad\qquad\qquad \Gamma = V \times 2\pi r,$

but for a free vortex $\qquad\qquad V = \dfrac{C}{r},$

whence $\Gamma = 2\pi C$ and is independent of r and the circulation does not depend on the size of the control volume.

For the control volume ABCD which does not include the vortex the circulation in an anti-clockwise direction must be calculated for each control surface and added. If θ is the angle subtended at the centre of the circle:

$$\Gamma_{AB} = V_A r_A \theta \quad = C\theta$$
$$\Gamma_{BC} = \Gamma_{DA} \quad = 0$$
$$\Gamma_{CD} = -V_c r_c \theta = \underline{-C\theta}$$
$$\text{Adding } \Gamma_{ABCD} = 0$$

Therefore the circulation for any control volume not enclosing the vortex is zero.

11.8 Secondary circulations

An important secondary effect is produced when flows contained between solid boundaries are bent round a corner (see example, p. 169). Water moving round a bend in a river, and air round a bend in a wind tunnel or in a duct of a ventilating system, are examples. Against the boundary corresponding to the bottom of a river, or the sides of a pipe bend, the fluid is slowed down, producing a boundary layer (see Fig. 11.9). The whole flow is now *not* two-dimensional as was assumed in 11.1, because the rotation is less in the

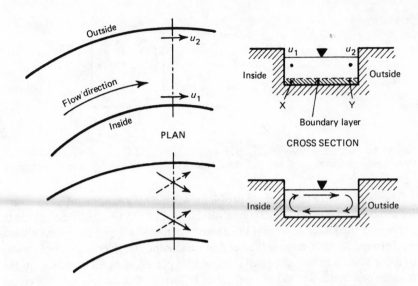

Fig. 11.9 The flow of water in a bend of a river. (Above) The velocities u_1 and u_2 form part of a free vortex motion caused by the bending of the streamlines and so cause a radial pressure gradient. The velocities in the boundary layer on the bottom are less than u_1 and u_2 on the surface so the radial pressure gradient on the bottom is less than that on the top. (Below) Consequently a spiral secondary motion is set up as shown in the cross-section, and also on the plan where the dotted arrows show the surface water movement, and the full arrows the bottom movement.

boundary layer than in the centre of the stream. The bending of the streamlines causes a free vortex so that there is a radial increase of pressure everywhere of

$$\Delta p = (u_1{}^2 - u_2{}^2)/2g,$$

where u_1 is the velocity at the inner wall and u_2 is that at the outer wall (see Fig. 11.9). But the velocities $u_1{}'$ and $u_2{}'$ in the boundary layer at X and Y are less than u_1 and u_2 in the main stream above, so that the pressure rise $\Delta p' = (u'_1{}^2 - u'_2{}^2)/2g$ in the boundary layer is less than the rise $\Delta p = (u_1{}^2 - u_2{}^2)/2g$ in the main stream. If the pressures in boundary layer and main stream are the same at the middle of the river, then at the outer wall the pressure in the boundary layer will be less than the pressure higher in the stream by an amount $\frac{1}{2}(\Delta p - \Delta p')$. At the inner wall the pressure difference is reversed. A current is therefore set up by these pressure differences that gives velocities down the outer wall, inwards across the bottom, up the inner wall and outwards across the top. This current, together with the main stream, gives a spiral combined flow, as shown in Fig. 11.9, where full arrows show bottom boundary layer flow, dotted arrows show surface flow. The effect is very noticeable in a cup of tea, for the tea-leaves are brought to the centre of the cup when it is stirred. Very similar spiral or *secondary* flows occur in closed ducts, where there are usually two spirals in opposite directions (see Fig. 11.10).

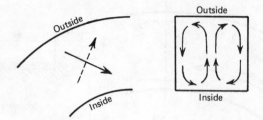

Fig. 11.10 Secondary flows in a bend of a rectangular pipe or duct. Dotted arrows show centre movement, full arrows the movement against the sides.

These secondary flows create problems for the hydraulic engineer in rivers, for the bottom flow will be carrying sand and silt from the outside of bends to the inside; the inside of a bend therefore becomes shoal and the outside bank tends to erode, thus making the bend more abrupt. If a river is to supply water to a canal or pipeline, the intake should be placed, if at all possible, on the outside of a bend, where the sand tends to move inwards, so as to reduce the quantity of sand taken in. See Fig. 11.11. In pipes and wind tunnels secondary

Fig. 11.11 The photo above shows a small straight river, with a canal at right-angles taking about thirty per cent of the river flow. Notice on the river bank opposite the canal an area of low speed and eddies: and in the canal entrance a separation (= breakaway) of the flow, causing a large permanent eddy on one side. As well as aluminium dust, blobs of heavy water-colour paint have been put on the floor of the river to show the direction of the flow there—which guides the movement of sand and other bed material. Due to secondary currents (see page 176) the bottom current is curved far more than the surface; thus all the bottom sediment in the river (coming from the left) goes into the canal.

Downstream, near the canal, there are only very small bed-velocities, and this is where siltation easily occurs. A canal in this situation always has a higher concentration of silt than exists in the river. (Photo by J. Gurr.)

flow is objectionable because the spiral persists well downstream of the bend that produces it. If it is desired to have uniform straight-line flow (as in the working section of wind tunnels), then it is essential to suppress this effect, or to work an unacceptably large distance downstream of the bend. One way of suppressing secondary flows is to guide the main flow around corners by closely spaced turning vanes (Fig. 11.12). Small-scale secondary flows are set up between the vanes, but these are dissipated in a much shorter distance than is a large secondary flow from a bend without guide vanes.

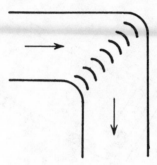

Fig. 11.12 Turning vanes in a pipe bend or corner of a wind tunnel.

11.9 The lift of aerofoils

In Chapters 2 and 6 the lift generated by an aerofoil has been explained by the pressure distribution on the aerofoil and the change of momentum flow rate in a vertical direction caused by the presence of the aerofoil. The change of direction of flow to produce the downward flow of momentum of necessity requires curvature of the streamlines which is associated with vortex motion. In streamline plotting it is possible to produce this type of flow pattern by combining the flow round an object obtained from a combination of sources and sinks with a free vortex of circulation Γ, Fig. 11.13. The lift on the body has been shown to be $\rho U \Gamma$, and in effect the object of an aerofoil is to generate the circulation from which the lift originates.

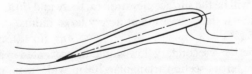

Fig. 11.13 The theoretical flow pattern around an aerofoil with a vortex providing circulation.

When a three-dimensional wing of limited span is considered as being represented by a free vortex there is the problem that it is not possible for a vortex to end in space. This can only be resolved by the continuation of the vortex from the ends of the wing in the familiar *wing tip vortices* which finally close with a vortex equal in magnitude but opposite in sign which was formed when the wing started to generate lift (Fig. 11.14).

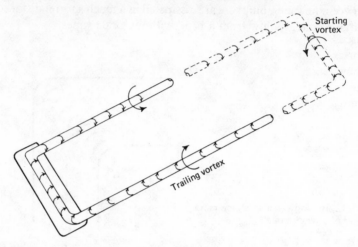

Starting vortex

Trailing vortex

Fig. 11.14 The representation of a lift producing wing as a vortex with the trailing and starting vortices.

Problems

1. A tube 50 cm long and closed at one end is half full of mercury ($\rho = 13.6 \times 10^3 \, \text{kg m}^{-3}$). If the tube is rotated about the open end in a horizontal plane at a constant rate of 180 rev min^{-1}, what pressure will exist at the closed end? *Ans.* 458 kN m^{-2}.

2. Find the maximum speed at which an open canister 25 cm diameter and 25 cm high can be rotated without spilling, if at rest it has been half filled with water.
   Hint. The volume of a paraboloid of revolution is half the volume of the circumscribing cylinder. *Ans.* 168 rev min^{-1}.

3. If the tangential velocity of a free vortex in water is 10 cm s^{-1} at a radius of 0.5 m, what will be the surface elevation (*a*) there and (*b*) at a radius of 3 cm, relative to the surface elevation at a very large radius?
   Ans. (*a*) 1.02 mm; (*b*) 28.3 cm.

4. Estimate whether a secondary flow would be prevented in a bend of a river, radius 600 m, cross-section triangular 100 m wide, max. depth 7.5 m, if a layer of heavily silt charged water, s.g. 1.04, existed on the bottom. The mean speed is 1 m s^{-1}.

Ans. Secondary prevented: free vortex gives press. difference less than that required to create a slope of the heavy fluid equal to the bed slope.

5. Eddies often cause temporary local depressions in the surface of a stream. The form of these eddies may be approximated by a forced vortex of radius R within a concentric free vortex. Compute the depth of the depression, H, of the surface if the central angular velocity of the vortex is ω. Show that the flow of angular momentum across a radial plane between $r = 0$ and $r = \infty$ is $4/3\ \rho R^3 \omega^2 D$, where D is the depth of the stream and $D \gg H$. Briefly discuss the effect on the central angular velocity if the stream carries the vortex into a region where the depth is $D/2$.　　　*Ans.* $H = R^2 \omega^2/g$.

6. Show that for a forced vortex of rotational velocity ω, the vorticity is 2ω, if the vorticity is defined as the circulation around a two dimensional control volume divided by the area of the control volume.

12

The production of shear forces in boundary layers

12.1 The basis of boundary layer theory

In several preceding chapters reference has been made to the effects that boundary layers have on other characteristics of a fluid flow. It is now appropriate to describe the properties of the fluid flow inside the boundary-layer and, in particular, to find the shear stress which has caused the boundary layer to form on the solid surface.

The concept of the boundary layer is due to Prandtl who, in 1904, realized that the effect of the shear stress on flow past a solid surface can be confined within a relatively thin layer of fluid adjacent to the surface, unless the flow breaks away and leaves a wake or region of separated flow (p. 46). Outside this thin boundary layer the shear stress has no effect, and any changes in velocity there are due solely to the distortion of streamlines by the solid surface, as given by the theoretical methods of Chapter 5. This idea now seems simple and obvious, and it has enabled a link to be made between theoretical ideal fluid hydrodynamics, which ignores shear forces, and experimental evidence with real fluids.

The essential assumption made in the theory of the boundary layer is that in a viscous fluid a shear stress at a point always produces a rate of strain there. As was shown in Chapter 1, a rate of strain is equivalent to a gradient of velocity in the stream, the direction of the gradient being across the direction of the stream (and of the stress). If there is no velocity gradient present, then there is no stress. If the flow is purely laminar, then there is a simple linear relationship between the stress τ and the velocity gradient du/dy: the constant of proportionality is called the coefficient of viscosity, μ, for the fluid, where $\tau = \mu \, du/dy$. More complicated relations hold good for turbulent flow but the shear stresses still depend on the velocity gradients across the flow. Solid objects, of course, distort the streamlines of flow around them as has been described in Chapter 5, and provide the boundaries which generate velocity gradients, with accompanying shear stresses, the shear stresses and velocity gradients due to boundary layer effect being additional to those caused by the distortions of the streamlines.

12.2 The boundary layer on a plane surface

A boundary layer on a plane surface parallel to the direction of the undisturbed flow is the simplest case to study. Such a surface, if exposed to a flow of ideal fluid, would not cause any velocity gradients at all. With real fluids, the flow is decelerated near the surface because of the shear stresses at the boundary, and the velocity gradients which result are due solely to these shear stresses.

The main properties of a boundary layer on a flat surface with a uniform pressure at every place on it are shown in Fig. 12.1. The boundary layer commences at the leading edge of the surface and grows thicker in a downstream direction. Within the first part, A in the figure, the flow is observed to be entirely laminar, whatever the roughness of the surface may be, or whether the oncoming, undisturbed, stream is laminar or turbulent. Apparently the random movements of turbulent motion cannot exist in a thin boundary layer which has a large mean value of the velocity gradient du/dy in it. After some distance in the laminar boundary layer the fluid motion becomes unsteady to some extent, and if sensitive velocity measuring instruments are placed there they show a periodic wavering of the velocity. These oscillations quickly degenerate into irregular motions, that is, fully developed turbulence. The short region B in which this change takes place is called the *transition boundary layer*. The thickness of the layer rapidly increases in this region. At all

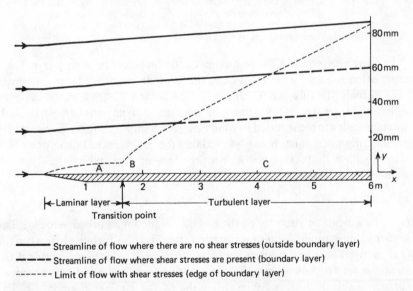

Fig. 12.1 Streamlines of the flow over a flat plate in a water flow of $1\,\mathrm{m\,s^{-1}}$. Note the exaggeration of the vertical scale and the displacement of streamlines outside the boundary layer.

places downstream of the transition the flow in the boundary layer is turbulent, with the thickness of the layer increasing, part C. If there is no other boundary to the flow there is no limit to the thickness of the layer as there will be no breakaway on a flat surface parallel to the flow. The *turbulent boundary layer*, continues to grow indefinitely, providing the surface continues downstream. Fig. 12.2 compares the shapes of the laminar and turbulent velocity distributions in the boundary layer.

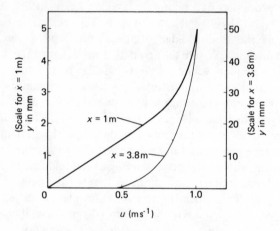

Fig. 12.2 A comparison of the velocity distributions in a boundary layer on a flat plate in water at 1 m s⁻¹. The laminar boundary layer velocity distribution occurs at 1m from the leading edge and the turbulent distribution at 3.8 m. (see Fig. 12.1) Note that the turbulent boundary layer thickness is ten times the laminar.

For engineering work the properties of the turbulent boundary layer C are most often required as in most cases the laminar boundary layer exists over only a small part of a surface. However, the laminar layer A, which has quite different properties, is sometimes of predominant importance, particularly when models are being tested in wind tunnels, in ship-testing tanks, or for civil engineering work. It has been observed that the laminar flow breaks down into the transition B at a distance X from the leading edge, where the Reynolds number based on X has a value

$$XU/v = 5 \times 10^5 \quad \text{to} \quad 2 \times 10^6$$

(v is the kinematic viscosity of the fluid; U is the undisturbed velocity). The higher value is applicable if the oncoming stream is non-turbulent, or when the surface itself is travelling into still fluid; the lower value is applicable if the stream is very turbulent, but XU/v is never smaller than 5×10^5 on a smooth surface. Table 12.1 shows typical lengths of the laminar boundary before transition at different flow velocities in water and air assuming that the transition Reynolds number is 1.58×10^6.

Table 12.1 Approximate thickness of boundary layers on a flat plate. Note how the thickness rarely exceeds 1 % of the length of the plate.

Stream speed typical of river or tidal flow

Water, $v = 10^{-6}\,m^2\,s^{-1}$ Stream speed $1\,m\,s^{-1}$	End of laminar boundary layer	Turbulent boundary layer		
Distance from leading edge $x/$(m)	1.58	2	4	16
Boundary layer thickness $\delta/$(mm)	6.3	18	53	206
Displacement thickness $\delta_1/$(mm)	2.17	2.25	6.65	25.8
Momentum thickness $\delta_2/$(mm)	0.83	1.75	5.17	20.07

Flow speed typical of ship

Water, $v = 10^{-6}\,m^2\,s^{-1}$ Stream speed $10\,m\,s^{-1}$	End of laminar boundary layer	Turbulent boundary layer		
Distance from leading edge $x/$(m)	0.16	4	16	256
Boundary layer thickness $\delta/$(mm)	0.63	45	139	1280
Displacement thickness $\delta_1/$(mm)	0.22	5.6	17.3	160
Momentum thickness $\delta_2/$(mm)	0.08	4.4	13.5	124

Airspeed of fresh breeze

Air, $v = 1.5 \times 10^{-5}\,m^2\,s^{-1}$ Stream speed $10\,m\,s^{-1}$	End of laminar boundary layer	Turbulent boundary layer		
Distance from leading edge $x/$(m)	2.3	4	16	256
Boundary layer thickness $\delta/$(mm)	9.3	46	216	2180
Displacement thickness $\delta_1/$(mm)	3.2	5.8	27	273
Momentum thickness $\delta_2/$(mm)	1.2	4.5	21	212

Airspeed of large aircraft climbing or small aircraft cruising

Air, $v = 1.5 \times 10^{-5}\,m^2\,s^{-1}$ Airspeed $= 100\,m\,s^{-1}$	End of laminar boundary layer	Turbulent boundary layer		
Distance from leading edge $x/$(m)	0.23	2	4	16
Boundary layer thickness $\delta/$(mm)	0.93	26	48	150
Displacement thickness $\delta_1/$(mm)	0.32	3.3	6.0	19
Momentum thickness $\delta_2/$(mm)	0.12	2.6	4.7	14

If a model bridge or structure is tested in a wind tunnel at the same speed U as the full-size prototype is expected to operate in the atmosphere, then the distance X to the transition point will be the same for both model and prototype. But in the model, X may be a large proportion of the total length, whereas in the prototype it may be only an insignificant fraction. The drag force due to a turbulent boundary layer is quite different from that due to a laminar one, so a misleading result may be obtained from such tests. As the atmosphere is usually in a less turbulent state than the air in a wind tunnel, the value of UX/v at transition will be rather higher for the prototype than for the model, that is, X will be rather greater in the prototype; but it would be purely fortuitous if the increase of X for this reason put the transition point B at the same relative place on the prototype as was observed on the model. An early transition may be artificially produced on a model by putting roughness on the surface, such as a series of wires stretched across the flow. Eddies are produced behind each wire which produce drag forces more nearly that of the naturally formed turbulent boundary layer.

12.3 Growth of boundary layers

As a boundary layer grows thicker, more fluid is decelerated from its original undisturbed velocity, so that the momentum of the fluid in a direction parallel to the solid surface is steadily decreased. Such a decrease of momentum can only be caused by retarding forces, and these are the tangential drag force on the surface and the pressure forces in the direction parallel to the surface caused by changes of pressure intensity (if any). But there are also important changes of the velocity component in the direction normal to the surface. If a rectangular control volume XX_1YY_1 (Fig. 12.3) is drawn to enclose a length of a boundary layer, since the boundary layer grows thicker downstream, the total quantity of fluid entering the control volume through XX_1 is greater than the quantity leaving across YY_1, where the mean speed of the fluid is less. There can be no velocity component normal to the surface X_1Y_1, so fluid must leave through XY. Thus there will be a small velocity component normal to XY and a consequent flow of momentum in this direction. An outward acting pressure force must therefore exist to provide this momentum so that there is a small outward pressure gradient in the fluid.

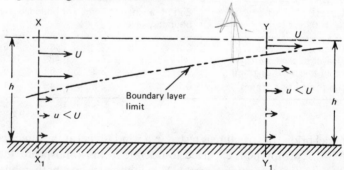

Fig. 12.3 Growth of a boundary layer. Since the mean speed across YY_1 is less than that across the same height XX_1, more fluid enters the boundary layer at XX_1 than leaves it at YY. Thus there must be an outward component of velocity across XY. Pressures at X, X_1, Y, Y_1 are not all necessarily the same.

The outward pressure gradient interacts with the pressure gradients along the surface to form a complicated system of forces and velocities which can only be explained by more advanced mathematical methods than will be given in this volume (they are however given in more detail in Boundary Layer Theory, H. Schlichting, McGraw-Hill). A simpler (and far less rigorous) description will be given here, which ignores the outward components of velocity and momentum, and assumes that the pressure is hydrostatic at right angles to the flow direction. These assumptions are tantamount to restricting the argument to fluids of low viscosity, which produce relatively thin boundary layers growing but slowly in thickness. These assumptions were implicit in the simple force–momentum analysis of the boundary layer in Section 6.5 page 93.

12.4 Properties of the laminar boundary layer

Though in normal circumstances the laminar boundary layer is so short, Table 12.1, its properties will be discussed in detail to demonstrate the general method of calculation for all boundary layers.

In any laminar flow the relation between stress τ and velocity u is given by the equation $\tau = \mu \, du/dy$, where y is measured in the direction at right angles to the direction of u. A prediction of the velocity distribution involves knowing or predicting how τ varies from the value τ_0 at the solid surface to zero at the outer edge of the boundary layer. In turn, it is τ_0, the drag of the fluid on the surface, that the engineer essentially desires to know. There is no known method whereby τ can be measured directly so that an indirect approach must be made. A distribution of τ will be assumed, the corresponding velocity distribution calculated and then compared with experiment.

The simplest and crudest measurements within a boundary layer on a flat surface indicate that the velocity decreases towards the surface with a smooth distribution curve, see Fig. 12.4(a)(ii); there is little evidence of a point of inflexion, which would imply that the gradient du/dy, and therefore τ, first increases to a maximum, then decreases again, Fig. 12.4(b)(iv). Such a behaviour of τ is unlikely: it is far more likely that τ should increase to a maximum, τ_0, at the surface, though not necessarily at a uniform rate. It is also unlikely that τ should be constant over the whole of the boundary layer thickness δ, falling off to zero suddenly at $y = \delta$, Fig. 12.4(b)(i). As a first approximation, it will be assumed that τ is varying at a constant rate throughout the layer, Fig. 12.4(b)(ii), so that $\tau = \tau_0(1 - y/\delta)$. As the condition $\tau = \mu \, du/dy$ is inherent to any wholly laminar flow, the stress equation is then

$$\mu \frac{du}{dy} = \tau_0(1 - y/\delta).$$

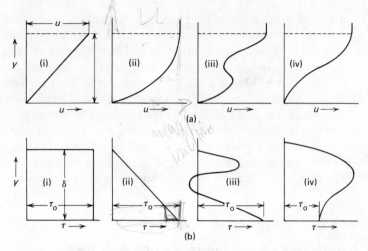

Fig. 12.4 (a) Suggested velocity distributions at a place in a laminar boundary layer at constant pressure and (b) corresponding distributions of shear stress. Distribution (ii) is similar to measured velocities.

Integrating each side, $u = \dfrac{\tau_0}{\mu}\left(y - \dfrac{y^2}{2\delta}\right) + \text{Constant.}$

The boundary conditions $y = 0$ when $u = 0$ gives Constant $= 0$; and the condition $y = \delta$ when $u = U$ gives $U = \dfrac{\tau_0}{\mu}\dfrac{\delta}{2}$.

Thus $U - u = \dfrac{\tau_0 \delta}{2\mu} - \dfrac{\tau_0}{\mu}\left(y - \dfrac{y^2}{2\delta}\right)$

or $(U - u) = \dfrac{\tau_0}{2\mu\delta}(\delta - y)^2 = \dfrac{U}{\delta^2}(\delta - y)^2.$ (12.1)

This equation shows that the velocity distribution at a given cross section of a laminar boundary layer is a parabola, *if the stress varies linearly.* Since the boundary condition gave $U = \dfrac{\tau_0}{\mu}\dfrac{\delta}{2}$, the all-important stress at the surface is

$$\tau_0 = 2\mu U/\delta$$ (12.2)

Considering the control volume shown in Fig. 6.9, the velocity distribution and thickness δ can be used now to determine the drag force on the flat surface, which has caused the slowing of the fluid. The reduction of momentum flow rate of the fluid passing through the control volume is equal to the shear force. The full reasoning has been given in Chapter 6, where it is shown that the force

F per unit width of the surface acting on the fluid in the direction of flow is

$$F = \int_0^\delta \rho u(u - U)\,dy. \tag{6.7}$$

Substituting for u from equation 12.1, see Example on page 190, it is found that

$$F = -\frac{2}{15}\rho U^2 \delta \tag{12.3}$$

which is the force on the fluid.

The reaction $-F$ is applied to the boundary layer along the surface from the leading edge back to the control surface where the boundary layer thickness is δ. It should not be thought that F is uniformly distributed along this distance X; in fact, the shear stress τ_0 changes along X so that F must be written as the integral of the elementary forces $\tau_0\,dx$, from $x = 0$ at the leading edge to $x = X$.

That is

$$-F = \frac{2}{15}\rho U^2 \delta = \int_{x=0}^{x=X} \tau_0\,dx$$

Substituting $\tau_0 = 2\mu U/\delta$ this equation can be solved by separation of the variables δ and x giving

$$X = \rho U \delta^2/30\mu$$

or

$$\delta^2 = 30\frac{\mu}{\rho}\frac{X}{U}$$

δ is variable with X.

or

$$\delta/X = \sqrt{30}(UX/v)^{-1/2} = 5.48(UX/v)^{-1/2} \tag{12.4}$$

where $v = \mu/\rho$ is the kinematic viscosity of the fluid, see Chapter 1.

An expression has thus been derived for the thickness of a laminar boundary layer as depending on the distance from the leading edge, the undisturbed velocity, and the kinematic viscosity of the fluid. An experimental check of this theoretical equation, by determining δ at a number of places X, would show whether or not the initial assumption of the linear distribution of τ with y is correct. This is a difficult experiment to do because it involves finding where the velocity first becomes less than U. The parabola of the velocity distribution curve shows that the slope du/dy is small near $y = \delta$ so that small errors in determining u give large errors in y.* However, careful experiments have at least shown that the above expression is not gravely in error, a value of 5.0 being generally accepted for the numerical constant.

* Not only is the accurate measurement of δ made difficult by the small slope of the velocity curve at the outside of the boundary layer, but δ itself is so small that instruments for measuring u take up so much space that inaccuracies result, see Table 12.1. Very fine instruments, such as small-diameter Pitot tubes, or hot wire anemometers, must be used. One simplification is to use the displacement thickness, δ_1, or the momentum thickness, δ_2, in place of the less easily measured δ. The definitions of δ_1, and δ_2 are given on pages 192 and 197.

Example

Estimate the thickness δ of a boundary layer 0.3 m from the nose of a thin aerofoil at zero incidence in an air flow of $U = 9\,\mathrm{m\,s^{-1}}$ with $\mu = 1.8 \times 10^{-5}\,\mathrm{kg\,m^{-1}\,s^{-1}}$ assuming the velocity in the boundary layer is given by:

$$u/U = (2y/\delta) - (y/\delta)^2.$$

How far away from the aerofoil are the streamlines outside the boundary layer displaced? What is the total shear force over the first 0.3 m of the aerofoil?

Air density $= 1.3\,\mathrm{kg\,m^{-3}}$

Solution

As the aerofoil is thin and at zero incidence the pressure will be assumed to remain constant along the surface.

Reynolds number at 0.3 m from nose

$$Re = \rho U x/\mu = 1.3 \times 9 \times 0.3/1.8 \times 10^{-5}$$
$$= 1.95 \times 10^5.$$

As transition does not generally occur on a flat plate until $Re > 5 \times 10^5$, it can be assumed that the flow is laminar.

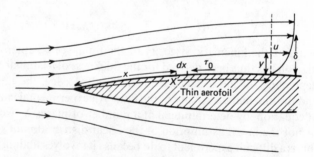

Fig. 12.5 The flow past the leading part of a thin aerofoil.

Using the momentum theorem, equate the total shear force to the rate of change of momentum in the boundary layer up to the distance X.

$$\text{Total shear force} = \int_0^X \tau_0\,\mathrm{dx},$$

but with laminar flow $\tau_0 = \mu(\mathrm{d}u/\mathrm{d}y)_{y=0}$

or
$$\tau_0 = \mu U \left(\frac{d}{dy} [(2y/\delta) - (y/\delta)^2] \right)_{y=0}$$
$$= 2\mu U/\delta.$$

So total shear force $F = \displaystyle\int_0^x \frac{2\mu U}{\delta}\, dx$, which cannot be evaluated as δ is an unknown function of x.

Now rate of change of momentum in the boundary layer

$$= \rho \int_0^\delta u(U - u)\, dy \text{ per unit width of the aerofoil,}$$

$$= \rho U^2 \delta \int_0^1 \frac{u}{U}\left(1 - \frac{u}{U}\right) d\left(\frac{y}{\delta}\right)$$

$$= \rho U^2 \delta \int_0^1 (2(y/\delta) - 5(y/\delta)^2 + 4(y/\delta)^3 - (y/\delta)^4)\, d(y/\delta)$$

$$= \rho U^2 \delta \left[\left(\frac{y}{\delta}\right)^2 - \frac{5}{3}\left(\frac{y}{\delta}\right)^3 + \left(\frac{y}{\delta}\right)^4 - \frac{1}{5}\left(\frac{y}{\delta}\right)^5 \right]_0^1$$

$$= \frac{2}{15} \rho U^2 \delta.$$

This rate of change of momentum also equals the shear force at the boundary, so:
$$F = \int_0^x \frac{2\mu U}{\delta}\, dx = \frac{2}{15} \rho U^2 \delta.$$

Differentiating with respect to x

$$2\mu U/\delta = \frac{d}{dx}\left[\frac{2}{15} \rho U^2 \delta \right] = \frac{2}{15} \rho U^2 \frac{d\delta}{dx}$$

or
$$15\mu/\rho U = \delta \frac{d\delta}{dx} = \frac{d}{dx}\left(\frac{\delta^2}{2}\right).$$

Now integrating, $\displaystyle\int_0^x (15\mu/\rho U)\, dx = \delta^2/2$

or
$$\delta^2 = 30\mu X/\rho U.$$

Hence
$$\frac{\delta}{X} = \sqrt{(30\mu/\rho U X)} = (30/Re)^{1/2}$$

So when $X = 0.3\,\text{m}$, $\delta = 0.3(30/1.95 \times 10^5)^{1/2} = 3.72\,\text{mm}$.

The distance the streamlines are displaced is the displacement thickness of the boundary layer δ_1. Applying the mass-continuity equation between a streamline just enclosing the boundary layer and the surface of the aerofoil then

$$\rho(\delta - \delta_1)U = \int_0^\delta \rho u \, dy,$$

or

$$U\delta_1 = \int_0^\delta (U - u) \, dy.$$

$$\delta_1 = \delta \int_0^1 \left(1 - \frac{u}{U}\right) d\left(\frac{y}{\delta}\right)$$

$$= 3.72 \, \text{mm} \int_0^1 (1 - (2y/\delta) + (y/\delta)^2) \, d\left(\frac{y}{\delta}\right)$$

$$= 3.72 \left[\frac{y}{\delta} - \left(\frac{y}{\delta}\right)^2 - \frac{1}{3}\left(\frac{y}{\delta}\right)^3\right]_0^1 \, \text{mm}$$

$$= 1.24 \, \text{mm}.$$

Substituting in the expression for total shear force:

$$F = \frac{2}{15} \times 1.3 \times 81 \times 3.72 \times 10^{-3} \frac{\text{kg}}{\text{m}^3} \frac{\text{m}^2}{\text{s}^2} \, \text{m}$$

$$= 0.052 \, \text{N} \, \text{m}^{-1}.$$

This demonstrates how small the shear forces are due to laminar flow over short surfaces.

A much better test of the validity of the stress assumption is the experimental determination of the drag force F, which is then compared with the theoretical values. It is convenient, as was described in Chapter 8, to express drag forces as Coefficients of Drag, which in this case of tangential stress on surfaces is commonly called the *Coefficient of friction* C_f. This is defined as

$$C_f = \frac{\text{Measured drag}}{\frac{1}{2}\rho U^2 \times \text{Area}} = \frac{F}{\frac{1}{2}\rho U^2 X}.$$

The theoretical values of C_f are determined by substituting

$$F = \frac{2}{15}\rho U^2 \delta \quad \text{and} \quad \delta = 5.48 \, X \left(\frac{UX}{v}\right)^{-1/2}$$

which give

$$C_f = 1.46 \left(\frac{UX}{v}\right)^{-1/2} \tag{12.5}$$

This equation is shown plotted in Fig. 12.6. Note how the value of C_f depends solely on the Reynolds number UX/ν, which also appears in the expression for δ and hence F.

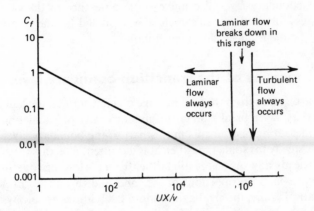

Fig. 12.6 Variation of Coefficient of Friction C_f with UX/ν for a laminar boundary layer. *Note the logarithmic scale.* Laminar flow will occur when $UX/\nu < 5 \times 10^5$ and may exist for $5 \times 10^5 < UX/\nu < 2 \times 10^6$, see p. 184.

It is comparatively easy to measure F (and therefore C_f) experimentally, even if it is difficult to make direct measurements of the thickness of the boundary layer. A flat thin plate with a sharpened leading edge is suspended in an air or water flow in such a way that the drag force F can be measured directly by some sort of weighing machine. The kinetic pressure $\frac{1}{2}\rho U^2$ is measured by a Pitot tube exposed to the undisturbed stream. Providing the plate is so short in the direction of the flow that $UX/\nu < 5 \times 10^5$, the laminar boundary layer entirely covers the plate and the values of C_f obtained from the experiment are immediately comparable to the computed ones of equation 12.5. Such experiments show that C_f is indeed very close to the calculated value $1.46(UX/\nu)^{-1/2}$. This good agreement might be taken as indicating that all the assumptions made in the analysis are correct, notably that the linear distribution of stress through the layer $\tau_0(1 - y/\delta)$ is true. Unfortunately, however, the situation is not so clear cut as this: other assumptions can be made about the way in which τ varies, and the resulting expressions for C_f are so close to the one already derived that the inevitable experimental errors in the observed values of C_f completely obscure the difference between the theoretical values on the several assumptions. For example, the unlikely assumption that τ is constant throughout a cross section of the boundary layer (i.e. $\tau = \tau_0$) gives a value of $C_f = 1.155(UX/\nu)^{-1/2}$, which is close to the value already derived for the much more likely linear increase of τ. Thus measurements of C_f cannot determine critically how the stress varies: any more or less likely stress distribution gives nearly the same value of C_f. From the engineer's point of view, however, this is a fortunate occurrence, for though it is not conclusive evidence, it is probable that the stress distribution in the more important turbulent boundary layer also makes little difference to the value of C_f. A simple stress distribution can therefore be assumed for the computations of turbulent boundary layer flows that follow.

12.5 Properties of the transition boundary layer

At some distance from the leading edge of a flat surface, the laminar flow in the boundary layer breaks down to give unsteady or oscillating velocities. Velocity measuring instruments in this part of the boundary layer show that the flow is steady for some time, then at intermittent intervals variations occur to give unsteady flow. At places still farther from the leading edge, the unsteady flow occupies larger proportions of the time, and this unsteadiness becomes more erratic. Finally, the erratic or random fluctuations are always present in all parts of the boundary layer, which is then said to be *turbulent*. The reasons for the breakdown from laminar to turbulent flow are still rather poorly understood, and are the subject of a good deal of research.

The properties of the transition boundary layer are partly those of laminar and partly of turbulent flow, and fluctuate from one to the other. The stresses and coefficients of friction for the boundary layer at this stage seem to be dependent on the turbulence of the undisturbed stream. The difficulties of measurement are such that no general values of C_f can be quoted. However, the point is of small importance in most engineering work, for the transition boundary layer occupies only a small part of the length of a surface exposed to a fluid flow. In fact, its length is considerably less than that of the laminar boundary layer. The drag due to this small length can therefore be neglected unless the surface is very short in the direction of the motion.

12.6 Properties of the turbulent boundary layer

The part of the boundary layer wherein the flow is entirely turbulent is by far the most important part to the engineer, because most surfaces subject to fluid flow are covered with it. As has already been described, after transition the turbulent boundary layer extends downstream to the full length of any surface to which the flow remains attached. A return to laminar flow is unusual unless strong pressure gradients are present to accelerate the flow. The thickness of the turbulent boundary layer increases and may eventually be many times greater than that for the laminar boundary layer, see Table 12.1. Because δ is so much greater in the turbulent than in the laminar boundary layer, it is easier to determine the velocity distribution. Experiments have shown that the velocity distribution can be represented, with minor errors only, as a power law

$u/U = (y/\delta)^n$, n varying from about 1/5 near the transition from laminar flow, to 1/7 farther downstream. The effect of the eddies in the turbulent boundary layer may be seen by comparing a plot of the above velocity distribution curve with that of the laminar layer, Fig. 12.2. Near the surface the velocity gradients of the turbulent layer are greater than those of the laminar layer, but over most of the boundary layer the velocity distribution is more nearly uniform.

As in the laminar boundary layer, the force that has caused the slowing down of the fluid is

$$F = \rho \int_0^\delta (U - u)u \, dy \qquad (6.7)$$

Substituting $u/U = (y/\delta)^{1/7}$ and integrating,

this becomes $F = \dfrac{7}{72} \rho U^2 \delta.$

As before, this force also equals the summation of the varying stresses τ_0 all along the surface from the leading edge. If the surface is long compared with the extent of the laminar layer near its leading edge, then an easy approximation to make is to ignore the laminar layer and assume that the *turbulent* layer starts at the leading edge, so that τ_0 everywhere is that applicable to a turbulent layer.

Thus $$F = \int_0^X \tau_0 \, dx$$

or $$\frac{7}{72} \rho U^2 \delta = \int_0^X \tau_0 \, dx.$$

To solve this equation another relationship is required to eliminate τ_0. In the laminar layer it was derived from the viscosity and velocity gradient at $y = 0$. This cannot be used for a turbulent flow where τ and du/dy are connected by the eddy viscosity, see Section 1.3, a coefficient that varies with the strength of the turbulence and with the position in the flow. Hence further information is required. Aid can come from data found from experiments on circular pipes. As will be described in Chapter 13, well downstream of its entrance, a pipe flowing full of fluid, is wholly occupied by the boundary layer on the walls. There is no central core of constant velocity fluid, but only a single point of maximum velocity (see Fig. 12.7). Thus, the flow within the pipe may be regarded as an approximation to a boundary layer on a flat plate which has been wrapped round an axis at a distance δ from the plate equal to the radius r of the pipe: and the axis velocity U_{max} is equivalent to the undisturbed stream velocity U of the flat surface boundary layer. It has to be assumed that the effects of the pressure gradient in the pipe and the radial symmetry are small. It is easy in the case of the pipe to measure τ_0, for this drag causes a measurable drop in pressure along the pipe. In Chapter 13 the connection will be developed between pressure drop

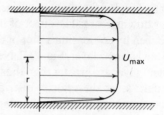

Fig. 12.7 Typical velocity distribution within a circular pipe.

and τ_0, but it suffices here to say that one empirical equation is

$$\tau_0 = 0.023 \, \rho \, U_m^2 \, (v/Ur)^{1/4}$$

for smooth-bore pipes of all sizes. Since $r = \delta$ for the corresponding boundary layer on a flat surface, the expression may be used with the force—momentum equation, that is

$$\frac{7}{72}\rho U^2 \delta = \int_0^X 0.023 \, \rho \, U^2 (v/U\delta)^{1/4} \, dx.$$

This may be integrated in the same way as was done for the laminar boundary layer and it gives

$$\delta/X = 0.376 \, (UX/v)^{-1/5}.$$

Measurements of δ are just as difficult to make in the turbulent layer as they were in the laminar layer. But the foregoing equation for δ/X can be used as before to determine the resultant coefficient of drag, C_f, giving

$$C_f = 0.073 \, (UX/v)^{-1/5} \qquad (12.6)$$

and this equation is plotted in Fig. 12.8. Hence, as with the laminar boundary layer, the main properties of a turbulent boundary layer depend on the Reynolds number UX/v.

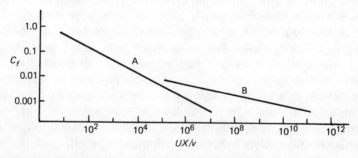

Fig. 12.8 The friction coefficient C_f for a smooth plate when the boundary layer is turbulent (curve B) compared with its value for laminar flow (curve A). Note logarithmic plotting.

Example

Show that if δ_2 is the momentum thickness at a control surface normal to the flow in a boundary layer on a flat plate the shear force, F, per unit width up to the control surface is given by

$$F = \rho U^2 \delta_2,$$

where δ_2 is defined as the thickness of a layer of the free stream flow which has a momentum flow rate equal to the deficit of momentum flow rate in the boundary layer.

Hence, assuming $\delta/X = 0.376(UX/\nu)^{-1/5}$ estimate the value of F for a flat surface 2 m long in a water flow of $9\,\mathrm{ms}^{-1}$ parallel to the surface if the boundary layer is assumed turbulent from the leading edge.

$$\text{For water } \nu = 10^{-6}\,\mathrm{m^2\,s^{-1}}.$$

Solution

Equating the deficit of momentum flow rate in the boundary layer with the momentum flow rate in a width δ_2 of the free stream gives

$$\rho U^2 \delta_2 = \rho \int_0^\delta u(U - u)\mathrm{d}y$$

$$= -F \text{ from equation 6.5 as there will be no pressure forces.}$$

Hence $\qquad \dfrac{\delta_2}{\delta} = \displaystyle\int_0^1 \dfrac{u}{U}\left(1 - \dfrac{u}{U}\right)\mathrm{d}\left(\dfrac{y}{\delta}\right)$

Now when $u/U = (y/\delta)^{1/7}$

$$\delta_2/\delta = \int_0^1 \left(\frac{y}{\delta}\right)^{1/7}\left(1 - \left(\frac{y}{\delta}\right)^{1/7}\right)\mathrm{d}\left(\frac{y}{\delta}\right) = \left[\frac{7}{8} - \frac{7}{9}\right] = \frac{7}{72}.$$

In the present case $Re = \dfrac{Ux}{\nu} = 9 \times 2 \times 10^6 = 18 \times 10^6$,

and $\delta = 0.376 \times Re^{-1/5} = 0.376 \times 2 \times (18 \times 10^6)^{-1/5} = 0.0266$ m

So $-F = \rho U^2 \dfrac{7}{72}\delta = 10^3 \times 81 \times \dfrac{7}{72} \times 0.0266 \dfrac{\mathrm{kg}}{\mathrm{m^3}}\dfrac{\mathrm{m^2}}{\mathrm{s^2}}\,\mathrm{m} = 210\,\mathrm{N\,m^{-1}}.$

These theoretical values of C_f can be checked, as before, by experiments where the drag is directly measured. Since C_f is partly based upon an empirical relation for τ_0 it is not surprising that the agreement between the results of the drag experiments and equation 12.6 is good: the only error lies in the approximate nature of the velocity distribution curve $u/U = (y/\delta)^{1/7}$. Up to $UX/\nu = 20 \times 10^6$ the experimental values of C_f all lie within about 2 per cent of the values given by equation 12.6. At higher values of UX/ν the errors become unacceptably large for engineering work concerned with large surfaces in the X

direction exposed to high fluid velocities U, thus giving large values of UX/v. (A 300 m ship sailing at 30 knots $(15\,\mathrm{m\,s^{-1}})$ gives $UX/v = 4 \times 10^9$ for the boundary layer at the stern.) Reliable information about the drag under these conditions requires a theory based more solidly upon rational explanations than upon experimental data. A simplified version of such a theory will therefore be given.

12.7 Transfer of momentum by turbulent eddies

To build up a rational theory for the drag caused by a turbulent boundary layer, it is necessary to consider the action of the irregular motions or eddies that are superimposed on the mean velocity. Consider a turbulent flow, such as a cross section of a boundary layer, where there is a gradient across the flow du/dy of the mean velocity u. The mean velocity in this context is the velocity at a point averaged over a long period: that is, the fluctuating velocities of turbulence are 'averaged out'. The drag force of the surface τ_0 would eventually stop the adjacent layer of fluid completely, unless this layer had exerted on it another force by the layer next above. Viscous forces given by the viscosity equation $\tau = \mu du/dy$, which act in a laminar boundary layer, are not nearly great enough, so that other ways must be found whereby the momentum of the layers nearer the surface is continually refreshed from layers farther away. One of these ways is by reason of the random and erratic motions that are superimposed on the mean velocity in turbulent flow, and are sometimes in a direction across the mean direction of flow. In this way a small quantity of faster moving fluid can be moved across the main flow into a layer of slower moving fluid, increasing the momentum of the latter. At the same time, the mass continuity principle requires a corresponding motion to take place somewhere in the opposite direction, taking slower fluid into the fast layer where it is in turn accelerated. But such a system would result in an accumulation of fluid at two opposite corners of a small control volume in the flow (AA in Fig. 12.9) unless there were corresponding motions along the direction of mean flow. Thus a rotary set of complementary motions is set up as shown in the figure and these are called *eddies*. It will be seen that the complementary motions change the speed of the upper and lower layers by the same amount. Thus, if u is the mean speed of the centre of the eddy relative to

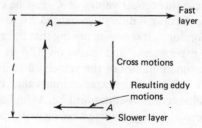

Fig. 12.9 The motions comprising an eddy connecting two layers of fluid travelling at different speeds.

the boundaries of the flow, and $\partial u/\partial y$ the instantaneous velocity gradient of the flow, then the speed of the upper layer is $u + \frac{1}{2} l \, \partial u/\partial y$ and of the lower layer $u - \frac{1}{2} l \partial u/\partial y$. The eddy motions across the flow are, therefore, $\frac{1}{2} l \partial u/\partial y$ relative to the mean motion (see Fig. 12.10).

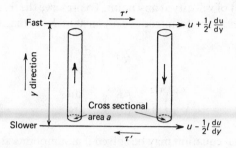

Fig. 12.10 Transfers of momentum occur through each of the cylinders of fluid shown, which are the same distance $l/2$ on opposite sides of an eddy. A shear stress τ', has produced a gradient $\partial u/\partial y$ of the mean velocity u so that the upper edge of the eddy is faster than the lower edge.

This simplified diagram of the motions in an eddy can be used now to determine the momentum transferred from one layer to another. In the cylinders of cross-sectional area a shown on each side of the eddy, a velocity $\frac{1}{2} l \partial u/\partial y$ is occurring, transferring a mass of $a\rho \frac{1}{2} l \partial u/\partial y$ in unit time from the fast layer to the slow, and vice versa. Every unit mass of fluid changes its momentum by an amount $l \partial u/\partial y$, so that the total rate of change of momentum by both cylinders is

$$2\left(a\rho\tfrac{1}{2}l\frac{\partial u}{\partial y}\right)\left(l\frac{\partial u}{\partial y}\right)$$

A momentum transfer of this sort across the flow would eventually mix the faster and slower layers together to form a uniform stream at the mean velocity, unless there are forces acting which preserve the velocity gradient. In a steady stream of real fluid with a velocity gradient there is a shear stress acting which is constantly forcing one layer to slide over the next. In the eddy of Fig. 12.10 the instantaneous value of this stress is τ' which, acting on the ends of the cylinder, just balances the tendency of the momentum transport to eliminate the velocity gradient.

Thus

$$\tau'2a = 2\left(a\rho\tfrac{1}{2}l\frac{\partial u}{\partial y}\right)\left(l\frac{\partial u}{\partial y}\right)$$

or

$$\tau' = \tfrac{1}{2}\rho\left(l\frac{\partial u}{\partial y}\right)^2.$$

Not all the eddies in the fluid are of course of this size l, and not all of any one eddy is necessarily giving the same momentum exchange. The stress τ' may therefore vary from place to place and from one moment to the next, depending upon the eddies that happen to be active. However, it is convenient to write l as the *mean* size of the eddy that occurs, τ as the *mean* stress and therefore du/dy as the *mean* gradient of velocity at any point. To preserve the truth of the equation it is therefore necessary to introduce a constant K so that now

$$\tau = K\tfrac{1}{2}\rho\left(l\frac{du}{dy}\right)^2$$

or

$$\left(\frac{\tau}{\rho}\right)^{1/2} = \left(\frac{K}{2}\right)^{1/2} l\frac{du}{dy}.$$

This differential equation may be solved if assumptions are made about the way τ and l change with y. A very simple one is that $\tau = \text{constant} = \tau_0$: in the laminar boundary layer it was shown that this simple assumption gives a remarkably good approximation to the results of more elaborate stress distributions. Also experiments tend to show that $(K/2)^{1/2} l$ is proportional to the distance y from the solid surface on which the boundary layer is formed. This is confirmed by simple observations of a river in flood: large eddies do not exist near the sides because of the lack of room for such large cross motions there. In fact, the experiments show that

$$\left(\frac{K}{2}\right)^{1/2} l = 0.4y.$$

Thus, substituting these assumptions into the eddy differential equation,

$$du = \left(\frac{\tau_0}{\rho}\right)^{1/2}\frac{1}{0.4y}dy$$

and integrating

$$u = \left(\frac{\tau_0}{\rho}\right)^{1/2}\frac{1}{0.4}[\log_e y] + \text{Constant}.$$

The constant of integration may be written as

$$C = \left(\frac{\tau_0}{\rho}\right)^{1/2}\frac{1}{0.4}\log_e\frac{1}{C_1} \quad \text{where } C_1 \text{ is another constant,}$$

so that the velocity equation is

$$u = \left(\frac{\tau_0}{\rho}\right)^{1/2}\frac{1}{0.4}\left[\log_e y + \log_e\frac{1}{C_1}\right]$$

or

$$u = 2.5\left(\frac{\tau_0}{\rho}\right)^{1/2}\log_e\frac{y}{C_1}.$$

Changing to common logarithms

$$u = 2.5 \times 2.303 \left(\frac{\tau_0}{\rho}\right)^{1/2} \log_{10} \frac{y}{C_1}$$

$$= 5.75 \left(\frac{\tau_0}{\rho}\right)^{1/2} \log_{10} \frac{y}{C_1}. \qquad (12.7)$$

This logarithmic velocity distribution has been confirmed by experiment. Values of u are taken at a number of distances y from a plate whereon there is a turbulent boundary layer. A graph is made of u against $\log_{10} y$ and it is found that all points lie in a straight line, as in Fig. 12.11: if τ_0 is measured as well, then the slope of the line is found to be $5.75\,(\tau_0/\rho)^{1/2}$. This logarithmic curve does not give a value for δ, where the velocity becomes constant: it gives a velocity increasing indefinitely with y, though at a decreasing *rate* of increase as in Fig. 12.12, page 207. Thus the experimental difficulty of finding δ to confirm a theoretical formula does not arise. But the place where there are large and therefore measurable changes of velocity is at small values of y, and here the very presence of the surface creates errors in the readings of Pitot tubes and other devices. Experimental verification of the logarithmic velocity formula at small values of y cannot be made. However, extrapolation to $u = 0$ of the straight line of the logarithmic velocity profile (Fig. 12.11) gives an intercept on the $\log y$ axis of $y = C_1$. Such a finite intercept is of course inevitable with a logarithmic equation. But it is now necessary to explain why apparently the

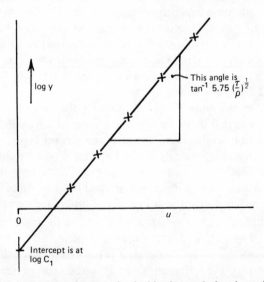

Fig. 12.11 Experimental determination of velocities in a turbulent boundary layer shows $u \propto \log y$. Slope and intercept therefore give $(\tau/\rho)^{1/2}$ and C_1 respectively.

fluid seems to be stationary at a finite distance C_1 above the surface. This seems hardly possible and in fact does not occur.

12.8 Mixing processes

Turbulent eddy motions, as an action producing a shear (friction) stress, can also be viewed more generally as a process of mixing. Consider one point in a stream of fluid (whether in a boundary layer or in the 'main stream') where the flow is turbulent, and so where the instantaneous velocity can be decomposed into components u and v at right-angles to each other. Each component then consists of the mean, steady velocity $\bar{u}$ (or $\bar{v}$) which does not vary, and an instantaneous velocity fluctuation u' (or v') which varies erratically from instant to instant. Thus $u = \bar{u} + u'$ and $v = \bar{v} + v'$. Notice that u' and v' can be either positive or negative. By choosing suitable axes, then if u is along the direction of total mean motion, $\bar{v} = 0$. The fluctuating components u' and v' are thus non-steady and could be measured by sufficiently sensitive and rapid-acting instruments.

Now consider the situation at an instant where $u = \bar{u} + u'; v = +v'$. In a unit of time the transverse velocity v' will carry across the mean total motion a mass of fluid $\rho v'$ per unit area: this fluid has temporarily a surplus u' of u-momentum above the steady value $\bar{u}$ (both values per unit mass). So the amount of fluctuating u-momentum transferred sideways in the v-direction per unit of time and of area is $\rho u'v'$. This surplus mixes instantaneously with the u-momentum already in the new layer to which the transverse velocity has brought it; the fluctuation would speed up the u-velocity in the new layer and so would lead to a non-steady situation, with gradual speeding up. However, this situation does not arise because of the existence of a shear stress which restrains the acceleration. In fact, a steady turbulent flow with a fluctuating transfer of momentum is inevitably accompanied by a shear stress $\tau = -\rho u'v'$ usually referred to as a Reynolds stress.

Now at another subsequent instant, due to a change in the turbulence, u' might be negative so that there is a *deficit* of u-momentum available. If v' is still positive, the motion would slow down in the second layer, unless it were reversed for the instant. Similarly, if the negative u' were accompanied by a negative v', there would be a slowing of a layer on the opposite side of the original layer, with the same effect as the original positive u' and v'. Over a long period, with many separate instants in which u' and v' are measured, the mean value $\overline{\rho u'v'}$ of all the separate values of $\rho u'v'$ will be the mean stress over that period. (The bar over $\rho u'v'$ implies a simple arithmetical average of a number of values of the product $\rho u'v'$.) Of course, it is possible to have this mean value zero (i.e. no stress), even though there is turbulence present; then, such turbulent velocities are called 'uncorrelated' with each other. On the other hand, if positive u' are *always* accompanied by positive v' (and negative u' with negative v'), there is a *relatively* large stress since the mean value $\rho u'v'$ is large: such turbulence is called 'correlated'. The stress in a boundary layer is accompanied by correlated turbulence: the correlation falls towards the outer edge of the boundary layer; and outside the boundary layer in the main stream the turbulence is uncorrelated (so there is no stress there).

A further point can now be considered. There is always present a wide variety of frequencies at which u' and v' change. There are slow changes of both mixed with fast changes. If the instants already referred to are widely spaced (in time), then the influence of the faster changes will be entirely missed by the instrument concerned. Instants close together in time sample both fast and slow changes and so take into account more turbulence than observations taken at instants further apart. So, in general, rapid and very frequent sampling of u' and v' gives a higher $\tau = \overline{\rho u'v'}$ than infrequent sampling. The type of instruments and their rapidity of following the velocities thus controls the apparent value of τ. It should be noted that this method of finding shear stress is the only one where the variation of τ with distance y across the flow can be measured. The method of Section 13.3 in a pipe where the pressure gradient along it is measured by small holes in the pipe wall, gives only τ_0, the particular value at the wall. Generally, experiments show that τ increases towards a maximum at τ_0. The tacit assumption of the mixing length theory that τ is constant through a boundary layer is therefore strictly incorrect, but more refined analyses show that the error is not great, and can be tolerated for many engineering purposes.

A very similar approach can be made to consider the mixing of a pollutant in a stream by the action of turbulence. Such a 'pollutant' may be a chemical, or heat, or merely a deficiency of a gas (oxygen for example). Consider a situation where polluted fluid exists at a mean strength $\bar{s}$ (units of polluted matter per unit volume of fluid) in a particular layer. In another layer there is less pollution because the pollutant is being taken away (for example clean water running over polluted water, or heat going away to the atmosphere). If a steady situation has been set up, the pollutant must be steadily transferred between the layers, by turbulent action of eddies. Thus, if a sensitive instrument is used to measure pollution, the instantaneous pollution is $\bar{s} + s'$, the bar and the prime having the same meaning as for the velocity u. So the amount of pollutant carried sideways is $s'v'$ at the one instant; if this is immediately accepted into the new layer, the pollution there would rise unless the transport away just balances it. So, if many instants are considered, the mean transport over them must be *exactly* balanced by the mean transferred pollution $\overline{s'v'}$. This transfer of pollution is an exact analogy of the stress caused by the transfer of u-momentum.

The method of determining a transport rate from turbulence measurements is always in principle possible, although there are sometimes severe instrumental problems to overcome so as to find u', v' or s' precisely and rapidly. It is then often helpful to relate the rate to the *gradient* of the mean value of the polluting substance across the stream—in the same way that in the mixing length theory τ is associated with the gradient of the mean value $d\bar{u}/dy$. Thus for a chemical pollutant

$$\overline{s'v'} = K_1 \frac{d\bar{s}}{dy}.$$

For heat

$$C_p\overline{T'v'} = K_2 \frac{d\bar{T}}{dy}$$

where T is the temperature and C_p is the specific heat at constant pressure. For u-momentum

$$\tau = \overline{\rho u'v'} = K_3 \frac{d\bar{u}}{dy}$$

The K terms, sometimes called exchange coefficients, have a direct analogy with the coefficient of viscosity (for the situation where there is *no* turbulence) for which in Chapter 1 it is shown

$$\tau = \mu \frac{du}{dy}$$

and K_3 is called the eddy viscosity.

It is relatively easy to measure gradients of mean properties such as $d\bar{s}/dy$, $d\bar{u}/dy$, for all that is necessary is a well-damped, averaging, instrument; but turbulence measurements need a rapid-acting instrument that will detect small changes of a property in the presence of a large average value of the property. Much effort has been devoted to finding K values in different situations. Unfortunately, K varies widely, and although in the same situation changes of K give a useful estimate of changes of transport rates, a K found in one situation cannot be relied upon for another. Generally, K for large-scale motions (ocean currents, and the atmosphere) will be larger than K for small scale motions. It would also be useful if in one particular situation $K_1 = K_2 = K_3$, for then a simple experiment to find K_3 (by measuring pressure gradients, for example) could then be directly used to find the more difficult transports of pollution or heat. Again, rather unfortunately, $K_1 = K_2 = K_3$ only in rather special circumstances, when the pollution does *not* change the density of the fluid: usually it does so (heat changes density by thermal expansion; pollution of a solid makes fluid heavier).

Example

At a point in a boundary layer, instantaneous and simultaneous measurements are made at intervals of 1 second, of the velocity components u and v, and of the concentration c of a pollutant. The fluid is water and pollutant is measured in milligrams/litre. Also, a well-averaged reading of u and c is taken at a place 1 m above the former point, giving $u = 1.09\,\mathrm{m\,s^{-1}}$; $c = 9.900\,\mathrm{mg\,l^{-1}}$.

A set of ten readings is given below: estimate the shear stress there, the upward mixing of pollutant and the exchange coefficients of momentum and pollution.

u	$\mathrm{m\,s^{-1}}$	1.00	1.02	1.04	0.90	0.97	1.01	1.02	0.96	1.00	0.98
v	$\mathrm{m\,s^{-1}}$	-0.02	$+0.03$	-0.01	$+0.05$	-0.01	$+0.01$	-0.05	$+0.02$	0	-0.02
c	$\mathrm{mg\,l^{-1}}$	10.05	10.00	9.97	10.08	10.04	9.99	9.98	10.02	10.00	9.94

A tabulation is necessary to find each value of u', v', c', and the products $u'v'$, $v'c'$, see Table 12.2. Care must be taken to ensure the calculations preserve the units of measurement. Work downwards in the u, v and c columns, then derive

$$u' = u - \bar{u}; \quad v' = v - \bar{v}; \quad c' = c - \bar{c}$$

Then multiply up $u'v'$ and $v'c'$ and work downwards in these columns.

Mean value $\overline{u'v'} = -\frac{1}{10} \times 58 \times 10^{-4} = -5.8 \times 10^{-4}\,\mathrm{m^2\,s^{-1}}$.

Thus stress $\qquad \tau = 1000 \times 5.8 \times 10^{-4}\,\mathrm{N\,m^{-2}}$
$$= 0.58\,\mathrm{N\,m^{-2}} \qquad\qquad\qquad \textit{Ans.}$$

Mean value $\qquad \overline{c'v'} = \frac{1}{10} \times 54 \times 10^{-4}\,\dfrac{\mathrm{mg}}{1}\cdot\dfrac{\mathrm{m}}{\mathrm{s}}$

$$= 5.4 \times 10^{-4}\,\frac{10^{-3}\mathrm{g}}{1000\,\mathrm{cm^3}}\cdot\frac{100\,\mathrm{cm}}{\mathrm{s}}$$

$$= 5.4 \times 10^{-8}\,\mathrm{g\,cm^{-2}\,s^{-1}}$$
$$= 5.4 \times 10^{-4}\,\mathrm{g\,m^{-2}\,s^{-1}} \qquad\qquad \textit{Ans.}$$

Exchange coefficient for momentum ('eddy viscosity') K_3

$$= \tau/\rho\,\mathrm{d}\bar{u}/\mathrm{d}y$$

Gradient of mean velocity $= \dfrac{1.09 - 0.990}{1.0} = 0.1\,\mathrm{s^{-1}}$

So $\qquad K_3 = \dfrac{0.58}{1000 \times 0.1} = 5.8 \times 10^{-3}\,\mathrm{m^2\,s^{-1}} \qquad\qquad \textit{Ans.}$

Notice how the eddy viscosity, and hence the turbulent shear stresses are much greater (5,800 times) than the kinematic viscosity (at $10^{-6}\,\mathrm{m^2\,s^{-1}}$) and the corresponding laminar shear stresses.

Exchange coefficient for pollution K_1
Gradient of mean pollution

$$\mathrm{d}c/\mathrm{d}y = \frac{10.007 - 9.900}{1.0} = 0.107 \times 10^{-3}\,\frac{\mathrm{g}}{1\,\mathrm{m}}$$

Table 12.2 Turbulent fluctuations in a flow

u m s⁻¹	u' m s⁻¹	v m s⁻¹	v' m s⁻¹	c mg l⁻¹	c' mg l⁻¹	$u'v'$ m² s⁻² ×10⁻⁴	$c'v'$ m mg s⁻¹ l⁻¹ ×10⁻⁴
1.00	+0.01	-0.02	-0.02	10.05	+0.043	-2	-8.6
1.02	+0.03	+0.03	+0.03	10.00	-0.007	+9	-2.1
1.04	+0.05	-0.01	-0.01	9.97	-0.037	-5	+3.7
0.90	-0.09	+0.05	+0.05	10.08	+0.073	-45	+36.5
0.97	-0.02	-0.01	-0.01	10.04	+0.033	+2	-3.3
1.01	+0.02	+0.01	+0.01	9.99	-0.017	+2	-1.7
1.02	+0.03	-0.05	-0.05	9.98	-0.027	-15	+13.5
0.96	-0.03	+0.02	+0.02	10.02	+0.013	-6	+2.6
1.00	+0.01	0	0	10.00	-0.007	0	0
0.98	-0.01	-0.02	-0.02	9.94	-0.067	+2	+13.4
Sum 9.90		0		100.07		$\left\{\begin{array}{c}+15\\-73\end{array}\right\}$	$\left\{\begin{array}{c}+69.7\\-15.7\end{array}\right\}$
Mean $\bar{u}=0.990$		$\bar{v}=0$		$\bar{c}=10.007$		$= -58 \times 10^{-4}$	$= +54 \times 10^{4}$

As $\bar{v}=0$, it turns out that the axes of measurement of velocity are precisely along and normal to the mean fluid motion.

So
$$K_1 = \frac{5.4 \times 10^{-4} \ \mathrm{g \, m^{-2} \, s^{-1}}}{0.107 \times 10^{-3} \ \mathrm{g \, l^{-1} \, m^{-1}}}$$

$$= 5 \times 10^{-1} \, \mathrm{m^{-1} \, l s^{-1}} = 5 \times 10^{-1} \ \mathrm{m^{-1}} \frac{\mathrm{m}^3}{1000} \, \mathrm{s^{-1}}$$

$$= 5 \times 10^{-4} \, \mathrm{m^2 \, s^{-1}} \hspace{3cm} \textit{Ans.}$$

It is quite common to find K_1 much smaller than K_3, particularly if the polluted water has a different density to that of the unpolluted stream.

12.9 The viscous sub-layer

The apparent paradox of the logarithmic velocity profile giving zero velocity at a finite distance C_1 can be explained by the presence next to the surface of a thin layer of fluid where the flow is controlled by the laminar shear stresses, even if flow in the rest of the boundary layer is turbulent. Apparently if the velocity gradient is great, as it is if y is small, the viscous forces tend to oppose the formation of eddies, and eventually prevent them from forming at all. Consequently the logarithmic profile of velocity is only valid for the part of the boundary layer outside this *viscous sub-layer*: inside it, the velocity increases approximately linearly with y (see Fig. 12.12). The sub-layer (which must not, of course, be confused with the wholly laminar boundary layer near the leading edge of the surface) is, however, usually thin: experiments have shown that its thickness δ' is given by

$$\delta' = 11.5 \nu \left(\frac{\tau_0}{\rho} \right)^{-\frac{1}{2}}. \tag{12.8}$$

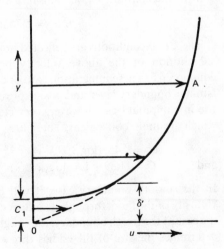

Fig. 12.12 The velocity distribution curve for a turbulent boundary layer on a smooth surface. A is the logarithmic velocity curve with intercept C_1, but in the layers of fluid nearest the surface, the flow is controlled by laminar shear stresses, the velocity decreasing linearly, as shown by the dotted line.

Thus for water at a speed of 3 m s^{-1}, at a point 3.6 m from the leading edge of a smooth plane surface (i.e. $UX/v = 10^7$ approx), the viscous sub-layer of the turbulent boundary layer has a thickness δ' of only 0.1 mm. The turbulent layer thickness above the viscous sub-layer is $\delta = 5.6$ cm, approximately as given by the empirical formula $\delta/X = 0.376 \, (UX/v)^{-1/5}$ already derived.

A viscous sub-layer then exists near a smooth surface, and the stress τ is transmitted finally through the distance δ' to the surface wholly by the molecular forces of viscosity. However, smooth surfaces are rather rare in engineering work, where surfaces usually have projections on them (roughnesses) whose size may be many times greater than δ'. Under these *rough boundary* conditions, the extrapolated value of the constant C_1 has been found to depend on the shape and size of the roughness. If the roughnesses are grains of sand, all just passing through a sieve of size k, experiments have given $C_1 = k/33$: but it must be emphasized that other shapes of roughness (perhaps, say, the hemispherical heads of rivets on steel plates) may give constants other than $k/33$.

From equation 12.8 it will be seen that for a given surface and fluid, δ' decreases if τ_0 increases. The stress τ_0 is largely controlled by the velocity U, for $\tau_0 = C_f \frac{1}{2}\rho U^2$, so that at high speeds the viscous sub-layer becomes thinner. Consequently if the surface is a rough one, it is quite possible that at low speeds the roughnesses are submerged in the viscous sub-layer, and the surface acts as if it were smooth. The velocity distribution in the turbulent part of the boundary layer is then quite independent of k and is

$$u/u_* = 5.75 \log_{10}(u_* y/v) + 5.5 \qquad (12.9)$$

Where $u_* = (\tau_0/\rho)^{1/2}$. By contrast the same surface at higher speeds has on it a thinner sub-layer so that the roughnesses may now project through it and control the constant C_1, leading to a velocity distribution

$$u/u_* = 5.75 \log_{10} 33y/k. \qquad (12.10)$$

These two conditions are reflected in the coefficient of friction found by the substitution of the above velocity profiles into the momentum integral equation 6.5. These integrations are of course more difficult than those of the laminar boundary layer and give expressions that are rather inconvenient for use in computations. However, very close approximations have been found which are more convenient. These are

$$C_f = 0.455(\log_{10} UX/v)^{-2.58} \text{ for the smooth boundary}$$

and $\qquad C_f = (1.89 + 1.62 \log_{10} X/k)^{-2.5}$ for the rough boundary.

In the smooth boundary case, it will be seen that C_f is controlled by the viscosity of the fluid, through its influence on the thickness of the sub-layer. In the case of the rough boundary, the roughness size alone controls C_f, because it determines the size of the eddies thrown off in the wake of each individual roughness element (i.e. grain of sand or rivet head). There is an intermediate stage between the true smooth and the rough boundary cases, when the viscous sub-layer is about the same thickness as the height of the roughnesses: the eddies thrown off by the top of the roughnesses are now weak and are strongly affected by the proximity of the viscous flow: C_f now lies between the smooth and rough boundary values, and depends on both UX/v and also X/k. A graph can therefore be made of C_f against UX/v, which gives a single line for all smooth surfaces. Coefficients for rough surfaces at low UX/v also follow this line, but gradually diverge from it in the intermediate stage, until at the higher values of UX/v they become independent of it, being then only dependent on X/k (see Fig. 12.13).

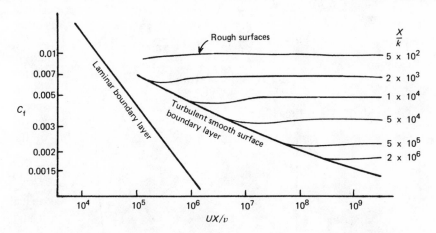

Fig. 12.13 The friction coefficient C_f for a rough surface. The left-hand line is for the laminar boundary layer at low values of UX/v, which is exactly the same as for a smooth surface. The lowest right-hand line is for the smooth surface equation which the rough surface also gives when there is a viscous sub-layer thicker than the size k of the roughness on the surface. The upper right-hand lines are for C_f when the roughness elements are causing eddies directly, each line being for a different value of X/k: C_f in this case is independent of UX/v but changes only with the ratio X/k. X is the distance from the leading edge of the surface.

Example

A long flat surface with roughness elements equivalent to sand of 5 mm size has an air flow past it and at a cross-section of the flow the velocity is 10 m s^{-1} at 20 mm from the surface. Estimate the shear stress on the plate at this cross-section. Find and plot an expression for the velocity profile if $\rho = 1.3 \text{ kg m}^{-3}$.

Solution

The logarithmic velocity distribution near a rough surface gives the relationship:–

$$u = 5.75 \, (\tau_0/\rho)^{1/2} \log_{10} y/y_0,$$

where $y_0 = k/33$ for sand, and hence at 20 mm

$$10 \, \text{m s}^{-1} = 5.75(\tau_0/\rho)^{1/2} \log\left(\frac{20 \times 33}{5}\right) = 12.2 \, (\tau_0/\rho)^{1/2}$$

so

$$\tau_0 = \rho\left(\frac{10}{12.2}\right)^2 \frac{\text{m}^2}{\text{s}^2} = 0.874 \, \text{N m}^{-2}.$$

As

$$\left(\frac{\tau_0}{\rho}\right)^{1/2} = \frac{10}{12.2} = 0.82 \, \text{m s}^{-1},$$

$$u = 5.75\left(\frac{\tau_0}{\rho}\right)^{1/2} \log_{10}\left(\frac{y \times 33}{5 \, \text{mm}}\right)$$

$$= 4.71 \log\left(\frac{6.6y}{\text{mm}}\right) \text{m s}^{-1}.$$

See Table 12.3 and Fig. 12.14.

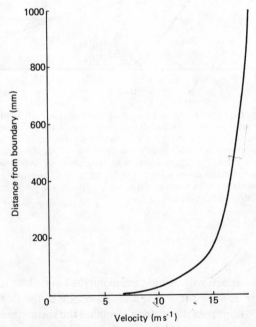

Fig. 12.14 The logarithmic velocity distribution in a turbulent boundary layer over a rough surface.

Table 12.3 Logarithmic velocity distribution

Distance from surface $y/$(mm)	5	10	20	50	100
Velocity $u = 4.71 \log (6.6y)/$(m s^{-1})	7.15	8.57	10.0	11.86	13.28

Distance from surface $y/$(mm)	200	350	500	700	1000
Velocity $u = 4.71 \log (6.6y)/$(ms^{-1})	14.70	15.84	16.57	17.26	17.99

12.10 Application of flat plate friction coefficients

The curve of C_f against UX/v for flat surfaces, both rough and smooth, may be also used, with caution, for finding the tangential friction forces on gently curved surfaces, providing that breakaway does not occur. If breakaway occurs the estimates so found will be greatly in error. A frictional drag force on a gently curved surface may be expressed in the form of a coefficient of drag so that

$$C_{\text{friction drag}} = \text{Frictional drag force}/\tfrac{1}{2}\rho U^2 A,$$

where A is, as in Chapter 8, the cross-sectional area of the object parallel to the flow direction. A distinction must be made between this frictional force and the drag force caused by the summation of all the pressure forces around the object, the pressures having arisen because of the distortion of streamlines around the object. The integration of an experimentally or theoretically determined pressure distribution to give a lift or drag force has already been described. The drag obtained in this way is called the *form drag*, and is quite independent of the friction drag. It is also usual to express form drag as a coefficient thus,

$$C_{\text{form drag}} = \text{Form drag force}/\tfrac{1}{2}\rho U^2 A.$$

The total drag, as found by direct measurement or by the wake traverse method of Chapter 6, is the sum of the form and friction drags, and gives a *total drag coefficient*. The relative magnitude of the coefficients of form and friction drags, as well as their absolute values, are related to the shape of the solid body. Well-streamlined shapes have the form drag and friction drag of the same order of magnitude with both small: poorly streamlined shapes with breakaway occurring near the leading edge and large wakes behind them have the friction drag only 1 per cent or 2 per cent of the form drag and the total drag coefficient is relatively large.

12.11 Effect of pressure changes on the boundary layer

The boundary layer on a solid surface has been shown to depend predominantly on the shear stress of the surface on the fluid, and to be nearly independent of the properties of the main stream. But if the pressure increases in the direction of motion, the boundary layer may affect the main stream by producing *breakaway* (sometimes called *separation*).

Consider a surface OABC, Fig. 12.15, along which there is a fluid flow and on which a boundary layer has formed on the leading portion OAB (the part OA will be wholly laminar flow). As has been described, the thickness δ increases downstream and at the point B the velocity distribution would be given by a similar curve to Fig. 12.15(*a*) if the pressure is constant along the surface. If, however, the pressure has been increasing in the direction of motion (i.e. $p_a < p_d$) there is a pressure force opposing the fluid motion. This opposing force, though spread uniformly over the whole cross section of the boundary layer BB_1, does not produce a uniform change of velocity on all the fluid passing across BB_1. The low-speed fluid in a stream tube of a particular area near the surface has a lower mass flow rate than one farther away so that the former has a greater reduction of velocity due to the opposing force. Thus the velocity distribution is modified to that shown in Fig. 12.15(*b*), there being little change in velocity near the outer limit of the boundary layer but much more change near the surface.

Farther downstream however, at C, the normal development of the boundary layer would have produced a thicker layer than that at B. Consequently there is more low-speed fluid close to the surface than at B and this is more sensitive to the opposing pressure force. A greater retardation occurs and an appreciable layer of fluid is brought to rest, Fig. 12.15(*c*). Still farther downstream, at D, the opposing force has actually reversed the flow and giving the profile of velocity shown in Fig. 12.15(*d*). The point where the flow is first reversed is called the *breakaway* point. It will be seen from Fig. 12.16 that the reversed flow causes a large slow eddy which is permanently present, and

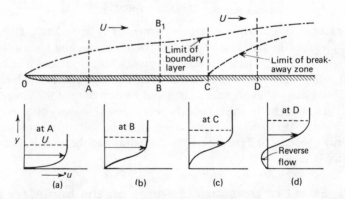

Fig. 12.15 Growth of a turbulent boundary layer on a flat surface along which the pressure rises in the downstream direction. Velocity distribution at A, B, C and D also shown, D being at a higher pressure than A. The breakaway (or separation) point is at C.

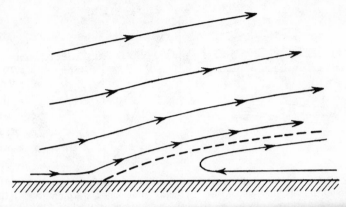

Fig. 12.16 The flow in the vicinity of a breakaway point. Arrows show fluid velocities. The effect is to make a generally rotatory motion centred in the breakaway zone (shown dotted).

which has large velocity gradients at its boundaries. A large degradation of energy into heat occurs in such an eddy. Thus breakaway is caused by a combination of the reduced velocities in a boundary layer and a pressure gradient opposing the flow (an *adverse* pressure gradient). Breakaway cannot occur unless there is an adverse pressure gradient: on the flat plate at constant pressure there is never breakaway; and breakaway can only occur with a real fluid which produces a boundary layer. Do not however confuse the permanent slow eddy due to breakaway with the smaller temporary eddies of turbulent motion.

Adverse pressure gradients are produced whenever there is a tendency for the fluid to decelerate, for by Bernoulli's equation,

$$\frac{u_1{}^2}{2g} + \frac{p_1}{\rho g} = \frac{u_2{}^2}{2g} + \frac{p_2}{\rho g}$$

if the energy is constant. Thus, if $u_1 > u_2$, $p_2 > p_1$, so that the pressure increases as velocity decreases. Such a deceleration occurs where the streamlines of a theoretical pattern tend to diverge, say, in the rear of solid objects travelling relative to a fluid. For instance, a cylinder, exposed to a stream of ideal fluid, causes a distortion of the streamline that can be predicted (Chapter 5), and there is no boundary layer. The velocity at different places round the periphery can be predicted from the streamline spacing and, by Bernoulli's equation, the pressure p at any point found as

$$(p - p_0)/\tfrac{1}{2}\rho U^2 = 1 - 4\sin^2\theta.$$

This distribution is shown in Fig. 12.17 and it will be seen that for $90° < \theta < 180°$ the pressure gradient is opposing the flow. In a real fluid a laminar boundary layer is formed on the leading side of the cylinder. The layer grows

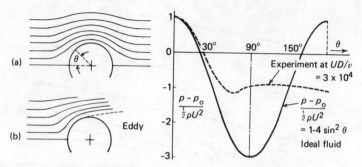

Fig. 12.17 Flow round a cylinder with axis normal to a stream of velocity U. (a) Theoretical ideal fluid streamlines. (b) Actual streamlines, showing breakaway and eddies in the wake. (Right) Theoretical ideal fluid, and actual experimental curves of pressure distribution on the surface of such a cylinder. Refer also to Fig. 10.1 for further details of flow patterns.

thicker as the fluid progresses towards $\theta = 90°$, so that there is retarded fluid near the surface of the cylinder. As soon as this boundary layer approaches the adverse pressure gradient, breakaway occurs and a large eddy is formed. The main flow is diverted to the outside of the eddy and no longer tends to revert to the streamline pattern which it had upstream of the cylinder (compare Fig. 12.17(a) and Fig. 12.17(b)): there is now no diverging flow to return the fluid to its original velocity, and consequently there is no appreciable rise of pressure in the eddy. The pressure distribution found experimentally under these conditions of breakaway is shown also in Fig. 12.17. Such an experimental pressure distribution can be integrated to find the form drag, and form drag coefficient, of a cylinder.

If the transition from a laminar to a turbulent boundary layer occurs before the breakaway commences, then the pressure distribution, form drag and flow pattern are all changed. The transition occurs when UX/v reaches a critical value (5×10^5 to 2×10^6 for a flat plate, but not necessarily the same for a cylinder), X now being the circumferential distance from the leading, stagnation point. Thus an early transition occurs if U is large or v small. The turbulent boundary layer following the transition has a velocity distribution with much higher speeds close to the surface than does a laminar boundary layer (see Fig. 12.2). Thus a turbulent layer has more momentum available at places near the surface to oppose the pressure gradient, and so is *less* susceptible to breakaway. On the cylinder the breakaway point therefore moves farther downstream to about $\theta = 130°$ or so. The eddy in the wake is much smaller than that when the boundary layer was wholly laminar, and the form drag coefficient is considerably reduced. As the velocity past a cylinder is increased the change between the two régimes occurs quite suddenly, as soon as the transition occurs before the breakaway. It is sometimes worth while deliberately to cause an early transition so as to have a small, low drag wake, see Fig. 12.20.

It will therefore be seen that the properties of the boundary layer are of the greatest importance in determining not only the frictional drag but also the wake and so the form drag of solid bodies. Although clever streamlining can delay the breakaway considerably it cannot always prevent it. Other ways of controlling the breakaway can be used if additional complications can be tolerated: for example, a narrow slot can be made in the surface across the flow and fluid drawn inwards by a suitable suction pump or fan as shown in Fig. 12.18(*a*). The nearer, slower, fluid which is most susceptible to adverse pressure gradients is thus removed. Breakaway can be entirely inhibited by this method on poorly streamlined shapes which will then have a very low drag. It is

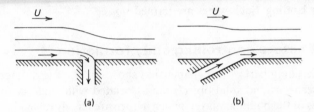

(a)   (b)

Fig. 12.18 Two ways of reducing the sensitivity of a boundary layer to an adverse pressure gradient, and therefore delaying the breakaway point. (a) Drawing-off the slow fluid through a slot; (b) speeding up the near-surface layers by a jet of fluid coming out of an inclined slot.

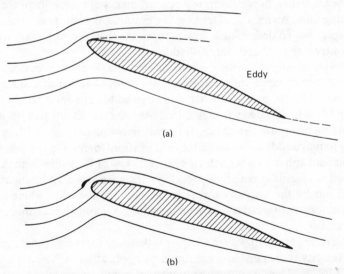

Fig. 12.19 (a) An unslotted aerofoil at a large angle of incidence has breakaway on upper surface. This stalling reduces the lift force considerably. (b) Slotted wing at the same incidence has no breakaway and continues to give a large lift force.

theoretically possible to apply boundary layer suction to aerofoils of any shape, reducing the drag, and thereby saving power, but the complications in the structure of an aircraft's wing to provide the slots and the internal passages is usually regarded as intolerable. Another method of control of breakaway is by fitting a slot as in Fig. 12.18(b), through which a jet of high-speed fluid is forced in the original direction of motion. The momentum of the jet speeds up the slow fluid near the surface and therefore makes it more resistant to adverse pressure gradients. This method has been successfully used on aerofoils, as shown in Fig. 12.19: passages are opened through an aircraft's wing at low speeds to convey the higher pressure air below the wing into the boundary layer on the upper side. A flow of air passes through the passage, which delays breakaway sufficiently to allow the aerofoil to be used at a much greater angle of incidence before stalling occurs, thus allowing the aircraft to fly at a lower speed for landing. Such wings are termed *slotted*.

12.12 Factors controlling eddy formation

The preceding part of this chapter has shown how quite different types of flow patterns around solid objects are associated with slight changes of the properties of the boundary layer, which in turn are largely dependent upon the turbulent motions therein. In fact, the type of flow pattern round a solid boundary depends greatly upon the eddy formation both in the boundary layer and outside of it: if the right conditions exist for a particular sort and size of eddy, then a certain flow pattern will appear, unless of course there are effects also acting other than those of friction, for example, the effect of a free surface, see Chapter 14. To determine a flow pattern it is therefore necessary to know what controls the life and death of eddies.

Eddies are greatly in evidence when one part of a fluid is accelerated relative to the rest. For instance, the water behind the blade of an oar of a rowing-boat is accelerated by the pull on the oar, and large eddies are immediately formed when the flow breaks away from the blade-tip: the harder the pull on the oar, the more vigorous are the eddies. It is reasonable to suppose that the forces tending to make eddies are inertia (i.e. acceleration) forces. On the other hand, eddies are not apparent when velocities are very low or viscosities high. An eddy produced in a highly viscous fluid, treacle for instance, is rapidly attenuated and damped out by the viscosity forces in the parts of the eddy where there is relative motion between it and surrounding fluid. Viscosity forces, then, tend to reduce eddies.

Inertia forces are dealt with in Chapter 6: these are the forces generated when a fluid velocity is changed and are found by application of Newton's Second Law of Motion. They are always proportional to $\rho u^2 a$, where u is a velocity of the fluid measured at some appropriate place, and a is a cross-sectional area. The particular inertia force per unit area producing eddies, at a certain place in a fluid flow, can therefore be written $K\rho U^2$. K is a constant for the given configuration of boundaries, chosen simply so that U may be taken as the velocity at a convenient point and not necessarily at the place where the eddies are produced. Changing the velocity at this convenient point, without changing the boundaries, will change the velocity at all places in the same proportion (including the place of the eddy). The inertia force per unit area is thus proportional to ρU^2.

Viscosity forces are considered in Chapter 1, the definition of a viscous stress being the product of coefficient of viscosity and the velocity gradient. For one given configuration of boundaries, the velocity gradient at any one place is proportional to U/d, where d is some length measurement of the particular boundaries concerned. Increasing the size of the boundaries (without changing their shape) decreases all velocity gradients in the same proportion, whether they occur locally in eddies or over large areas of the flow. Thus the viscous force per unit area at any place is proportional to $\mu U/d$.

Now it is conventional to express the relative importance of the two sorts of force by their ratio, so that

$$\text{Inertia force/Viscous force} = \rho U^2/\mu\, U/d = \rho Ud/\mu = Ud/\nu,$$

where ν is the kinematic viscosity of the fluid $= \mu/\rho$. This ratio is the Reynolds number derived by dimensional analysis in Chapter 10. When the number is small there will be little tendency towards eddy production, because the viscous forces are large compared with the inertia ones: when the ratio is large there will be a great tendency for eddies to occur.

The expression UX/ν which appeared often in the boundary layer analyses is a Reynolds number which expresses the sort of eddy present in the layer. The property of these eddies to produce a momentum change, and therefore a drag force on a smooth surface, is solely dependent upon the Reynolds number at the place considered, and this is shown by the unique character of diagrams such as Fig. 12.8. For rough plates both the relative roughness X/k and the Reynolds number fix the coefficient of friction C_f.

It should not be thought that the flow pattern changes continuously as Re increases: there may be ranges of Re when the pattern does not change at all, and there may also be sudden changes of pattern. A good example is the flow pattern around a circular cylinder with its axis normal to the flow. The pattern can be seen by the experimental methods described in Chapter 5, by dust particles or other tracers. Fig. 12.20 is divided into several sections, each of which shows the flow pattern for a certain range of $Re = Ud/\nu$ where U is the undisturbed velocity well upstream, d is the diameter of the cylinder, and ν is the kinematic viscosity of the fluid. If $Re < 2$ approximately, the flow pattern is that shown in (*a*) and there are no eddies at all: the pattern is very similar to that given by the theoretical streamline pattern (Chapter 5). If Re lies between about 2 and 40, there is breakaway on each side of the cylinder, with two simultaneous eddies in the wake, which do not, however, detach themselves periodically and create turbulence in the stream. If Re lies between 40 and 10^5 a breakaway occurs alternately on each side of the cylinder, at about $\theta = 85°$, and a wake eddy alternates on one side and the other. These eddies periodically detach themselves and create additional turbulence in the stream. And if $Re > 10^5$ the breakaway is delayed to about $\theta = 130°$ and a much narrower wake is caused. Thus a succession of several different types of flow pattern exist, dependent only upon the value of $Re = Ud/v$. Sometimes the change from one type to another is sudden (as is the one at $Re = 10^5$ with the cylinder), and sometimes gradual (as is the case at about $Re = 2$). Changes such as these occur with most flow patterns though it must not be assumed that they occur at the same values of Re as occurred with the cylinder, or even that they are of the same sort.

Since the type of flow pattern affects the pressure distribution and therefore the drag, it is not surprising to find a unique connection between the coefficient of drag and the Reynolds number, a result which was obtained by Dimensional Analysis (Chapter 10). The curve of Re against C_{drag} for smooth cylinders is given in Fig. 12.20, showing both the form drag coefficient and the total drag coefficient. The difference is the part of the drag due to the shear force. Similar

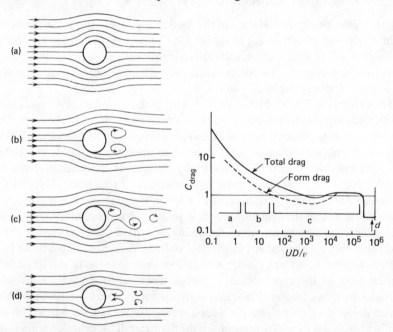

Fig. 12.20 (Left) Four different sorts of flow pattern around a circular cylinder, each with its own range of UD/v. See Fig. 10.1 (Right) A logarithmic plot of C_{drag} against UD/v for a smooth circular cylinder, showing also the ranges in which the flow patterns occur. Both Total drag and Form drag coefficients are shown the difference being the Friction drag coefficient.

experimental curves can be drawn for any other shape. These experiments can be done with a wide variety of sizes, velocities and viscosities of fluids, *providing the shape under test and the flow conditions are always the same.* The Reynolds number correlates all tests to a common basis.

The wake behind a solid 'bluff' body, as has been described above, is an area where fluid has been slowed down by the drag forces on the body. Outside the wake, the velocity is only changed by a small amount from the original upstream value—(usually increased, particularly if there are constricting walls each side of the body). Thus a cross-section of the flow just downstream of the body will show a non-uniform velocity distribution in the wake (Fig. 12.21(*a*)). This non-uniformity gradually fades out further downstream, until the original velocity of the stream is regained. A similar situation of a non-uniformity dying out arises in the neighbourhood of a *diffusing jet*. Such a jet occurs, for example, when a duct discharges air into a large room; when a chimney discharges into the atmosphere; or when a water flow from a pipe enters a tank well below water surface level (Fig. 12.21(*b*)).

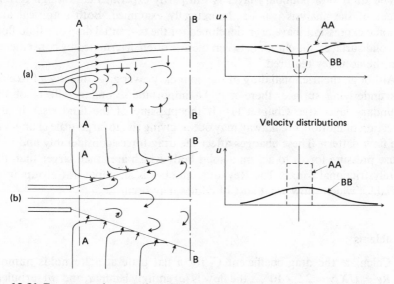

Fig. 12.21 Two cases of non-uniform flow gradually becoming less non-uniform by diffusion processes: (a) The wake behind a cylinder gradually disappearing downstream and a uniform flow re-appearing; (b) The spreading of a jet from a pipe which gradually changes from a high speed, concentrated flow to a much more widespread one. In both cases cross-section AA, taken as a whole, gives pronounced non-uniformity, and BB shows much less non-uniformity. Entrainment shown by small arrows at right angles to main flow.

In both diffusion situations there are no solid boundaries affecting the flow. Clearly there is no external force on the wake or the jet, so no change of the momentum flow can take place along the jet or wake axis; the observed changes of speed must be accomplished by a different mechanism. This is done by the effect of the eddies which steadily draw in, or *entrain*, fluid across the originally sharply defined boundary between fast and slow fluid. In a jet, the quantity of fluid in motion increases at successive cross-sections downstream as fluid, originally still, is entrained into the moving fluid. In the wake, the quantity of slow moving fluid decreases at successive cross-sections as fast fluid is entrained into it. This change of mass in motion, while there is no change of momentum, is characteristic of diffusion systems. At the same time, because of the high local shear stresses inside eddies, there is a degradation of kinetic into a thermal form of energy. Any situation where high and low speeds are adjacent in a fluid gives rise to turbulent entrainment and so to energy degradation, while the momentum flow remains constant at all cross-sections.

12.13 Summary

The study of a boundary layer is still partly experimental, though certain stages of the analysis can be theoretically examined. Both empirical and rational expressions have been developed for the tangential drag of a fluid flow on solid surfaces: the choice between these is often determined by convenience and the accuracy required.

Although the frictional drag of the boundary layer gives directly a force on the underlying surface, there may be important secondary effects of the boundary layer (see Chapter 11). If the pressure of the fluid rises in the direction of motion, breakaway may occur, giving rise to large-scale changes in the flow pattern. These changes affect the drag forces considerably and may cause pressure forces to act on a solid body which are much larger than the purely frictional forces. The Reynolds number is an important grouping of variables which show what sort of eddies may occur.

Problems

1. Calculate the drag coefficient C_f for a flat plate at a Reynolds number $Re = UX/v = 2.5 \times 10^5$, if the flow is (a) entirely laminar, and (b) turbulent assuming the power law $u/U = (y/\delta)^{1/7}$ and $\delta \propto Re^{-1/5}$. What are the practical consequences of any difference of (a) from (b)?

    Ans. (a) 0.00265. (b) 0.00616.

2. A flat plate is placed in a very turbulent stream of water at 40°C with the flow parallel to its surface. The water speed is 3.57 m s^{-1}. Where is the transition point and how thick is the boundary layer just before this point?

    Ans. 0.09 m; 0.069 cm.

3. Find the thickness δ of the boundary layer 30 cm from the leading edge of a flat plate held edgewise to a water current of 0.5 m s^{-1} and viscosity

0.01 poise, assuming that the velocity distribution in a laminar layer is

$$u/U = 2y/\delta - 2y^3/\delta^3 + y^4/\delta^4.$$

What is the distribution of shearing stress τ?
Calculate also the drag force per metre width. *Ans.* 0.142 N

4. Develop an expression for the coefficient of drag and δ for the laminar boundary layer on a flat plate assuming that the velocity in it increases to its maximum as the sine of the distance from the plate, i.e. $u/U = \sin \pi y/2\delta$.
Ans. $C_f = 1.310 \, Re^{-1/2}$.

5. Find the discharge through a boundary layer of thickness δ, in which the velocity distribution is $u/U = (y/\delta)^{1/5}$. Then find the *displacement thickness* of the boundary layer. *Ans.* $\delta_1 = \delta/6$.

6. Find the ratio of the friction drags on the front and rear halves of a flat plate set parallel to a uniform stream. Assume that the boundary layer is turbulent over the whole plate and that the velocity profile can be represented by the seventh root approximation. *Ans.* 1.35.

7. Find the ratio of the drag in water to the drag in air on a thin flat plate of chord 1 m at zero incidence in a stream flowing at $5 \, \text{m s}^{-1}$.

   The drag coefficient is $2.656/Re^{1/2}$ or $0.148/Re^{1/5}$ for a laminar or a turbulent boundary layer respectively. The critical Reynolds number is 0.5×10^6. Assume that the effective origin of the turbulent layer after transition is at the leading edge of the plate.

   The density and kinematic viscosity are to be taken as $1000 \, \text{kg m}^{-3}$ and $1.11 \times 10^{-6} \, \text{m}^2 \, \text{s}^{-1}$ for water, $1.21 \, \text{kg m}^{-3}$ and $1.47 \times 10^{-5} \, \text{m}^2 \, \text{s}^{-1}$ for air.
Ans. About 1100:1.

8. A rectangular thin flat plate is held in a uniform airstream with its surface parallel with, and one edge perpendicular to, the direction of the stream. The boundary layer is entirely turbulent, and the drag coefficient, based on the wetted area, can be taken as $0.074/Re^{1/5}$.

   Re is the Reynolds number based on the dimension parallel with the stream direction.

   Obtain a general expression for the drag in terms of the area and aspect ratio of the plate, the density and kinematic viscosity of the air and the air speed.

   Hence compare the drag of a rectangular plate which has its sides in the ratio 4:1 with the drag of a square plate of the same area in the same airstream, for both the possible orientations of the rectangular plate. Outline briefly the physical reasons for the difference in drag.
Ans. 1.15: 0.87

9. Distinguish between the *Reynolds Number* and the *Reynolds stresses* in a fluid flow.

   At one point, distant 0.1 m from the wall of a channel with a water flow in it, a sensitive instrument measured two components u and v of the velocity.

Eleven consecutive measurements, at constant time intervals are listed below:

u mm s^{-1} +104 +109 +83 +88 +101 +93 +110 +100 +86 +94 +88

v mm s^{-1} -2 -15 +10 +24 -5 -19 -19 +3 +20 -2 +5

Determine the mean velocity of flow in the u-direction and find the local value of the Reynolds shearing stress.

At another point, distant only 0.05 m from the wall, the corresponding stress was 0.35 N m^{-2}. Estimate the stress at the wall.

Ans. $\bar{u} = 96$ mm s^{-1}; Stress 0.09 N m^{-2}; 0.61 N m^{-2}.

13

Flow through pipes and closed conduits

13.1 The need to calculate pipe flows

The engineer is often engaged in designing works to convey fluid from place to place by a closed conduit or pipe wholly filled with the fluid. In general, because of the relative motion, a drag force is exerted by the walls on the fluid so that flow work has to be done to preserve the motion. The pressure intensity of the fluid is therefore higher at points upstream than at points downstream in the pipe. It is helpful to know this pressure drop accurately so that if it is desired to use a pump a suitable size may be chosen (Chapter 15). Alternatively, if the pressure is being applied by gravity (for example, when a high level reservoir is forcing water down a pipe to a lower level), then it may be essential to forecast the size of pipe necessary to supply the required quantity of fluid. There is a great economic importance in high accuracy for this sort of calculation, and much research effort has been put into the problem. As an example, an error of $7\,\mathrm{kN\,m^{-2}}$, (a pressure-head of only 0.715 m) in forecasting the pressure drop along a length of 1.5 m diameter water main in which the mean water speed is $3\,\mathrm{m\,s^{-1}}$ causes an error of the estimated pump power of 37 kW.

13.2 The inlet to a pipe

The conditions near the inlet of a pipe from a large container of nearly static fluid can be investigated using the energy equation. If, over a short length of the pipe adjacent to the inlet, the energy degradation into heat is small, the conditions are therefore rather like those of orifice flow (see Chapter 8). Consider a reservoir of water with the surface at a height h above the centre-line of the pipe, the entry of which has a well-rounded, 'bellmouth' shape, Fig. 13.1(a). Water is being made to flow through the pipe at a speed u. In the reservoir, however, the water is all static, so that the total energy per unit weight, $H = p/\rho g + z$, is the same everywhere in it. It is convenient to take a datum level for the head at the level of the centre-line of the pipe, so that $H = h$ everywhere. Since there is little or no degradation of energy in the initial short length of pipe, the total head H is preserved to the point where a vertical pressure tube (piezometer) has been drawn. The water stands static in this tube

223

224   Flow through pipes and closed conduits

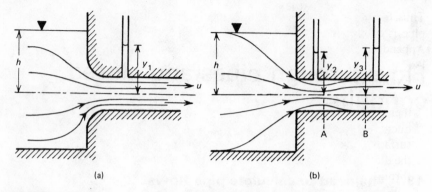

(a)   (b)

Fig. 13.1 (a) Flow out of a reservoir into a well-rounded entry of a pipe. The pressure just within the pipe is measured by a vertical piezometer. (b) Flow out of a reservoir into a sharp-edged pipe. Notice how there is breakaway at the edge, causing an eddy to form, and the jet reduces in diameter, subsequently expanding to pipe diameter with an accompanying energy degradation.

to a height y_1 above the centre-line so that the total head in the pipe at this place is

$$H = u^2/2g + y_1 = h$$

or
$$(h - y_1) = \frac{u^2}{2g} \tag{13.1}$$

Thus at a well-designed inlet to a pipe there is a drop of pressure head $h - y_1$ due solely to the flow work required to accelerate the fluid from rest to a velocity u. Such a pressure measurement is often a convenient way of finding the flow. It must be emphasized that this drop of pressure is reversible and quite unconnected with the irreversible pressure drop due to shear stresses at the walls, which will be discussed later.

If the entry to the pipe is not curved to a bellmouth, but is sharp edged as in Fig. 13.1(b), then the conditions are different, being rather like those in a sharp-edged orifice. The streamlines show a vena contracta at A and the pressure head there, y_2, is given by Bernoulli's equation i.e. $h - y_2 = u'^2/2g$, where u' is the velocity in the vena contracta. Downstream, however, the flow diverges until it occupies the whole of the pipe cross section, causing eddies as it does so. There is a consequent degradation of energy E in these eddies which must be included in the energy equation if it is to be applied to points downstream of the vena contracta. For example, at the second piezometer tube, where the water rises to a height y_3,

$$h = y_3 + \frac{u^2}{2g} + E$$

or
$$(h - y_3) = \frac{u^2}{2g} + E \tag{13.2}$$

Thus, for a particular flow rate, the drop of pressure at a sharp-edged entry of a pipe is always greater than that at a rounded entry, by an amount E (see Appendix at the end of this chapter for values of E).

Example
A ventilation duct has a well rounded entry, just downstream of which there is a static pressure tapping point. When there is air flow through the duct, a manometer filled with alcohol, specific gravity 0.8, has a difference in surface levels of 0.23 m when one tube is connected to the tapping point. If the duct has a cross-section of 0.1×0.3 m^2, what is the mass flow rate of air through the duct?

What extra power would have to be provided to give the same flow rate if a sharp edged entry were used?

Solution
Assuming that there is no loss of energy in the well rounded entry then writing the energy equation in terms of gauge pressures

$$0 = \frac{P_1}{\rho_1 g} + \frac{u^2}{2g}.$$

But from the manometer, $p_1 = -\rho_2 g h$

$$= -800\, g \times 0.23 \text{ kg m}^{-2}.$$

If the air is considered incompressible and to have a density of 1.23 kg m^{-3},

$$\frac{u^2}{2g} = \frac{800}{1.23} \times 0.23 \text{ m} = 150 \text{ m},$$

and $\qquad\qquad u = \sqrt{2 \times 9.81 \times 150} = 54 \text{ m s}^{-1}.$
Mass flow rate $= 1.23 \times 54 \times 0.03 = 2 \text{ kg s}^{-1}.$

If the entry were sharp edged, the head loss would be $\alpha u^2/2g$, where $\alpha = 0.5$ from the table on page 265.

Hence $E = \alpha \dfrac{u^2}{2g} = 0.5 \times 150 = 75 \text{ m}.$

Then extra power $= \rho g Q E$
$$= 2 \times 9.81 \times 75 \text{ kg s}^{-1} \text{ m s}^{-2} \text{ m}$$
$$= 1.5 \text{ kW}.$$

13.3 The middle part of a pipeline

A rounded entry to a pipe will generally produce a flow of fluid whose velocity is nearly uniform over the cross section. As this flow passes over the inner surface of the pipe, the shear stresses at the wall produce boundary layers which grow thicker downstream.

The flow in the boundary layer is retarded and so mass continuity requires that the velocities near the centre of the pipe increase, thus more flow work has to be done on the fluid and the pressure continues to fall as the flow outside the boundary layers is accelerated. Eventually after a distance of about 30 to 50 pipe diameters the boundary layers from opposite sides meet at the centre-line: thence forward no further growth of the boundary layer can occur, and the velocity distribution in a pipe of uniform cross-section is the same at all subsequent cross sections. There is no centre core of constant mainstream velocity, all the flow is subjected to shear stresses, and the pipe is completely full of boundary layer (see Fig. 13.2). In the common case of a pipe cross section which is symmetrical and which has uniform roughness everywhere on it, the maximum speed is at the centre.

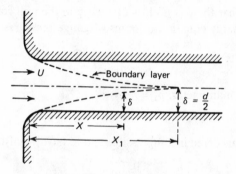

Fig. 13.2 The growth of boundary layers at the inlet to the pipe. At a distance X_1 from the inlet the boundary layers have grown to a width equal to the pipe radius: thenceforwards the whole pipe is full of boundary layer and there is no central core of fluid with uniform velocity.

The drag of the fluid on the walls which has caused the boundary layers is due to exactly the same mechanics as was the case for the flow over a flat surface, discussed in Chapter 12. In the initial stages of the development of the layers, before they have met, they grow and produce a stress in accordance with the same sort of laws which apply to the flat plate, the parameter UX/v (where X is the distance from the entry to the pipe) governing both the shear stress τ_0 and the layer thickness δ. The numerical values of the constants in the equations differ in the pipe case from the flat plate boundary layer case due partly to the drop of pressure in the flow direction. In the final, fully developed, stage of the flow when the layers from each side have met, δ is constrained to be equal to the radius of the pipe r, for the thickness cannot go on increasing indefinitely as was the case on the flat surface. The shear stress τ_0 therefore remains constant at a value fixed by r, U_{max}, v, and the roughness of the walls.

There is a convenient connection between τ_0 and the pressure drop along the pipe. Consider a short length δx of a pipe through which fluid is passing with a shear stress τ_0 on the walls. The resultant pressure drop is the difference between a pressure p downstream and a pressure $p + \delta p$ upstream, as shown in Fig. 13.3. The pressure force balancing the shear force is therefore $a\delta p$, where a is the cross-sectional area of the pipe. The shear force is the product of τ_0 and the surface area of the walls over the length δx, that is $\tau_0 P \delta x$, where P is the perimeter of the cross-sectional area of the pipe.

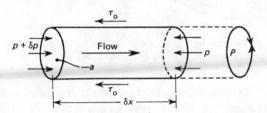

Fig. 13.3 The forces on the fluid flowing steadily in a short length δx of a pipe. The excess of pressure intensity δp acting over the cross-sectional area a just balances the shear stress τ_0 which is spread all over the inside of the pipe (perimeter P).

Equating the shear and pressure forces

$$\tau_0 P \delta x = a\delta p$$

or
$$\delta p/\delta x = \tau_0 P/a.$$

In this expression, p and τ_0 are measured in the same units of force per unit area. Engineers often express pressure intensity however as the head h of fluid causing the pressure p. Thus

$$\delta p = \rho g \delta h$$

Substituting above,
$$\rho g \frac{\delta h}{\delta x} = \tau_0 \frac{P}{a}$$

or
$$\delta h/\delta x = \tau_0 \frac{P}{a} \frac{1}{\rho g}$$

The ratio a/P is termed the *hydraulic radius* and is denoted by m: for a circular pipe of diameter d, $m = d/4$. Also the ratio $\delta h/\delta x$ is termed the *hydraulic gradient*; it is the angle made with the horizontal by the line joining the fluid surfaces in a succession of piezometers along the pipe, Fig. 13.4. From above, for a pipe of constant cross-section without bends,

$$\frac{\delta h}{\delta x} = \tau_0/\rho gm \qquad (13.3)$$

so that the hydraulic gradient is constant where τ_0 is constant, that is, where the

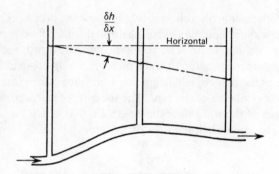

Fig. 13.4 The hydraulic gradient of a pipe is the gradient of a line joining the fluid levels in vertical piezometers at intervals along the pipe.

pipe has a constant cross-section, the flow is fully developed, and has no central core of constant velocity. In the entry length (about 30–50 diameters long) neither τ_0 nor $\delta h/\delta x$ are constant. However, this length is usually small compared with the remainder of the pipe and provided the pipe cross-section is constant and there are no sudden changes in the flow direction, it is usually good enough to regard $\delta h/\delta x$ as constant throughout the whole length of the pipe.

13.4 Pipe outlets

The outlet of a pipe into a large reservoir full of nearly still fluid (or a duct discharging air into a large room) presents the problem of high-speed fluid being decelerated. If the pipe finishes abruptly as in Fig. 13.5(*a*) a jet of high-speed fluid is projected into the still fluid and the steady streamline flow pattern breaks down into eddies (Chapter 12, p. 219). Near the outlet pipe the boundary, between moving and still fluid is well marked, and there is a pronounced velocity gradient. Vigorous eddies are formed by this gradient, tending to mix the moving and still fluids, the mixture having an intermediate speed. The mixing process occurs all along the jet so that the boundary becomes less distinct and the velocity gradient negligible at places far from the pipe. The incoming momentum of the jet is preserved, the reduced speed farther from the outlet being compensated by the increased quantity of fluid in motion. For constant momentum there must be no forces acting in the direction of motion, so the pressure remains constant throughout this dissipating jet, and the surface of water in a reservoir, for instance, remains horizontal above the jet. The kinetic energy $u^2/2g$ per unit weight of the incoming fluid is, however, completely degraded into heat by the eddy motions, so that there is a drop of total head of $u^2/2g$ as the fluid leaves the abrupt outlet.

The energy so wasted in low-grade heat must have been supplied to the fluid somewhere, in a ventilation duct by the fan and motor, in a water pipe by gravity or a pump. For a flow of Q units of volume per second, or of $Q\rho$ mass per second, the total power so wasted at an abrupt outlet is $\rho Q u^2/2g$. This power may be considerable and it is often desirable to reduce it and so decrease the energy required for the whole system. For example, in a 1.5 m diameter water main with a mean speed $u = 3\,\mathrm{m\,s^{-1}}$, the wasted power would be 24 kW. The waste of kinetic energy can be reduced by decelerating the fluid gradually in a slowly diverging pipe, so that only a low-speed jet is released into the reservoir. Such a diverging pipe is called a *diffuser*, Fig. 13.5(*b*). The difficulty of designing an efficient diffuser lies in avoiding breakaway of the main flow from the walls, for there is a pressure gradient opposing the flow (see Chapter .12). If breakaway does occur, then eddies are set up on one or other side of the central jet (see Fig. 13.6), and the energy degradation is increased nearly to that which occurs with an abrupt outlet. A conical diffuser of about 6° semi-vertex angle will avoid breakaway and will regain about 80 per cent of the inlet kinetic energy in the form of a rise in pressure, only about 20 per cent is wasted. If a diffuser is used on a pipe outlet, the total energy in it is nearly constant, and the pressure *rises* as the fluid passes through and is decelerated.

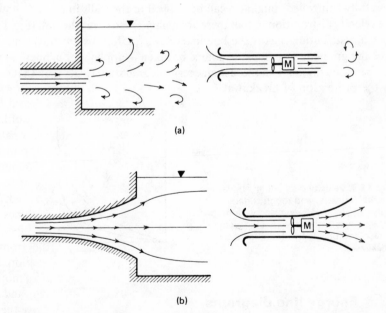

(a)

(b)

Fig. 13.5 (a) Abrupt outlets from pipes cause vigorous eddy formation and consequent large energy degradation: (left) the outlet of a water pipe into a reservoir; (right) the outlet of a ventilation duct; the energy so dissipated must be supplied by the fan and motor. (b) Gradual outlets (diffusers) reduce the eddy formation and energy degradation considerably, and exhaust the fluid at low speed.

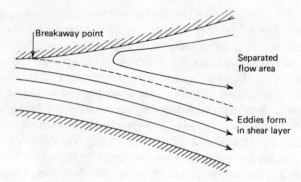

Fig. 13.6 A poor diffuser allows breakaway to take place, the asymmetric jet causes eddies, and there is considerable energy degradation.

A great deal of research has gone to produce efficient diffusers that are not unduly narrow angled and long. As in the case of aerofoils and other surfaces, methods of boundary layer control can be used with success to delay or inhibit breakaway altogether (see Chapter 12). Suction is used to reduce the boundary layer thickness if the complications of the equipment required is tolerable. Carefully controlled roughness can be applied to the walls in order to modify the velocity distribution so that there are higher speeds near the wall (Fig. 13.7); such a distribution makes the boundary layer more resistant to the adverse pressure gradient. Care must of course be taken in applying roughness so that the increased drag does not cause an energy degradation larger than that saved by the prevention of breakaway.

Fig. 13.7 Velocity distribution above smooth (curve a) and rough surfaces (curve b). At any given place y_1, a rough surface gives higher speeds than a smooth surface and thus is less likely to cause breakaway.

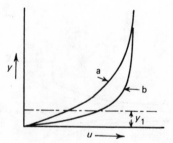

13.5 Energy line diagrams

A convenient way of showing graphically the degradation of energy in the several parts of a pipe system is by using an *energy line diagram*. Examples are shown in Fig. 13.8 for water pipelines and a ventilation duct. On a vertical control surface across the flow, the total energy of the fluid is set off at any point

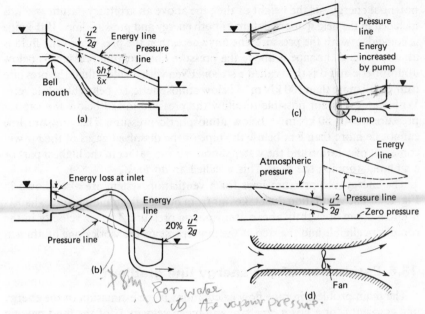

Fig. 13.8 Four examples of energy line diagrams. (a) A pipeline between two reservoirs. Notice the vertical drop in the energy line at the abrupt outlet. (b) A pipeline which rises above the pressure line has a pressure in it *below* atmospheric, and the arrangement is called a *siphon*. Notice the drop in the energy line at the abrupt entry. (c) A pump in a pipeline gives a concentrated increase to the energy in a pipe. (d) An energy line diagram can be made for a ventilation duct, but an atmospheric pressure datum line must be shown. There may be an additional decrease of energy and pressure as the air flows through filters or coolers, bends, changes of section etc.

by a vertical ordinate above the centre-line of the pipe. The line joining the tops of the ordinates is called the *energy line*. Thus at reservoirs of still liquid, the energy line is at the level of the liquid surface: at places where there is a concentrated degradation of energy there is a step down in the energy line, as occurs at abrupt inlets and outlets: at places where there is an energy input, such as at a pump or fan, there is a step up in the direction of motion of the fluid: and along pipelines where there is a uniform degradation of energy, due to fluid friction, the energy line is inclined to the horizontal at an angle equal to the energy line gradient i. It is usual to assume that i is the same in the inlet length (about 50 diameters long) as in the remainder of the pipe where the flow has developed its full boundary layers to the pipe axis. The error is small, for most pipes are much longer than 50 diameters.

As well as the energy line it is also useful to plot the *pressure line*, equivalent to the hydraulic gradient line of Fig. 13.4. Since the total head H is equal to the sum of kinetic, potential and pressure heads, $H = u^2/2g + p/g\rho + z$, it will be seen that the pressure line is always $u^2/2g$ vertically below the energy line. The potential energy z is the height of the pipe above an arbitrary datum and it is included in the height above datum of both energy and pressure lines. If the pipe is coincident with the pressure line anywhere, then the pressure of the fluid is atmospheric: if the pipe is above the pressure line then the pressure is below atmospheric and it is then called a *siphon*. Even with an ideal fluid the pressure cannot fall more than $100 \, kN \, m^{-2}$ below atmospheric, i.e. below absolute zero. With water it is not possible to allow the pressure to fall below the vapour pressure about $80 \, kN \, m^{-2}$ below atmospheric pressure. The pressure line cannot lie more than 8 m below the pipe or the dissolved gases of the air will come out of solution and the water vapourize and gather in the highest part of a siphon, stopping the flow. This is called an *air lock.*

The energy and pressure lines for a ventilation system are also shown in Fig. 13.8. The ventilation duct takes air from outside the building and exhausts it into a room at slightly above atmospheric pressure. The air velocity in the room is negligible and the step at the fan represents the work done on the air.

13.6 Estimation of the energy line gradient

The main problem in pipe flow calculations is the estimation of the energy line gradient i for a given pipe size and mean velocity $\bar{U}$ of the fluid passing through it. Alternatively, i may be given and it is desired to find which size pipe is required for a certain discharge Q to pass through it. The problem is of such great economic importance that few other subjects in fluid mechanics have attracted so much research effort, and so many formulae have been proposed that they may appear confusing.

The obvious way to solve the relationship between i, Q, and pipe size is entirely empirical. Experiments are made on lengths of uniform pipes, both circular and non-circular in cross section, and both Q and i measured. Usually i is found from the drop in pressure δh for a certain length δl of the pipe so as the pipe is uniform, equation 13.3 can be rewritten as

$$i = \frac{\delta h}{\delta l} = \frac{\tau_0}{\rho gm}.$$

It is found that a law of the form

$$\bar{U} = K m^x i^y \tag{13.4}$$

gives a very fair approximation to the experimental data if the variables do not extend over a wide range. In this equation $\bar{U} = Q/a$ is the mean velocity in the pipe, $m = a/P$ is the hydraulic radius of the pipe, K is a coefficient and x, y are indices. a is the pipe cross-sectional area.

Unfortunately, there is considerable divergence of opinion as to the numerical values of K, x and y. Not only does it appear that they vary according to the roughness of the pipe walls, but they change with the pipe size and the mean velocity, even if the roughness remains constant. Also the equations seem to be accurate only in the restricted range of $\bar{U}$ in which the experiments have been conducted; at greater or smaller velocities the equations give unacceptably inaccurate predictions. However, such formulae are often used in the present state of engineering knowlege, though a good deal of experience and guess-work is required to estimate K, x and y for pipes which have not been subject to experiment and, indeed, may not have been made.

Of the formulae of the above sort, those due to Chezy and Manning are most often used. Chezy's formula is $\bar{U} = C \sqrt{(mi)}$ and the coefficient C depends on the roughness of the walls and on m. In Manning's formula, $\bar{U} = n^{-1} m^{2/3} i^{1/2}$, the coefficient n is much less dependent on m. Both equations are simple to use, and there is an ever-increasing body of experience with which to estimate C or n for a pipe of a given construction and roughness. There are several formulae (also empirical) which purport to express C and n in terms of m, and of parameters typical of the roughness. It is important to remember when using either Chezy's or Manning's equation that neither are dimensionally correct so that the numerical value of C or n is dependent on the system of measurement in use. For example, C for a large, very smooth pipe would be of the order of $66 \, \text{m}^{1/2} \, \text{s}^{-1}$ while a corresponding value of n is $0.015 \, \text{s} \, \text{m}^{-1/3}$.

Alternatively a rational approach may be used to establish suitable equations. If the shear stress in equation 13.3 is expressed in terms of the shear stress coefficient, C_f, (page 192) then, for a uniform pipe,

$$\tau_0 = \rho g m i = \tfrac{1}{2} \rho \, \bar{U}^2 C_f.$$

This may be expressed as

$$\bar{U} = \left(\frac{2g}{C_f} m i \right)^{1/2},$$

which, if $C = (2g/C_f)^{1/2}$, becomes the Chezy equation. Similarly if $n = m^{1/6} (C_f/2g)^{1/2}$ the Manning equation is produced. However the most common form of the equation is that attributed to Darcy or Weisbach which substitutes h/l for the energy line gradient i, Fig. 13.9, and gives

$$h = C_f \frac{l}{m} \frac{\bar{U}^2}{2g} \tag{13.5}$$

The dimensionless shear stress coefficient, C_f, is often written as f and termed the *pipe friction coefficient*. The term $\bar{U}^2/2g$ is the kinetic energy per unit weight of fluid corresponding to the mean velocity and it is a length measurement, or head of fluid in motion. Notice that it is *not* the mean kinetic energy of the fluid in the pipe (see Chapter 4).

For a circular pipe of diameter d the equation is $h = 4f \dfrac{l}{d} \dfrac{\bar{U}^2}{2g}$ because now

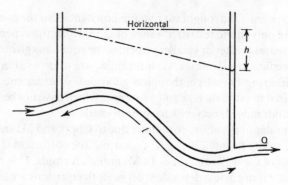

Fig. 13.9 The hydraulic gradient gives a drop of pressure *h* in a length *l* of pipe. The length *l* may be greater than the straight line distance between pressure tapping points.

$m = d/4$. Confusion is sometimes caused by the figure 4 because many textbooks (notably American ones) prefer to incorporate it into the coefficient f, but still call the combined coefficient f. It should always be made clear whether the general coefficient (as $f = C_f$ in equation 13.5) is meant, or the particular value f' applicable to circular pipes $\left(\text{that is } f' = 4f \text{ and so } h = f' \dfrac{l}{d} \dfrac{\overline{U}^2}{2g} \right)$.

It requires just as much experience to estimate f as was required for the coefficients and indices of the other empirical equations. There is now no intention of trying to make an equation so that the coefficient f remains constant, as was attempted before, so that it is not surprising that f changes with $\overline{U}$ and d even if the same roughness is used on the pipe walls. Nevertheless, in restricted ranges of $\overline{U}$ and d, roughnesses on the pipe walls caused by certain materials give sufficiently constant values of f that tables of them can be made and used by engineers. These values have been found by experiments where $\overline{U} (= Q/a)$, m, and i are all measured, and f calculated from them. But large numbers of these values of f so found will only produce confusion unless they can be related to a common basis. One way of achieving such a basis is by the Method of Dimensions as given in Chapter 10. It is expected that the wall stress τ_0 will depend on $\overline{U}, \rho, d, \mu$ and k. The last variable is the size of the roughnesses that are on the walls.

Suppose then that the complete dependence of τ_0 can be written as an equation

$$\tau_0 = \overline{U}^a \rho^b d^c \mu^f k^g + \text{other terms with the same variables but different indices,}$$
every term being dimensionally the same as the first.

Equating the dimensions of τ_0 and those of the terms on the right-hand side

$$MLT^{-2}L^{-2} = L^a T^{-a} M^b L^{-3b} L^c M^f L^{-f} T^{-f} L^g.$$

Collecting indices, form three simultaneous equations

$$\text{for } M, \quad 1 = b + f$$
$$\text{for } L, \quad -1 = a - 3b + c - f + g$$
$$\text{for } T, \quad -2 = -a - f$$

One solution of the equations giving a final result which is generally known can be obtained by solving for a, b, and c. That is

$b = (1 - f):$ $c = -1 - a + 3b + f - g:$ $a = (2 - f)$

$$c = -1 - (2 - f) + 3(1 - f) + f - g$$
$$c = (-f - g)$$

Inserting into the original equation for τ_0

$$\tau_0 = \bar{U}^{2-f} \rho^{1-f} d^{-f-g} \mu^f k^g + \text{other terms}$$

or
$$\tau_0 = \bar{U}^2 \rho \left(\frac{\mu}{\bar{U}\rho d}\right)^f \left(\frac{k}{d}\right)^g + \text{other terms}$$

This may be simplified to

$$\tau_0 / \bar{U}^2 \rho = \phi\left(\frac{\mu}{\bar{U}\rho d}\right)\left(\frac{k}{d}\right) = \tfrac{1}{2} C_f = \tfrac{1}{2} f$$

where ϕ means 'a function of'.

Hence f is a function of $(v/\bar{U}d)$ and (k/d). The function can be found by plotting the experimentally found values of f against $\bar{U}d/v$; and several curves will appear, each for a particular (k/d). Fig. 13.10 shows the generally accepted curves for roughnesses that are grains of sand stuck onto the walls. Other types of roughnesses give slightly different curves but these roughnesses are more difficult to express as a single size k. Because all the values of f fall onto their correct curves of (k/d), it is seen that no important variables have been omitted from the dimensional analysis. If there had been an omission, a value of f found for a certain k/d might have fallen amongst values applicable to another k/d, if the omitted variable had not remained constant.

The parameter $\bar{U}d/v$ is the *pipe Reynolds number* and k/d the *roughness ratio* for the pipe. When $k/d = 0$, the pipe is smooth walled and f only depends on $\bar{U}d/v$. The curve for this condition has been experimentally determined as far as $\bar{U}d/v = 1 \times 10^6$, and in this range it is closely approximated by

$$f = 0.079 \, (\bar{U}d/v)^{-0.25}. \tag{13.6}$$

Rough-walled pipes always give higher values of f than those for a smooth pipe at the same $\bar{U}d/v$; but at sufficiently high values of $\bar{U}d/v$, f then becomes constant. At lower $\bar{U}d/v$ the curves converge upon the smooth pipe curve, so that in some circumstances a physically rough walled pipe acts as if it is hydraulically smooth. Notice how f has a similarity to the coefficient of friction

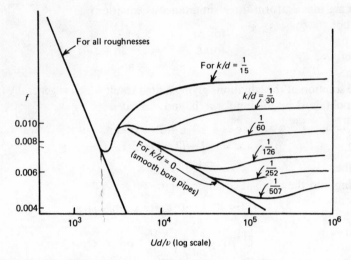

Fig. 13.10 The pipe flow shear stress coefficient diagram for circular pipes of all diameters and roughnesses and for all velocities and kinematic viscosities. The curves have been plotted from the results of many experiments with artificial roughnesses of uniform size. Slight changes are evident for commercial pipes where roughness is non-uniform. (After Nikuradse.)

C_f for a flat plate (Fig. 12.13). Both are functions of a Reynolds number composed of a velocity, a length and the kinematic viscosity.

In the case of the flat plate, however, both the Reynolds number and roughness ratio are expressed in terms of the length of the plate X which does not compare directly with the pipe diameter, d.

13.7 Pipe flow when Ud/v is small

A striking feature of the pipe flow diagram of Fig. 13.10 is the way in which the many branches of the curves converge into one line at about $\bar{U}d/v = 2300$ approximately. This line is straight, it overlaps the left-hand end of the branched curves, and it extends to the smallest values of $\bar{U}d/v$ with f increasing all the while. The single line is found to represent all values of f in the range of $\bar{U}d/v \leqslant 2300$, no matter what the roughness of the pipe may be. The equation $f = 16(\bar{U}d/v)^{-1}$ exactly fits the line.

The reason for this rather sudden change is that there is no turbulence in the fluid at all if $\bar{U}d/v$ is small enough. The flow is wholly laminar, and must be contrasted with the flow at the higher values of $\bar{U}d/v$ when the whole flow is turbulent and there is mixing throughout the pipe. The differences between the 'all laminar' and 'all turbulent' flow cases can be ascribed to the conditions in the boundary layers on the pipe walls at the entry to the pipe, before they have met at the pipe axis to form the fully developed pipe flow. The initial part of these boundary layers is laminar, precisely as was the upstream part of the layers on a flat surface (see Chapter 12). If the boundary layers unite at the pipe axis before they have reached their transition points (and so before they have become turbulent), then the whole flow in the pipe is laminar and $f = 16(\bar{U}d/v)^{-1}$. If the transition occurs before the layers unite then the flow in the pipe is turbulent and the right-hand, branched curves of Fig. 13.10 apply. The turbulence in the oncoming stream largely determines where the transition occurs, as was described in Chapter 12: with little or no turbulence in the oncoming stream entering the pipe, the transition is delayed to a high UX/v so that there is a correspondingly high $\bar{U}d/v$ before the flow in the pipe is turbulent. But however vigorous the eddies are made in the oncoming stream, UX/v never falls below a critical value (5×10^5 on a flat plate, but different in a pipe) and this corresponds to $\bar{U}d/v = 2300$. In other words, turbulent flow *cannot* exist in a pipe if $\bar{U}d/v < 2300$: laminar flow *may* exist if $\bar{U}d/v > 2300$, depending upon the turbulence present at the entry. With very still conditions indeed in a tank from which a bell-mouthed entry to the pipe extends, laminar flow has been known to exist in a pipe up to $\bar{U}d/v = 40,000$, though the slightest vibration then will start disturbances which grow and create fully turbulent flow in the pipe.

For the laminar flow case the stress τ_0 and consequently f can be calculated precisely and does not depend on experiment. The proof is similar in many respects to that for the laminar boundary layer on a flat surface. It can be shown that the velocity distribution is a parabola, that $\bar{U} = \frac{1}{2}U_{max}$, that τ decreases linearly from τ_0 at the pipe wall to zero at the centre, and that $f = 16(\bar{U}d/v)^{-1}$. Substituting this value of f in Darcy's formula it is found that

$$h = 32\mu l\bar{U}/\rho g d^2. \tag{13.7}$$

and this is sometimes called *Poiseuille's equation*. While it is satisfying to find a solution to a problem in fluid mechanics that does not depend on any experimental evidence, this problem of the laminar flow friction in pipes is not of great engineering importance. When air or water is passing through pipes of the usual sizes, $\bar{U}$ is very small to allow $\bar{U}d/v < 2300$. It is usually if viscous fluids, such as oil, are passing through the pipe that v is high enough to ensure $\bar{U}d/v < 2300$ and thus to have laminar flow.

It should be noticed that the value 2300 for the *lower critical pipe Reynolds number* is applicable to flow in pipes only, and has no relevance at all to other flow phenomena also depending on a Reynolds number Re (which is length × velocity ÷ kinematic viscosity). For instance in open channels, the critical Re based on the depth of flow for the change from laminar to turbulent flow is about 6000: the critical Re for the boundary layer to become turbulent before breakaway on a cylinder (and therefore cause a narrow wake with low drag) is about 1×10^5 based on cylinder diameter. The existence of a phenomenon at a certain numerical value of a Reynolds number in one particular set of boundaries applies only to that arrangement.

Laminar motion in a pipe can be demonstrated by injecting a thin stream of dye or of smoke. The stream remains a thin thread throughout the length of the pipe, not diffusing and becoming faint as it would do if there were eddies in the flow mixing the stream with the remainder of the fluid. If the velocity is carefully increased so that $\bar{U}d/v > 2300$ then the slightest disturbance at the entry causes turbulent flow to replace laminar flow, and the dye stream is immediately diffused.

13.8 The universal pipe flow law *No need for it I agree!*

The pipe flow diagram of Fig. 13.10 was presented as having been experimentally determined throughout. Parts of the diagram may, however, be derived rationally and then checked by experiment. In laminar flow, the coefficient f may be determined as described in the previous paragraphs; in the turbulent range, the logarithmic velocity profile derived for turbulent flow in Chapter 12 may be used, together with the properties of a laminar sub-layer near the walls to find f for a smooth-bore pipe as

$$1/\sqrt{f} = 4.0 \log_{10}(Re.2\sqrt{f}) - 1.6 \qquad (13.8)$$

where $Re = \bar{U}d/v$. The proof will not be given here. This equation fits the experimental data better than the empirical law $f = 0.079\,Re^{-\frac{1}{4}}$ at high Re (equation 13.6), but is not sufficiently superior at lower Re to make it worth while using for engineering purposes since it is more difficult to use (see Fig. 13.11).

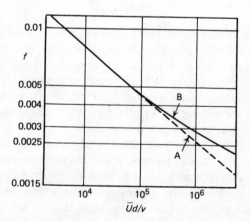

Fig. 13.11 A comparison between the values of f obtained by the Blasius empirical equation $f = 0.079\,Re^{-\frac{1}{4}}$ (curve A) and those found by experiment and the exact law (curve B) $1/\sqrt{f} = 4.0 \log_{10}(Re\,2\sqrt{f}) - 1.6$.

In a similar manner the exact solution can be made for f due to the logarithmic velocity distribution in a rough-walled pipe on the assumption that the roughnesses themselves project so far through the laminar sub-layer that their wakes throw off eddies into the main stream. The solution is

$$1/\sqrt{f} = 4.0\log_{10} r/k + 3.48. \tag{13.9}$$

showing that f is now solely dependent upon the roughness ratio r/k.

To compare effectively the equations for smooth- and rough-bore pipes it is necessary to find how f in the smooth-wall case depends on the thickness of the laminar sub-layer δ'. Now experiments show that approximately $\delta' = 10v/\sqrt{(\tau_0/\rho)}$ (p. 207).

But since $\qquad \tau_0 = g\rho id/4$ for a circular pipe

So $\qquad \sqrt{(\tau_0/\rho)} = \frac{1}{2}\sqrt{(idg)}$

or $\qquad \delta' = 20v/\sqrt{(idg)}$.

By definition $f = i\dfrac{d}{4}\dfrac{2g}{\bar{U}^2}$ (equation 13.5), and substituting for (idg)

$$\sqrt{f} = \frac{v}{\bar{U}\delta'}10\sqrt{2}$$

Consider first the rough-wall flow formula equation 13.9. This may be rewritten

$$1/\sqrt{f} - 4\log_{10} r/k = 3.48,$$

that is to say, the left-hand side of the equation is completely invariable.

Now consider the smooth-wall flow formula of equation 13.8; subtract from both sides $4\log_{10} r/k$, thus giving

$$1/\sqrt{f} - 4\log_{10} r/k = 4.0\log_{10} 2Re\sqrt{f} - 1.6 - 4\log_{10} r/k.$$

The first term on the right-hand side may be simplified by substitution for $\sqrt{f} = 10\sqrt{2}v/\bar{U}\delta'$ above, that is

$$4.0\log_{10} 2Re\sqrt{f} = 4.0\log_{10}\frac{2\bar{U}2r}{v}10\sqrt{2}\frac{v}{\bar{U}\delta'}$$

$$= 4.0\log_{10} 40\sqrt{2r/\delta'}$$

$$= 4.0\log_{10}\frac{r}{k}\frac{k}{\delta'}40\sqrt{2}$$

$$= 4.0(\log_{10} r/k + \log_{10} k/\delta' + \log_{10} 40\sqrt{2}).$$

Insert this term into the modified equation 13.8 above, and

$$1/\sqrt{f} - 4\log_{10} r/k = 4.0(\log_{10} r/k + \log_{10} k/\delta' + \log_{10} 40\sqrt{2})$$
$$- 1.6 - 4\log_{10} r/k$$

$$= 4.0\log_{10} k/\delta' + 5.4.$$

Thus for a smooth-walled pipe, or a rough pipe having a laminar sub-layer submerging the roughness elements, the expression

$$1/\sqrt{f} - 4\log_{10} r/k$$

is not constant but depends on k/δ'.

Both the 'smooth wall' and 'rough wall' equations can now be shown on the same graph (Fig. 13.12), where $1/\sqrt{f} - 4\log r/k$ is plotted against $\log k/\delta'$. The smooth-wall law shows a straight inclined line, and the rough-wall law a straight horizontal line. The experimental data always fit one or the other lines at high or low values of k/δ', but in a middle range of k/δ' the points form a transition curve from one line to the other. The shape of the transition depends on the sort of roughness but does not depend on its size which has already been taken into account by the variable k. Roughnesses consisting of a single layer of sand grains all of the same size give one transition which diverges from the smooth-wall line when k/δ' is a little less than 1.0, that is when the crests of the grains are just emerging from the laminar sub-layer. Commercial pipes have roughnesses which are a mixture of large and small grains, and these give another transition starting at a lower k/δ'. This can be explained if k is the mean size of the roughnesses which will be used to correlate the high k/δ' data to the rough pipe line of Fig. 13.12. But while the laminar sub-layer is still thicker than k, some of the larger roughnesses are projecting through it, producing eddies from what is now a partially rough surface. It will not be until δ' is a good deal smaller than k that all the roughnesses are exposed and creating the completely rough surface.

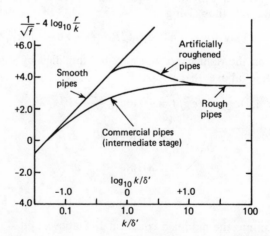

Fig. 13.12 The universal pipe curve for turbulent flow. The ratio k/δ' is the parameter showing if the boundary roughness project through the laminar sub-layer or not. The shape of the middle part of the curve depends on the type of roughness and its uniformity of size.

The success of the boundary layer analysis (much of which is too advanced and lengthy to give here) in correlating all the pipe flow data onto one curve (Fig. 13.12) may well be regarded as one of the triumphs of fluid mechanics. But great problems remain: how to find k by direct measurement without performing a hydraulic experiment in the high k/δ' or rough-surface law zone of Fig. 13.12? How to express the size of non-uniform roughnesses in terms of a single parameter k? How to select the appropriate transition curve for a particular roughness shape? It is not always possible, for instance, in large hydro-electric turbine supply pipes, to carry out experiments at a high enough $\bar{U}d/v$ for the roughnesses to protrude completely through the laminar sublayer and thus to give k. Yet the accurate estimation of f is often critically important, and improved methods along the foregoing lines will no doubt be attempted.

13.9 Effect of the pipe shape on the flow

The whole derivation of the flow formulae rests on the assumption that the radial distribution of velocity is the same as that over a flat plate; that in fact the pipe flow is the boundary layer of a plane surface wrapped around the pipe axis. It is therefore assumed that the flow is symmetrical about the axis and the shear force per unit area is the same at all places round the perimeter. This is undoubtedly true for a circular pipe, but not so for pipes of other sections. In Fig. 13.13 the contour lines of equal velocities are shown for three shapes of pipes. For a circular pipe they are symmetrical; those for a rectangular pipe are more crowded near the midpoints of the walls than they are near the corners. There is accordingly a greater velocity gradient at the midpoints than elsewhere, giving a greater τ_0 there. For a somewhat square section, Fig. 13.13 (b), the corner effects form an appreciable part of the whole flow: but for a long narrow section, Fig. 13.13 (c), the corner effects occupy less of the

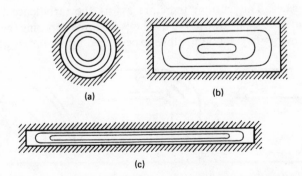

(a)   (b)

(c)

Fig. 13.13 Cross sections of three different pipes, showing typical contours of equal velocity.

total cross-sectional area. On the other hand, the velocity gradients at the short sides of pipe (c) are more different from those at the long sides, compared with the squarer pipe (b). The two effects are to some extent compensating so that it is found experimentally that f for non-circular pipes is not greatly different from that for a circular pipe, and is certainly near enough for most engineering purposes. The concept of the hydraulic radius (cross-sectional area divided by wetted perimeter) enables a close correlation to be obtained between friction data for many shaped pipes. If the pipe cross-section is very irregular, however, with many reversals of curvature, the concept may break down and there may be larger errors in using the circular pipe values of f. Experiment will be necessary for every separate case of this sort.

13.10 Effect of pipe bends and other non-uniformities

Pipelines are not always straight from end to end; they have bends in them, which may be abrupt or gentle, and they may have valves and changes of cross section in them. The effect of these non-uniformities is usually to increase the degradation of energy locally.

If the bend is sufficiently abrupt as in Fig. 13.14(a), breakaway may occur at the inside of the bend and will cause a large eddy formation downstream. These eddies degrade a good deal of energy in a short length of pipe, so that such bends cause a sudden downward step in the energy gradient line, in addition to the gradient caused by the normal pipe friction. The breakaway can be prevented, and the energy degradation reduced, by placing curved guide vanes across the corner, Fig. 13.14(b). These constrain the fluid to turn at the appropriate radius. If the bend is gentle, as in Fig. 13.14(c), no breakaway occurs, but secondary flow is caused by the interaction of boundary layer and radial pressure gradient due to the bending of the streamlines. The double spiral flow so resulting (see Chapter 11) acts across the centre of the pipe, giving a constant cross velocity which brings slow fluid to the centre and fast fluid to the inner sides. This current therefore assists the turbulence to transport momentum across the flow, and so increases the force opposing the motion. An increased energy degradation is thus experienced around even gentle bends.

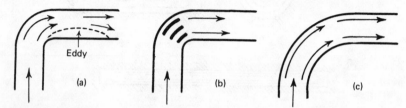

Fig. 13.14 Three different pipe bends. (a) A sharp elbow may create breakaway and consequent eddies which (b) may be eliminated by using turning vanes. (c) A gentle bend usually causes secondary spiral flows downstream of it (Chapter 11).

All other non-uniformities in pipes cause additional and concentrated degradation of energy. Expansions from one size pipe to another give eddy formations similar to those already described for the inlet and outlet of pipelines. Valves always cause some eddies even when they are fully open, the designs having a more intricate fluid passage giving a greater degradation than those of the straight-through flow variety. For any particular type of non-uniformity, experiments must be carried out to find the reduction of energy through it. It is found that if the flow in the pipes is turbulent (the common engineering case), then the energy degradation in the fittings is porportional to the kinetic energy $\bar{U}^2/2g$. In laminar flow a different relation may hold. Comparison with the usual pipe friction equation $h = 4f\dfrac{l}{d}\dfrac{\bar{U}^2}{2g}$ shows that the friction of every pipe bend or fitting can be expressed as an additional length of the pipe concerned. The sum of the real length and the additional length gives the effective length, which may then be used in the pipe friction equation to obtain the relation between i, d and Q. Typical valves of the loss coefficient, α, and the equivalent additional length of pipe, β, are given in Fig. 13.25.

Example
A smooth-walled pipeline 3500 m long is to connect two reservoirs 7 m different in elevation. Entry is abrupt, outlet is a conical diffuser. There are two globe valves, and five 60° bends are inevitable. If gravity is to cause a flow of water, what should be the diameter so that the discharge is 0.03 m³ s⁻¹? Water temperature 15°C.

Solution
First approximation—ignore valves, inlet, bends, etc. Flow is turbulent, most probably, so guess $\bar{U}d/v = 10^4$. Thus estimate f from Fig. 13.10 as $f = 0.0079$.

So
$$h = 4f\frac{l}{d}\frac{U^2}{2g} = 4f\frac{l}{d}\frac{Q^2 16}{\pi^2 d^4 2g}$$

or
$$d^5 = \frac{32}{\pi^2}\frac{flQ^2}{gh}$$

with $f = 0.0079$, $l = 3500$ m, $g = 9.81$ m s⁻², $h = 7$ m.

$$d^5 = 117 \times 10^{-5}\,\text{m}^5 \quad \text{or} \quad d = 0.259\,\text{m provisionally.}$$

As d is derived from a fifth root its value will not be very sensitive to changes in the other values in the equation. Use this provisional value of d to check that the guessed value of $\bar{U}d/v$ was correct ($v = 1.45 \times 10^{-2}\,\text{cm}^2\,\text{s}^{-1}$).

$$\bar{U} = \frac{0.03}{\pi/4 \times 0.259^2} = 0.568\,\text{m s}^{-1}\ \text{provisionally.}$$

So $\qquad \dfrac{\bar{U}d}{v} = \dfrac{0.568 \times 100 \times 0.259 \times 100}{1.45 \times 10^{-2}} = 1.01 \times 10^5$

Thus the first approximation of $\bar{U}d/v$ was underestimated by a factor of ten. *Second approximation*—try $\bar{U}d/v = 1.2 \times 10^5$. Thus $f = 0.0044$ from Fig. 13.10, a decrease of 40 % in f, which is small compared with the factor of ten change in the Reynolds number.

So $\qquad\qquad d^5 = 65.3 \times 10^{-5}$ or $d = 0.230$ m
and $\qquad\qquad \bar{U} = 0.717$ m s^{-1}

Check for $\qquad \dfrac{\bar{U}d}{v} = \dfrac{71.7 \times 23.0}{1.45 \times 10^{-2}} = 1.14 \times 10^5$.

This estimation is therefore much better than the first. Since d is changed so little by the change of $\bar{U}d/v$ from 1×10^4 to 12×10^4, no further approximation will be made. It is probably good enough for most engineering purposes.

Third approximation takes bends, valves, etc., into account.
From the appendix (p. 265)

an abrupt entry is equivalent to 25 diameters of pipe
a diffuser outlet $\qquad$ „ $\qquad$ 6 $\qquad$ „ $\qquad$ „
a globe valve $\qquad$ „ $\qquad$ 75 $\qquad$ „ $\qquad$ „
a 60° bend $\qquad$ „ $\qquad$ 22 $\qquad$ „ $\qquad$ „

Thus the total additional length of plain pipe equivalent to the fittings is

$$25 + 6 + (2 \times 75) + (5 \times 22) = 291 \text{ diameters}$$

or a length of $\qquad\qquad 291 \times 0.230 = 67$ m
Thus the effective length is $\qquad\qquad 3567$ m
Amending the second approximation

$$d^5 = 65.3 \times 10^{-5} \times \frac{3567}{3500} = 66.7 \times 10^{-5}$$

and d is barely affected. Thus $d = 0.23$ m is the diameter necessary and probably the nearest standard pipe size, perhaps 0.25 m diameter would be used. The conclusion above that the fittings in a *long* pipeline are of little importance in the determination of the diameter demonstrates the usual engineering convention of ignoring them.

The type of expression which occurs in this example where the variables cannot be completely separated, i.e. $f = \phi(Re) = \phi\left(\dfrac{\bar{U}d}{v}\right)$ and so d appears on both sides, is referred to as *implicit*. If the pipe diameter and flow were both given and the unknown was the head, it could be found directly and the solution would be *explicit*.

13.11 Changes of pipe flow with time

It is commonly found that in the course of time the shear stress coefficient of pipes carrying water increases, so that the flow decreases for a given pressure difference. This is due to the growth of rust and of lime nodules, especially if the water is hard. It has been found that to a first approximation the nodules grow always at the same rate, so that the roughness size k increases linearly with time. That is, $k = k_0 + \alpha t$, where k_0 is the initial roughness of the pipe. The constant α can be found from data previously obtained from other pipes in the same neighbourhood. Since for rough pipe

$$1/\sqrt{f} = 4\log_{10} r/k + 3.48,$$

the original roughness k_0 can be found by a hydraulic experiment with the new pipe: substitution of the future roughness k will give the future value of f which is then used in the friction formula $h = 4f\dfrac{l}{d}\dfrac{U^2}{2g}$ to give the future flow.

13.12 Non-steady flow in a pipe

Some important engineering problems arise when a fluid flow in a pipe is accelerated or decelerated. The pressure within the pipe may change temporarily so much that the pipe may burst or collapse if it is not strong enough. It is rarely economic to design a large pipe strong enough to take these pressures, and safety devices are usually essential.

Consider a long pipe from a reservoir to a consumer's valve. If the latter is fully open the mean speed in the pipe is u, and the pressure falls along the pipe. Suppose now that the valve is shut instantaneously, thus completely stopping the fluid near it. Farther away upstream, the fluid is still moving so that the near fluid is compressed, increasing its pressure and density. In the piece of pipe shown in Fig. 13.15, the fluid near the valve is stopped, its pressure being p above the pressure in the moving fluid, so that to a first approximation its density is raised from ρ to $\rho(1 + p/K)$. K is the coefficient of compressibility of the fluid (Chapter 1). In a short time t, the length of the stationary fluid increases,

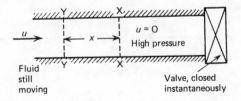

Fig. 13.15 Portion of a pipe upstream of a valve which has just been shut instantaneously. XX and YY are two positions of the junction between compressed, stationary fluid, and moving, low pressure fluid. The two positions are separated by a time t.

the junction between it and the moving fluid changing from XX to YY, a distance x. Due to the increase of density $\rho p/K$, the mass of fluid in the fixed volume XXYY has also increased, and this can only be achieved by a transport of mass of fluid through the section YY. During the whole of the time t, the fluid speed across YY is u, so that the transport of mass is $\rho u t a$, and so

$$ax\rho p/K = \rho u t a$$

where a is the cross-sectional area of the pipe.

That is. $$x/t = uK/p = c. \tag{13.10}$$

where c is the velocity of the junction of the moving and stationary fluid.

In order to obtain another expression for p and c, it is necessary to use the momentum theorem of Chapter 6. In Fig. 13.16(a) the junction between stopped and moving fluid in the pipe is shown moving at velocity c towards the fluid which still has a velocity u. If the whole system is given a velocity c in the opposite direction to this, as in Fig. 13.16(b), then the junction is stopped and the problem becomes one of steady flow into and out of the control volume shown. Now because K is large for most liquids, to a first order of approximation the density downstream of the now stationary junction is not

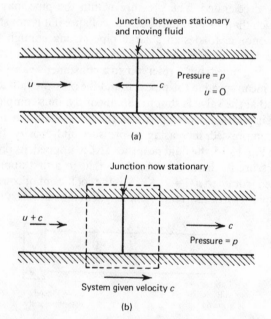

Fig. 13.16 A junction between moving and stationary fluid with velocity c, as in (a), thus giving non-steady motion. By giving the whole system an equal and opposite velocity, the motion becomes steady as in (b), and the momentum theorem may be applied to the control volume shown by the dotted lines.

greatly changed from the original value, so that the simplified momentum equation (6.6, page 88) may be used. That is, an incoming flow of $a(u+c)$ has its velocity changed by u so that the difference of the rate of flow of momentum into and out of the control volume is

$$\rho a(u+c)u$$

which is equal to the force imposed on the fluid, pa.

So substituting for c from equation 13.10

$$pa = \rho a(u+uK/p)u.$$

Again because K is very large compared with pressures usually experienced $(K = 2 \times 10^6 \,\text{kN m}^{-2}$ for water), so u may be ignored compared with uK/p.

That is $\qquad p = \rho u^2 K/p$

or $\qquad p = u\sqrt{(K\rho)}$ $\qquad\qquad\qquad\qquad$ (13.11)

Thus the pressure set up by the deceleration is independent of the size or length of the pipe. If water is the fluid concerned then by substituting numerical values for K and ρ it is seen that although p is small compared with K, it is large (4200 kN m^{-2}, or 40 atmospheres at $u = 3 \,\text{m s}^{-1}$) compared with the pressures usually imposed in steady flow. The pipes must therefore be thick walled for safety. Instantaneous closure of valves is clearly to be avoided if costs are to be kept low.

Substitution for p from 13.11 into 13.10 gives $c = \sqrt{(K/\rho)}$, from which it will be seen that c is always high though by no means infinite ($c = 1430 \,\text{m s}^{-1}$ for water). Thus the assumption of ignoring u compared with c is justified. It was mentioned in Chapter 1 that $c = \sqrt{(K/\rho)}$ is also the speed of sound in a substance, so that in the pipe the high pressure fluid grows outward from the valve at sonic speed. The junction is a pressure-wave which can be heard in the pipe as a knocking noise, sometimes called *water hammer*. Notice, however, that the expression for p involves the assumption that K is large; if it is small (as with a gas for example), then more elaborate analysis is required.

In the compressed fluid, energy is stored in the form of strain energy according to the usual law,

$$\text{Strain energy per unit volume} = \tfrac{1}{2}\,\text{Stress} \times \text{Strain}$$
$$= \tfrac{1}{2}p^2/K.$$
$$\text{Substituting for } p = u\sqrt{(K\rho)},$$

$$\text{Strain energy per unit volume} = \tfrac{1}{2}\frac{u^2}{K}(K\rho) = \tfrac{1}{2}\rho u^2.$$

But $\tfrac{1}{2}\rho u^2$ is the kinetic energy per unit volume due to the velocity u (observe $u^2/2g$ is the kinetic energy per unit *weight* of fluid). Thus all the original kinetic energy of the fluid is transformed to strain energy at the junction, and none is converted to heat or otherwise degraded by the water-hammer phenomenon though such a degradation will of course occur in the pipe due to the usual pipe friction.

The history of the pressure at the valve (or at any other point in the pipe) can now be found (Fig. 13.17). When the valve is shut the pressure wave at the junction of the moving and stationary fluid moves away from the valve until at a time l/c afterwards the whole pipe is filled with stationary high-pressure fluid (Fig. 13.17(a) and (b)). This situation is unstable, for the high pressure is now not balanced by any deceleration of the fluid. The high pressure therefore commences to accelerate the fluid in the opposite direction, starting from the open, reservoir end. The pressure wave moves back again towards the valve, the pressure remaining high there, Fig. 13.17(c). When the pressure wave reaches the valve, all the fluid is now moving away from the valve at the same speed as it originally moved in the opposite direction. To decelerate this reverse flow the pressure must therefore be reduced at the valve to give an inward force on the outward moving fluid. The conditions are exactly the same

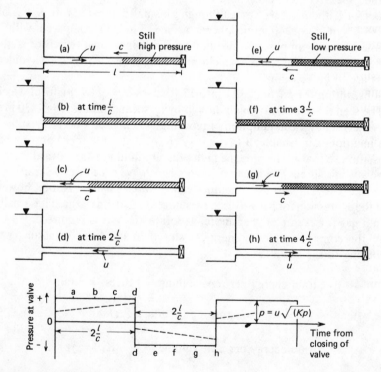

Fig. 13.17 The history of pressure changes in a pipeline after the valve at the end has been shut instantaneously. The pressure-time graph at the valve is shown below the eight sketches (a) to (h). The dotted line shows the pressure if shear stresses gradually degrade kinetic energy to heat.

as occurred when the valve was shut, but reversed in sign. The pressure therefore falls to $p = -u\sqrt{(K\rho)}$ below the pressure at the reservoir which could cause the liquid to cavitate and the pressure wave moves away from the valve again, at the same speed c (Fig. 13.17(e)). When the pressure wave arrives at the reservoir end, Fig. 13.17(f), the pipe is all filled with low pressure fluid, which is again in an unstable state: the higher pressure in the reservoir accelerates the fluid in the original direction and the pressure wave moves back again towards the valve. The fluid now has its original velocity, so as the pressure wave gets to the valve, the hydraulic conditions are exactly the same as occurred when the valve was first shut. The complete cycle of events is therefore repeated and goes on repeating if there is no degradation of the kinetic energy into other forms of non-available energy.

The cyclic nature of these pressure pulses is shown in Fig. 13.17 by a graph of the pressure at the valve, plotted against the time after the valve is closed. The full line shows the pulses as predicted in the preceding paragraph. The dotted lines show approximately how shear forces on the moving fluid in the pipe steadily degrade the kinetic energy so that the strain energy in the stationary fluid is also steadily reduced. The magnitude of the pressure pulses is therefore reduced with time until they are imperceptible. If the negative pressure pulse is greater than the absolute hydrostatic pressure that the fluid possessed when it was originally moving, a further modification is made to the pulses. The pressure cannot fall below zero absolute so that if larger pulses are generated the fluid cavitates, and temporarily fills part of the pipe with vapour. The low-pressure part of the pulses may thus be cut off short.

If the valve is not shut quite instantaneously but takes a short time, *less* than $2l/c$, then the pressure is gradually built up at the valve in this time. The maximum pressure remains the same, however, because eventually all the kinetic energy is converted to strain energy before the pressure wave arrives back at the valve. But if the valve is shut in a time rather greater than $2l/c$, only part of the kinetic energy has been converted into strain energy by the time that the pressure wave between stopped and moving fluid has returned to the valve as in Fig. 13.17(d). Only a part of the increased pressure $p = \sqrt{(K\rho)}$ is therefore experienced by the valve: the situation of Fig. 13.17(e) does not arise because some of the flow of fluid away from the valve is provided by fluid coming through the partly open valve. The flow is restricted to some extent by the valve so that a reduced negative pressure is experienced, compared with the case when the valve was fully shut.

13.13 Slow closing of a valve

If the valve is shut in a relatively long time compared to the time required for a pressure-wave to travel twice along the pipe, say $10 \times 2l/c$, the pressure pulse will be hardly measurable. But a pressure rise of a different nature may quite well occur, and although this rise is usually much less than that caused by the compressibility effect above, it may still be too great for the safety of the pipe. The pressure rise is now determined by considering the fluid to be incompressible and originally travelling at a speed u. If a deceleration α is caused by the slowly shutting valve, then a deceleration force must be applied to the whole mass of fluid. This force is the increase of pressure at the valve, above the pressure at the supply end of the pipe. If h' is the head of static fluid corresponding to this pressure increase, then the deceleration force is $\rho g h' a$ on the whole cross-sectional area a of the pipe. Since the mass of fluid being decelerated is $\rho a l$, where l is the length of the pipe, then by Newton's Second Law

$$\rho g h' a = \rho a l \alpha$$

or

$$h' = l\alpha/g.$$

If the valve is made to move in such a way that the deceleration is uniform in time t, then $\alpha = u/t$ and $h' = lu/gt$. This does not imply that the rate of closing the valve is uniform, as will be later explained. The increase of pressure due to the deceleration is therefore dependent on both the length of the pipe and on the time of closing the valve: a result which should be contrasted with that for the case of the valve shutting suddenly, when the pressure is independent of both these variables.

The effect of the deceleration pressure on the energy line gradient of a pipeline is shown in Fig. 13.18. The valve, X, is at the far end of a long pipeline from a reservoir O. The energy line gradient for a steady flow might be the line OA (the pressure at X being fixed by the downstream pressure, another reservoir, for example). In this diagram the usually small energy degradations at the inlet, and the kinetic energy $u^2/2g$ have been neglected. If X is now shut slowly to give a decelerating pressure h', the gradient line initially moves upward to OB, where $AB = h'$, before either the velocity u or the head loss

$$h = 4f\frac{l}{d}\frac{u^2}{2g}$$ has changed significantly. As the deceleration takes effect, u

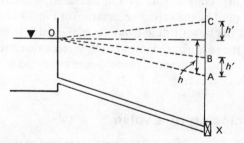

Fig. 13.18 Pressure changes in a pipe whose valve X is closed slowly. OA is the steady flow pressure line, and as soon as the fluid is decelerated a head h must have been applied opposing the motion. OC is the pressure line just before the motion ceases.

decreases and so does h, while h' stays constant if the deceleration is uniform. The gradient line therefore steadily rises, decreasing the energy line gradient. At the instant of complete closure of the valve, both u and h are zero, but h' still remains. The gradient line is therefore now sloping upwards to OC, and the pressure at the valve is greater than that due to the hydrostatic pressure. It is this additional pressure which may damage pipe or valve. An instant after complete closure, the pressure falls to hydrostatic.

It will now be seen how the pressures generated by a deceleration change radically according to whether the time of closure of the valve is much greater, greater, or smaller than $2l/c$. All three types of action can occur successively in one operation of the valve in the following way. A valve decelerates a flow by introducing a degradation of energy additional to that caused by friction in the pipe. In the case of a sluice valve (*gate* valve) the pipe is partly blocked so that the flow area is reduced to a_2 from the full bore of the pipe a_1. Downstream of the valve the flow expands again to a_1, and so there is a degradation which is approximately given by $E = (u_2 - u_1)^2/2g$. (For a proof, see problem 5 on p. 130).

If the valve closes very slowly the initial deceleration is negligible. Then the total degradation in the whole system is the sum of the pipe flow degradation and this valve degradation, and the sum equals the total change in pressure h between the ends of the pipe.

Thus
$$h = E + 4f\frac{l}{d_1}\frac{u_1{}^2}{2g}$$

where u_1 is the velocity in the pipe.

Substituting, $h - (u_2 - u_1)^2/2g - 4f\dfrac{l}{d_1}\dfrac{u_1{}^2}{2g} = 0$

or
$$K_1 - K_2 u_1{}^2\left(\frac{a_1}{a_2} - 1\right)^2 - K_3 u_1{}^2 = 0$$

where K_1, K_2, K_3 are constants for the particular system.

The equation may be solved for u_1 in terms of a_2/a_1 and a curve plotted. The shape is as shown in Fig. 13.19(a), the curvature depending on the relative values of the constants K_1, K_2, K_3. It will be seen that the initial part of the closing of the valve causes only a small change in u_1, but that later the same change of a_1 causes a much larger change of u_1 and the deceleration of the flow must be taken into account. If the valve is shut so that a_2 decreases uniformly with time, as in Fig. 13.19(b), then the deceleration in the pipe increases as the valve shuts. In the final period of time $2l/c$ before closing, the pressure at the valve is according to the water hammer equation $p = u\sqrt{(K\rho)}$, and if the velocity curve is sufficiently steep, p may be large and may cause serious damage. Consequently valves on long pipelines are frequently arranged so that

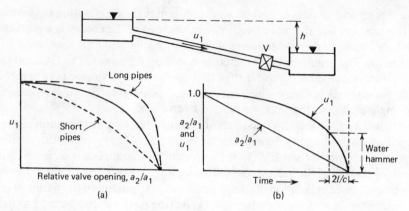

Fig. 13.19 (a) Typical curves showing the decrease of fluid speed in a long pipeline as the valve is shut. (b) The high rate of deceleration of the fluid when the valve is approaching its seating at a constant speed may give water hammer unless the valve movement is deliberately decelerated in the later stages of shutting.

they begin to shut quickly, but later the closing motion is slowed down to avoid water hammer.

In the case of long tunnels used in hydro-power schemes the low strength of the rock may limit the rate at which the valve can be closed to give an unacceptably long time for stopping the flow. A method of overcoming this difficulty is described in Section 15.9.

Example

A 0.3 m diameter pipe 3 km long with $f = 0.004$ has a head of 10 m across its ends. At the downstream end a short transition piece brings the section to 0.25 m square where there is a sluice valve, the coefficient of contraction of the flow through the valve being 0.6. If the valve is closed uniformly over a long time, plot the variation of flow rate with valve position. What would be the pressures at the valve if it were shut in 10 s?

Solution

If the valve is shut very slowly the deceleration in the pipe will be negligible. If T is the time the valve takes to shut then the flow area downstream of the valve will be

$$a_{2t} = 0.6 \times 0.25 \left(0.25 - 0.25 \frac{t}{T} \right) \text{m}^2$$

$$= 0.0375 \left(1 - \frac{t}{T} \right) \text{m}^2$$

at time t.

Hence head loss across the valve, h_v, is given by

$$h_v = \frac{Q^2}{2g}\left(\frac{1}{a_2} - \frac{1}{a_1}\right)^2 = \frac{Q^2}{2g}\left(\frac{26.7}{1-\dfrac{t}{T}} - 14.15\right)^2 \text{ m}^{-4}.$$

Similarly the head lost in the pipe, h_f, is

$$h_f = 4f\frac{l}{d}\frac{Q^2}{2ga_1^2} = 4 \times 0.004 \times 10^4 \times \frac{Q^2}{2g \times 0.0707^2 \text{ m}^4}$$

$$= 1632\,Q^2\,\text{s}^2\,\text{m}^{-5}.$$

Equating to the available head of 10 m

$$10\,\text{m} = Q^2\left[\left(\frac{6.0}{1-\dfrac{t}{T}} - 3.2\right)^2 + 1632\right],$$

and the values of Q computed for various values of t/T, Fig. 13.20.

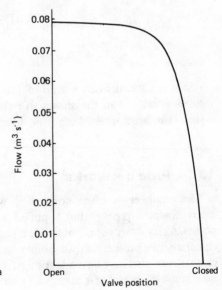

Fig. 13.20 The flow in a pipeline during a very slow valve closure.

If the speed of sound in water is $1430\,\text{m s}^{-1}$ then if the valve closure is to be sudden the time of closing must be less than or equal to $2l/c$.

$$\frac{2l}{c} = \frac{2 \times 3 \times 10^3}{1430} = 4.2\,\text{s}$$

The initial stages of shutting the valve will give a slow stoppage of the flow with a rise in head of $l\alpha/g$. As an approximation find the flow at this time assuming the valve closing is very slow and calculate the average deceleration.

With valve open $\qquad Q_1 = 0.0781 \, \text{m}^3 \, \text{s}^{-1}$.

After 5.8 s $\qquad Q_2 = 0.0754 \, \text{m}^3 \, \text{s}^{-1}$.

Whence $\qquad \alpha = \dfrac{u_1 - u_2}{t} = \dfrac{Q_1 - Q_2}{a_1 t}$,

and $\qquad h = \dfrac{l\alpha}{g} = \dfrac{3 \times 10^3}{9.81} \times \dfrac{0.0027}{5.8\,\pi} \times \dfrac{4}{0.3^2} \dfrac{\text{ms}^2}{\text{m}} \dfrac{\text{m}^3}{\text{s}\,\text{s}\,\text{m}^2}$

$$= 2.0 \, \text{m}$$

which is a small pressure even compared with the static pressure when the valve is shut.

However for the last 4.2 seconds the effect will be that of a sudden valve closure, hence

$$\Delta P = \rho u c$$

$$= 10^3 \times \frac{0.0754}{\pi \times 0.3^2} \times 4 \times 1430 \frac{\text{kg}}{\text{m}^3} \frac{\text{m}^3}{\text{s}\,\text{m}^2} \frac{\text{m}}{\text{s}}$$

$$= 1.5 \, \text{MN}\,\text{m}^{-2},$$

which is equivalent to a head of 156 m, that is, of the order of a hundred times greater than the slow valve closure, and fifteen times greater than the static head when there is no flow.

13.14 Pipe networks

The engineer is often concerned with the properties of a system of interconnected pipes being supplied with, or supplying, fluid at different points. A city water supply network is such a system, and it may be necessary to calculate the pressure at all points in the network for a given flow. The calculations also may be made for a particular network if it is desired to know how much the pressure is changed if a large flow (say for fire-fighting purposes) is taken from one point.

One sort of network, Fig. 13.21(*a*), is essentially a group of pipes all meeting in a common junction point J. Different pressures are applied to the outer ends of the pipes (by connecting to reservoirs or pumps) so that a flow occurs from the higher pressure ends towards the lower pressure ends. The problem is to find what the flow will be in each pipe. It is clear that the total fluid arriving at J must equal the total flow away from J; that is, if flow towards J is reckoned as

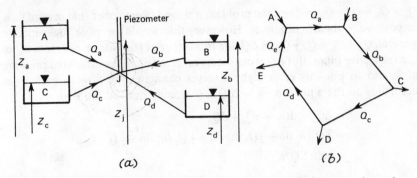

Fig. 13.21 Two sorts of pipe networks. (a) Several reservoirs feeding a junction point. (b) A ring main.

positive then the total flow towards J is zero, or

$$Q_a + Q_b + Q_c + Q_d = 0,$$

where Q_a is the flow through pipe A, etc. Now imagine an open-ended pipe erected vertically from J, wherein the fluid stands hydrostatically with its surface at z_J above some arbitrary datum level. If the surface of the reservoir connected to pipe A is at a height z_a, then

$$(z_A - z_J) = 4f_a \frac{l_a}{d_a} \frac{\overline{U}_a^2}{2g} = 4f_a \left(\frac{l_a}{d_a}\right) \frac{16Q_a^2}{\pi^2 d_a^4 2g}$$

or

$$h_a = K_a Q_a^2,$$

where $h_a = z_a - z_J$ and $K_a = 4f_a \dfrac{l_a}{d_a^5} \dfrac{16}{\pi^2 2g}.$

Similarly for all the other pipes,

$$h_b = K_b Q_b^2; \quad h_c = K_c Q_c^2; \quad h_d = K_d Q_d^2.$$

It will be seen that z_J appears in all these head loss equations so that if it changes all the flows will change. There are therefore 5 unknowns (z_J, Q_a, Q_b, Q_c and Q_d) and, using $Q_a + Q_b + Q_c + Q_d = 0$, there are 5 equations with which to solve them. The quickest and neatest way to do so is as follows.

Consider the effect of a small change dh of the head loss h between the ends of a pipe.

Since

$$h = KQ^2,$$

then, differentiating,

$$dh = 2KQ dQ,$$

or substituting for K

$$dh = 2dQ\, h/Q.$$

Thus the small change dh gives a change dQ in the flow through the pipe. Now suppose an estimate is made of the unknown z_J: then h_a, h_b, h_c, h_d are all estimated, and from the four head loss equations an estimate is made of Q_a, Q_b, Q_c, Q_d. If the original estimate of z_J happened to be correct then

$Q_a + Q_b + Q_c + Q_d = 0$ and the problem is solved (Remember that inward flow is positive, outward negative). However, this would be pure luck and, in general, $Q_a + Q_b + Q_c + Q_d = \Delta Q$, the total error in Q, made up of errors dQ in the respective pipes. But ΔQ is caused by an error dh in the estimated pressure at J; and to put this error right involves changing the flow in each pipe, according to the equation $dh = 2dQ\, h/Q$ already derived. So that

$$\Delta Q = dQ_a + dQ_b + dQ_c + dQ_d$$
$$= \tfrac{1}{2} Q_a/h_a\, dh + \tfrac{1}{2} Q_b/h_b\, dh + \tfrac{1}{2} Q_c/h_c\, dh + \tfrac{1}{2} Q_d/h_d\, dh$$
$$= \tfrac{1}{2}\, dh\, \Sigma Q/h$$

or $$dh = 2\,\frac{\Delta Q}{\Sigma Q/h}$$

Thus from an initial estimate of z_J, ΔQ and $\Sigma Q/h$ may be found and the correction dh determined by the equation. The calculation is best done in tabular form, as in the following example, which may be extended by further approximations until the desired degree of accuracy is obtained. This type of iterative solution is particularly appropriate when a computer is available.

Example

Four pipes from reservoirs meet at a point J, viz.:

Pipe	Reservoir level m above datum	Pipe length m	Diameter m	f
a	100	3000	1.5	0.004
b	110	6000	1.0	0.007
c	80	3000	1.0	0.006
d	40	10 000	2.0	0.004

Determine the flow in each pipe, and the pressure at J.

Solution

Clearly, z_J must lie between 40 and 110 m above datum, so for first approximation let a and b discharge to c and d. Therefore estimate $z_J = 90$ m above datum.

Pipe	$4f\dfrac{l}{d^5}\dfrac{16}{\pi^2}\dfrac{1}{2g} = K$	Estimated h m	$Q = \sqrt{(h/K)}$ m^3 s^{-1}	Q/h m^2 s^{-1}
a	0.52	$100 - 90 = 10$	4.4	0.44
b	13.9	$110 - 90 = 20$	1.19	0.06
c	5.95	$80 - 90 = -10$	-1.3	0.13
d	0.41	$40 - 90 = -50$	-11.0	0.22

$$\Delta Q = -6.7 \qquad \Sigma\frac{Q}{h} = 0.85$$

Thus
$$dh = 2\frac{\Delta Q}{\Sigma\frac{Q}{h}} = -2 \times \frac{6.7}{0.85} = -15.8\,\text{m}$$

Clearly the first estimate of z_J was badly in error, so make a second approximation of

$$z_\text{J} = 90 - 15.8 = 74.2\,\text{m}$$

Pipe	h new estimate	Q	Q/h
a	25.8	7.04	0.27
b	35.8	1.61	0.05
c	5.8	0.99	0.17
d	-34.2	-9.08	0.27

$$\Delta Q = +0.56 \qquad \Sigma\frac{Q}{h} = 0.76$$

New correction
$$dh = +2 \times \frac{0.56}{0.76} = +1.47\,\text{m}$$

Thus the third estimate of $z_\text{J} = 74.2 + 1.47 = 75.7\,\text{m}$.
Check this estimate by a third tabulation.

Pipe	h	Q
a	24.3	6.83
b	34.3	1.57
c	4.3	0.85
d	-35.7	-9.30

$$\Delta Q = -0.05\,\text{m}^3\,\text{s}^{-1}$$

This is small enough for most purposes, though further corrections could be made if necessary.
Thus
$$z_\text{J} = 75.7\,\text{m}$$
$$Q_\text{a} = 6.83; \quad Q_\text{b} = 1.57; \quad Q_\text{c} = 0.85; \quad Q_\text{d} = 9.30\,\text{m}^3\,\text{s}^{-1}$$

13.15 Doubling pipes

A special case of the pipe network just described is when a pipeline between two reservoirs is doubled by a parallel pipe for a part or the whole of its length. This situation occurs in water supply systems when a city outgrows the capacity of its supply mains, and it is desired to increase the flow.

The network is shown in Fig. 13.22, and, as before, it is convenient to use the junction pressure z_J as a variable.

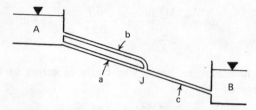

Fig. 13.22 A doubling pipe part way between two reservoirs increases the flow. Pipes *a* and *c* are usually the same size, being the original pipe. Pipe *b* is the new pipe.

So for pipe a,

$$4f_a \frac{l_a}{d_a{}^5} \frac{16 Q_a{}^2}{\pi^2 2g} = z_A - z_J$$

for pipe b,

$$4f_b \frac{l_b}{d_b{}^5} \frac{16 Q_b{}^2}{\pi^2 2g} = z_A - z_J$$

for pipe c,

$$4f_c \frac{l_c}{d_c{}^5} \frac{16 Q_c{}^2}{\pi^2 2g} = z_J - z_B$$

at the junction

$$Q_a + Q_b = Q_c$$

and

$$l_a = l_b.$$

These equations, together with the additional information that the left-hand sides of the first two are equal, may be solved directly by the usual methods of subtraction for simultaneous equations. Sometimes the increased flow $(Q_a + Q_b)$ and d_b is known and the correct length l_b for the doubling pipe calculated: sometimes $(Q_a + Q_b)$ and l_b are fixed and d_b is calculated: sometimes l_b and d_b are fixed and it is desired to know by how much the flow is increased. Usually the diameter of pipe c is the same as that of pipe a, the original pipe.

Doubling pipes are often laid in portions at intervals of time, as the water demand grows, until they are the same length as the original pipe. By the time this full development is required, the first pipe has usually become corroded, the roughness has increased, and *f* is larger than that for the newer pipe.

Example

A water main has a length of 1.07 km and delivers a flow of 0.08 m³ s⁻¹ with a head difference of 14.2 m. The flow required increases to 0.1 m³ s⁻¹ and it is proposed to add a length of doubling pipe of the same diameter to achieve this. Estimate the length of doubling pipe required and the head at the junction.

Solution

Before the doubling pipe is added the pipe flow equation can be reduced to

$$lkQ^2 = Z_A - Z_B$$

for the particular pipe, whence

$$k = \frac{Z_A - Z_B}{lQ^2} = \frac{14.2 \, \text{m s}^2}{1.07 \times 0.08^2 \, \text{km m}^6} = \frac{2.074 \, \text{s}^2}{\text{m}^6}.$$

After the doubling pipe has been added (Fig 13.22), if the losses at the junction are neglected and the pipes are identical then

$$Q_a = Q_b = \tfrac{1}{2} Q_c,$$

and

$$Z_A - Z_J = k l_a Q_a^2 = k l_a \frac{Q_c^2}{4}$$

$$Z_J - Z_B = k(l - l_a) Q_c^2.$$

Eliminating Z_J by adding the equations

$$Z_A - Z_B = k l_a \frac{Q_c^2}{4} + k(l - l_a) Q_c^2$$

$$= k Q_c^2 \left(l - \frac{3 l_a}{4} \right).$$

But $Z_A - Z_B$ can be expressed in terms of the original flow Q and so

$$k l Q^2 = k Q_c^2 \left(l - \frac{3 l_a}{4} \right),$$

re-arranging,

$$\left(\frac{Q}{Q_c} \right)^2 = 1 - \frac{3}{4} \frac{l_a}{l}$$

Whence

$$\frac{l_a}{l_c} = \frac{4}{3} \left[1 - \left(\frac{Q}{Q_c} \right)^2 \right] = \frac{4}{3} \left[1 - \left(\frac{0.08}{0.1} \right)^2 \right] = 0.48.$$

Then $l_a = 0.48 \times 1070 = 514 \, \text{m}$

and $Z_J - Z_B = 2.074 \times 556 \times 0.1^2 \, \text{m}$

$$= 11.5 \, \text{m}.$$

Notice how an increase of only 25 % in the flow rate requires half the pipe to be doubled.

13.16 Ring mains

The other sort of pipe network is shown in Fig. 13.21(b), and is essentially a ring of pipes supplied with fluid at some points and supplying fluid to outgoing pipes at others. The arrangement is a common one in town water supply systems, water being supplied at a few points and being taken off at many. The problem is to determine the pressures at every junction point, and the flow in every part of the ring, if the supply quantity and the several outlet quantities are given.

The method of solution of the problem is very similar to that for the junction network. A first guess is made of the discharge in each pipe, the guesses being made so that at each junction the flow arriving is the same as the flow leaving. The head loss for each pipe, h, is then calculated by the usual pipe flow formula $h = 4f\dfrac{l}{d}\dfrac{\bar{U}^2}{2g}$. In the ring one direction, say clockwise, is taken as the positive direction. A fall of head in this direction would be regarded as a positive h, and a rise in this direction a negative h. This can be achieved algebraically by placing $h = 4f\dfrac{l}{d}\dfrac{\bar{U}}{2g}|\bar{U}|$. Under this convention, the sum of all the values of h for the successive pipes of the ring should be zero if correct guesses have been made of the discharges. Usually, of course, an error Δh is found which is made up of the sum of the errors in the several pipes. This error is caused by the wrong estimation of the discharges. However, the only correction that can be made to the discharge is one that is applied to all the pipes in the ring so that the algebraic sum of the flows at every junction remains zero. For example, Fig. 13.23 shows one junction where a flow Q_x has been specified to be removed. The original guesses Q_a and Q_b have been found to be wrong, giving together with the other ring pipe flows an error Δh around the ring. If a correction δQ is applied as shown in the clockwise direction to *all* pipes, then the condition of flow into the junction equalling the flow out of the junction is preserved at each junction.

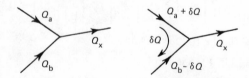

Fig. 13.23 One junction of a ring main showing how the balance of quantity can only be preserved if the same correction is made to both of the flows in the limbs of the ring. Pipes *a* and *b* belong to the ring. Pipe *x* is the outflow from the ring at the junction point.

The error in h due to an error in Q has already been established as

$$dh_a = 2dQ_a h_a/Q_a \text{ for pipe a}$$

and similar expressions for the other pipes.

The total error round the ring is therefore

$$\Delta h = dh_a + dh_b + dh_c + \ldots \text{ etc.}$$

$$= \left(2\frac{h_a}{Q_a} + 2\frac{h_b}{Q_b} + 2\frac{h_c}{Q_c} + \ldots\right)dQ$$

That is
$$dQ = \frac{\Delta h}{2\Sigma h/Q}.$$

The error dQ so found can now be corrected by applying a balancing discharge of the same magnitude but opposite in sign. The error in h should now be small, but the estimation may be repeated until the required accuracy is obtained, as shown in the following example.

Example
A four-sided ring main ABCD has a supply of $0.5 \text{ m}^3\text{ s}^{-1}$ of water at A and delivers to other pipes at B, C and D to the extent of $0.33, 0.1,$ and $0.07 \text{ m}^3\text{ s}^{-1}$ respectively. What are the flows in the ring, and the pressures at B, C and D if the pressure head at A is 30 m? The pipe characteristics are given below.

Pipe	Length km	Diameter m	f
AB	1.0	0.5	0.004
BC	2.0	0.3	0.007
CD	3.0	0.6	0.004
DA	1.0	0.3	0.006

Solution
Original estimates of discharges are made as shown in the following table. It was noticed that AB is a low resistance pipe (K low) so that the discharge through it must be higher than AD. Having thus fixed Q_{AB} the remaining discharges are also fixed because at each junction the flow arriving must equal the flow leaving.

Tabulate the first estimate as follows:

Pipe	$K = 32fl/\pi^2d^5g$ $\text{s}^2\text{ m}^{-5}$	Q estimated $\text{m}^3\text{ s}^{-1}$	$h = KQ|Q|$ m	h/Q s m^{-2}
AB	42.3	0.40	6.77	16.9
BC	1904.0	0.07	9.33	133.3
CD	51.0	-0.03	-0.045	1.5
DA	816.0	-0.10	-8.16	81.6

$$\Delta h = +7.89 \qquad \Sigma\frac{h}{Q} = 233.3$$

Thus error in flow $dQ = \dfrac{\Delta h}{2\Sigma h/Q} = \dfrac{+7.89}{2 \times 233.3} = +0.017 \text{ m}^3\text{s}^{-1}$.

The second estimate may then be made by applying a correction of $-0.017 \text{ m}^3\text{s}^{-1}$ to all discharges, which should then reduce the above error.

Pipe	K s^2m^{-5}	Q revised m^3s^{-1}	h m	h/Q s m^{-2}
AB	42.3	$0.40 - 0.017 =$ 0.383	6.21	16.2
BC	1904.0	$0.07 - 0.017 =$ 0.053	5.35	100.9
CD	51.0	$-0.03 - 0.017 = -0.047$	-0.11	2.4
DA	816.0	$-0.10 - 0.017 = -0.117$	-11.17	95.5

$$\Delta h = +0.28 \quad \Sigma\frac{h}{Q} = 215.0$$

The error in h ($+0.28$ m) is probably near enough for most engineering purposes, but a still nearer approximation can be made by applying a correction of

$$dQ = \frac{\Delta h}{2\Sigma h/Q} = 0.0006 \text{ m}^3\text{s}^{-1}.$$

Thus
$$Q_{AB} = 0.3824 \text{ m}^3\text{s}^{-1}$$
$$Q_{BC} = 0.0524 \text{ m}^3\text{s}^{-1}$$
$$Q_{CD} = 0.0476 \text{ m}^3\text{s}^{-1}$$
$$Q_{DA} = 0.1176 \text{ m}^3\text{s}^{-1}$$

Pressure heads are
$$h_A = 30 \text{ m}$$
$$h_B = 30 - 6.21 = 23.79 \text{ m}$$
$$h_c = 23.79 - 5.35 = 18.44 \text{ m}$$
$$h_d = 18.44 + 0.11 = 18.55 \text{ m}$$
$$\text{or } h_d = 30 - 11.17 = 18.83 \text{ m}$$

Where the two different values of h_d show the error Δh of 0.28 m.

The first estimate of the discharge was close so that the corrections were small.

13.17 More elaborate networks

Networks on water supply systems are usually more complicated than the foregoing examples, though they can often be broken up into a number of rings and junctions as, for example, Fig. 13.24. In this case the ring main method is used, first on ABC, say; then using the value of the flow in EC which has just been calculated, the flows in the ring DEC are then guessed and corrected: then using the flows in DE and AE, ring AED is guessed and corrected. But by now, the newly corrected flows for AE and EC have made ring ABC in error, so a further correction procedure must be made in ABC. Finally, the outer ring ABCD is tested for errors in *h*, using the computed flows given by the analyses for the inner rings, and if necessary a correction applied. This in turn puts errors into the inner rings so the whole computation is then repeated, when it will be found that the errors in the rings slowly decrease until an acceptably correct result is given.

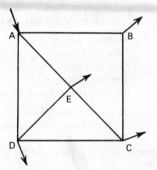

Fig. 13.24 A more elaborate pipe network which would involve lengthy approximations.

The complete procedure is tedious and slow for complicated networks so that other mathematical methods have been evolved. These are usually more easily adapted for use with computers than the above method. However, very high precision is not needed because of the uncertainty in practice of estimating *f*.

13.18 Conclusion

The flow of fluids in pipes is a most important branch of engineering fluid mechanics. Essentially the engineer wishes to know the energy degradation in the several parts of a pipe system. This degradation results in a decrease of pressure in a downstream direction, which must be made good by pumps or by gravity in order to preserve the flow. Changes of energy and of pressure occur at the inlets, outlets and at valves and other fittings in the pipe, but the major cause of a gradient of energy is the turbulence in the pipe caused by the shear stresses on the walls. There have been many empirical attempts to correlate pipe size and roughness, hydraulic gradient and flow in a simple yet accurate manner. These are successful within small ranges of the variables concerned. Much more successful attempts have been made to rationalize friction measurements using boundary layer theory, but they are only truly valid for uniform sand roughnesses on the walls of the pipe. There are still great difficulties in expressing the usual sort of roughnesses found in commercial pipes by one simple parameter. Systems comprising a network of pipes supplied by and supplying fluid at a number of points can be analysed to find the flow in each pipe, but the accuracy is no better than that of finding the friction coefficient for the pipes.

Appendix: Values of coefficients in the pipe flow formula

Straight uniform pipes.   If the roughness to diameter ratio k/d is known then the curves of Fig. 13.10 may be used to determine f for a given $\overline{U}d/v$. This always involves a successive approximation method of calculation if h, l and Q are known and it is desired to find $\overline{U}$ and d (see example in text). The following values of k are sometimes used.

Asbestos cement	0.0012 cm
New steel pipes	0.005–0.012 cm
New cast iron	0.025 cm
Smooth concrete	0.025–0.05 cm
Rock—depending on smoothness of finishing	

In general it is rare to find f smaller than about 0.0035, or much higher than 0.01 unless artificial roughness have been deliberately added to the pipe walls.

Chézy's formula, $\overline{U} = C\sqrt{(mi)}$.   One fairly adequate empirical formula for C is that due to Bazin

$$C = 87 \, \mathrm{m}^{1/2} \, \mathrm{s}^{-1} \Big/ \left(1 + \frac{\gamma}{\sqrt{m}}\right)$$

where γ is a constant depending on the roughness of the pipe (or of an open channel).

Planed timber, smooth plaster	$\gamma = 0.06$
Brick	$\gamma = 0.16$
Rubble masonry	$\gamma = 0.46$

Manning's formula $\overline{U} = n^{-1} m^{2/3} i^{1/2}$

Cast iron	$n = 0.010 \, \mathrm{s\,m}^{-1/3}$
Riveted pipes	$n = 0.013 \, \mathrm{s\,m}^{-1/3}$
Asphalted cast iron	$n = 0.009 \, \mathrm{s\,m}^{-1/3}$
Concrete	$n = 0.011 \, \mathrm{s\,m}^{-1/3}$

The Manning formula is attributed to Strickler in European countries outside Britain.

Fittings in pipes.   Nearly all fittings and irregularities in pipes cause an energy degradation. A selection of fittings is shown in Fig. 13.25, with the degradation shown as a proportion, α, of $\overline{U}^2/2g$. As well as the constant α, another way of presenting the data is to give the equivalent length of uniform pipe that would cause the same degradation. The constant β is given so that the equivalent length is βd, where d is the diameter of the pipe. It is assumed that f

for this pipe is 0.005. Note that if $\alpha = 1.0$, then $\beta = 50$, so that in this case the equivalent length is 50 diameters.

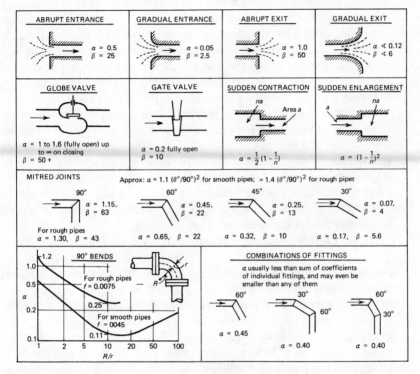

Fig. 13.25

Problems

1. A flow of 420 l min^{-1} of oil is pumped through a 7.5 cm pipe, 62 m, whose outlet is 3 m higher than the inlet. The oil density is 0.91 g cm^{-3} and viscosity 1.24 poise. Estimate the power required.

   Hint. Compute the pipe Reynolds number first.   *Ans. 675 W.*

2. A 75 cm bore water pipe carries 0.66 m^3 s^{-1}. At point B the pressure is 175 kN m^{-2} gauge and the elevation is 36 m. At point C, 1500 m from B, the pressure is 310 kN m^{-2} absolute and the elevation is 30 m. Determine the pipe friction coefficient f.   *Ans. $f = 0.0072$.*

3. Calculate the diameter of a pipe 800 m long to convey coal gas at 600 m^3 h^{-1} from a gasholder to a power station. Delivery is 15 m above the entrance to the pipe; pressure at holder is 10 cm water gauge and at station is 5 cm.

   Density of gas 0.7 kg m^{-3}; of air 1.25 kg m^{-3} pipe flow coefficient $f = 0.005$

What would be the diameter of the pipe if the delivery was 15 m below the inlet, all other conditions remaining the same?

Hint. Variation of atmospheric pressure is important.

Ans. 20.6 cm : 24.2 cm.

4. In a hydro-electric scheme the power house is to be 5 km from the dam and the available head between impounding level and tail race is 75 m. It is desired to generate 12 000 kW and the turbine and electrical efficiencies are 82 per cent and 96 per cent respectively. The water is to be conveyed by a double pipeline with the conservative value of $f = 0.006$.

Calculate the diameter of the pipes (a) if the energy degradation in the pipes is not to exceed 1 per cent of the available energy and (b) if this degradation is 5 per cent.   *Ans.* (a) 4.29 m: (b) 3.16 m.

5. If in question 4 the amount of degraded energy is not important, what are the maximum powers that can be developed by the scheme using the pipe diameters calculated above? Discuss the factors which would affect the final choice of diameter, including cost of pipeline, of water, of power-house and the future development of the plant.

Hint.   Establish that differentiation shows $h = \frac{1}{3}H$ for max power.

Ans. (a) 56 800 kW: (b) 26 600 kW.

6. Discuss cavitation and its effect on design of structures. A spillway pressure tunnel 6 m in diameter has a gate in it 10 m below the free surface of the lake supplying the tunnel. In a model study, the speed at the corners of the gate was found to be twice the mean speed in the tunnel. At what discharge will cavitation begin in the full-size tunnel? Assume that the water commences to boil at -10 m gauge pressure.   *Ans.* 280 m^3 s^{-1}

7. Given that the pipe flow coefficient f in a pipe which is roughened to an effective roughness size k is

$$1/\sqrt{f} = 4 \log_{10} r/k + 3.48,$$

show that Chézy's coefficient is $C = (18 \log_{10} 3.7 \, d/k) \text{m}^{1/2} \, \text{s}^{-1}$ where d is the pipe diameter.

C for a certain 30 cm water pipe has been observed to deteriorate from 65 m$^{1/2}$ s^{-1} to 47 between the years 1920 and 1950, due to the growth of roughnesses by corrosion on the bore. Estimate C for a 75 cm main at the beginning and end of a 20 years' service in the same district. Assume the roughnesses grow linearly with time.

Ans. 72 m$^{1/2}$ s^{-1}; 56 m$^{1/2}$ s^{-1}.

8. Find the sequence and magnitude of the events following the sudden closure of a valve at the downstream end of a rigid pipe 1500 m long, conveying at 3 m s^{-1} water of elasticity 20×10^5 kN m^{-2}.

9. Four reservoirs, A, B, C, and D, in which the water-levels are respectively 50, 30, 10 and 20 m above datum, are connected by pipe to a common junction point J.

Pipe	Length	Diameter
	m	cm
AJ	500	60
BJ	1000	30
CJ	1500	30
DJ	3000	45

Assuming $f = 0.005$, calculate the head at J and the flow in each pipe.

Ans. Head at J 45.6 m: AJ 0.63, JB 0.15, JC 0.19, JD 0.29 $m^3 s^{-1}$.

10. A quadrilateral network of pipes ABCD is joined across AC by another pipe. There is an inflow of 0.1 $m^3 s^{-1}$ at A and outflows of 0.03 and 0.07 $m^3 s^{-1}$ respectively at B and C.

Pipe	Length	Diameter
	m	cm
AB	1000	30
BC	1000	30
CD	1000	30
DA	1000	30
AC	2000	45

Assuming $f = 0.005$, calculate the flow in each pipe.

Ans. AB 0.275: CB 0.025: AC 0.532: ADC 0.193 $m^3 s^{-1}$.

11. At what radius should a Pitot tube be placed in a pipe in order to give a direct measure of the average velocity? Give two answers, one for turbulent and one for laminar flow. Assume the 1/7th root approximation for the velocity distribution in turbulent flow.

Ans. 70.7 and 76 per cent of the pipe radius.

12. A town is supplied with water through a pipe 2 m in diameter, $k = 0.2$ mm, of length 5×10^4 m. If the difference of levels between entry and exit reservoirs is 150 m, estimate the flow.

For many years, only half the flow will be required. Explain how the flow could be so limited, estimate the power to be dissipated, and show with diagrams how the position of the dissipator will affect total head and piezometric lines along the pipe. Indicate regions of low pressure.

Kinematic viscosity of water 0.01 $cm^2 s^{-1}$.

Ans. $\delta' < k_s$, so rough pipe; 9.9 $m^3 s^{-1}$; 5.5 MW.

14

Flow in open channels

14.1 The basis of open channel flow calculations

The flows of water with a free surface exposed to atmospheric pressure present the most common problems of fluid mechanics to a civil engineer. These will range from flows on a geographical scale, such as estuaries and rivers, to flows in man-made structures like drainage and irrigation canals and sewers. By applying the principles of fluid mechanics as given in preceding chapters, the changes of depth, velocity, pressure, force and energy of the flows may be found; the special feature of free surface flow being that along one boundary of the flow (i.e. the free surface) the pressure is the same irrespective of changes of depth and velocity. In this respect open channels are strikingly different from pipe-flows, where every change of velocity must be accompanied by a change of pressure. However, in channels just as in pipes, there are two distinct families of problems: the first where differences of depth (pressure in pipes) and velocity occur within short distances and shear forces are small (rapidly varied depth flow); and the second where appreciable differences of depth occur only over long distances and shear forces are significant (gradually varied depth flow).

Useful and simple approximations to many changes-of-depth problems in channels will be derived in this chapter by invoking, in general, the following simplifications.

(i) The fluid is incompressible and homogeneous—this excludes problems where air is entrained into water making a froth, as occurs in high speed flows.

(ii) At all control surfaces in the flow where calculations are made the stream-lines are straight, i.e. there are no pressure gradients due to curvature of the flow. Hence the pressure distribution on any control surface is hydrostatic and the free surface is horizontal in the direction at right angles to the flow.

(iii) That the flow is steady.

The effect of these assumptions is to change the general forms of the equations of motion to equations specifically for open channel flows. The first four sections of this chapter will be concerned with the derivation of the open

268

channel flow equations and their properties. The remainder of the chapter will demonstrate the application of the equations to open channel flows. Much of the information on which the analysis is based is the result of observations of open channel flows, but it is the principles of fluid mechanics which have enabled this empirical information to be organised in a form which is understandable and convenient for the use of engineers.

14.2 The general analysis of an element of a steady channel flow

Consider the control volume ABCD which contains part of an open channel flow as shown in Fig. 14.1 to which the assumptions of Section 14.1 can be applied. The control surfaces AD and BC represent respectively the free surface and the surface of the channel in contact with the flow, on which there is an average shear stress τ. The control surfaces AB, DC are cross sections of the channel normal to BC of area, a, which have a liquid surface width, b, a distance, d, above the bed and a perimeter, P, in contact with the flow. The level of the bed of the channel at any cross section is Z, and the horizontal distance x such that the slope of the channel bed, S, is given by

$$S = \tan^{-1} \frac{(Z_1 - Z_2)}{x}.$$

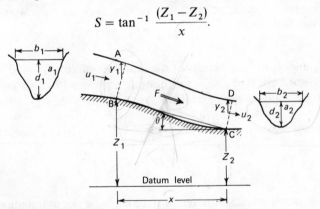

Fig. 14.1 A control volume containing a section of an open channel flow. The two control surfaces AB, CD are normal to the flow direction.

Note that downward slopes are considered positive and that for the small slopes usually encountered in open channel flow $S \simeq \theta \simeq (Z_1 - Z_2)/x$. The mass of water in the control volume is represented by m, and there may be an external force, F, acting on the flow. In Fig. 14.1 this force is drawn in the flow direction to conform to the convention that all positive forces are in the flow direction.

It is now possible to apply the equations of motion to the flow through control volume using the assumptions of Section 14.1.

14.2(a) The continuity equation

For steady flow the continuity equation is:

$$Q = a_1 \bar{u}_1 = a_2 \bar{u}_2 = \int_{AB} u_1 \, da_1 = \int_{CD} u_2 \, da_2$$

assuming that there is no inflow across the sides or free surface of the control volume.

14.2(b) The force–momentum equation

The forces in the flow direction which have to be considered are the resultant hydrostatic pressure force, R, the component of the weight force, the shear force, and the external force. The resultant of these forces acting on the liquid in the control volume in the flow direction gives the change of momentum flow rate between the two ends of the control volume, i.e.,

$$R + mg \sin \theta - \tau x \frac{(P_1 + P_2)}{2} + F = \rho \int_{CD} u_2{}^2 \, da_2 - \rho \int_{AB} u_1{}^2 \, da_1 \quad (14.1)$$

If $\bar{d}$ is the depth of the centroid of the area δa, where δa is the area of the sides of the channel projected in the flow direction, then the hydrostatic pressure force in the flow direction is:

$$R = \rho g \bar{d}_1 a_1 \cos \theta - \rho g \bar{d}_2 a_2 \cos \theta + \rho g \bar{d} \, \delta a$$

where $\bar{d}_1$ and $\bar{d}_2$ are the depths of the centroids of areas a_1 and a_2. As θ is generally small $\cos \theta \simeq 1$ so substituting in equation 14.1 and grouping the terms which apply to the control surfaces AB, CD gives:

$$\rho g \bar{d} \, \delta a + mg \sin \theta - \tau x \frac{(P_1 + P_2)}{2} + F = \left(\int_{CD} \rho u_2{}^2 \, da_2 + \rho g \bar{d}_2 a_2 \right)$$

$$- \left(\int_{AB} \rho u_1{}^2 \, da_1 + \rho g \bar{d}_1 a_1 \right) \quad (14.2)$$

It is conventional to refer to the terms of the left hand side of the equation as the 'side force', 'weight force', 'shear force', and 'external or applied force'. Then, by analogy, the terms in brackets on the right hand side of the equation which contain the momentum flow rate plus the hydrostatic pressure force at a particular cross section of the flow may be considered as 'flow forces' (or 'specific forces'), M, corresponding to the cross-section so:

$$M_1 = \int_{AB} \rho u_1{}^2 \, da_1 + \rho g \bar{d}_1 a_1$$

$$M_2 = \int_{CD} \rho u_2{}^2 \, da_2 + \rho g \bar{d}_2 a_2.$$

One inconvenience of this definition of the flow force is the integral form of the momentum flow rate due to the fact that the velocity may not be uniformly distributed. This can be simplified by comparing the actual momentum flow rate with that of a uniform stream of the same depth and flow rate and defining a coefficient, β, such that:

$$\int \rho u^2 \, da = \beta \rho \bar{u}^2 a.$$

As, by definition, the mean velocity, $\bar{u}$, is the volumetric flow rate divided by the area,

$$\int \rho u^2 \, da = \beta \rho Q \bar{u} = \beta \rho \frac{Q^2}{a}.$$

As the Reynolds number is large in open channel flows where the momentum flow rate is significant compared to the hydrostatic pressure force, the flows are turbulent with relatively uniformly distributed velocities. Hence the values of β are close to one ($\beta \simeq 1.03 \rightarrow 1.15$) and for a first approximation uniform flow may be assumed ($\beta = 1.0$) so that:

$$M = \beta \rho \frac{Q^2}{a} + \rho g \bar{d} a$$

$$\simeq \rho \frac{Q^2}{a} + \rho g \bar{d} a.$$

The general equation can now be stated as

change of flow force = $M_2 - M_1$
= resultant force in flow direction
= side force + weight force − shear force
+ external force. (14.2a)

This equation is general and allows one term to be evaluated if all the others are known and can be summarized by the statement that 'change of flow force is the resultant force on the stream in the flow direction'.

The equation can often be simplified and this is particularly helpful for illustrative purposes. Firstly, if the cross section of the channel remains unchanged, as is the case in most man-made channels, the side force is zero and hence:

$$M_2 - M_1 = \text{weight force} - \text{shear force} + \text{external force.} \quad (14.2b)$$

As mentioned in Section 14.1, external forces due to hydraulic structures are usually applied to the flow over a short distance in which the shear force is negligible. Hence for these rapidly varied depth flows the force–momentum equation becomes:

$$M_2 - M_1 = \text{weight force} + \text{external force,} \quad (14.2c)$$

and if the structure has a horizontal bed the weight force acting in the flow direction is zero so:

$$M_2 - M_1 = \text{external force.} \tag{14.2d}$$

Alternatively for flows with no external forces where the shear force is important, i.e. gradually varied depth flows:

$$M_2 - M_1 = \text{weight force} - \text{shear force.} \tag{14.2e}$$

In later sections rapidly varied depth and gradually varied depth flows will be considered separately, but it should always be remembered that the equations used to solve them are special cases of the general flow equation.

14.2(c) The steady flow energy equation

If a stream tube which crosses AB, in Fig 14.1, at a depth y_1 below the surface, is at a depth y_2 at CD, then the energy equation for the streamtube may be written as

$$\frac{u_1^2}{2g} + \frac{p_1}{\rho g} + Z_1 + d_1 - y_1 = \frac{u_2^2}{2g} + \frac{p_2}{\rho g} + Z_2 + d_2 - y_2 + E', \tag{14.3}$$

where E' is the hydraulic energy dissipation per unit weight of the liquid flow. But assuming the pressure distributions over the control surfaces AB and CD are hydrostatic (Section 14.1(ii)), then,

$$p_1 = \rho g y_1, p_2 = \rho g y_2$$

Substituting in equation 14.3 and simplifying gives

$$\frac{u_1^2}{2g} + \frac{\rho g y_1}{\rho g} + Z_1 + d_1 - y_1 = \frac{u_2^2}{2g} + \frac{\rho g y_2}{\rho g} + Z_2 + d_2 - y_2 + E'$$

and

$$\frac{u_1^2}{2g} + Z_1 + d_1 = \frac{u_2^2}{2g} + Z_2 + d_2 + E' \tag{14.4}$$

It is usual to rearrange the energy equation and define the total head H of the flow in a streamtube as:

$$H = \frac{u^2}{2g} + d + Z,$$

so that equation 14.4 becomes:

$$H_1 = H_2 + E'.$$

In most cases it is necessary to consider the energy of the whole flow at a control surface in terms of the sum of the energy flow rates for each stream tube. As the presence of shear forces will cause the velocity distribution in any open channel to be non-uniform it is more convenient to compare the actual kinetic energy flow rate at a control surface to that of a uniformly distributed flow of the same depth and flow rate by defining a coefficient, α, such that:

$$\int \rho \frac{u^3}{2}\, da = \alpha \rho \frac{\bar{u}^3}{2}\, a.$$

The average total head for the whole flow is then:

$$\bar{H} = \alpha \frac{\bar{u}^2}{2g} + d + Z$$

$$= \alpha \frac{Q^2}{2ga^2} + d + Z$$

Although the non-uniformity of the velocity profile may cause values of α as great as 1.3, a value of one may often be used as a first approximation, as with the coefficient β for the momentum flow rate.

As the first two terms in the equation for the total head of the flow depend on the flow depth relative to the channel bed, it is possible to define the *specific head*, E, the total head relative to the channel bed, as:

$$E = \alpha \frac{Q^2}{2ga^2} + d$$

whence
$$H = E + Z,$$

and the steady incompressible flow energy equation becomes:

$$E_1 + Z_1 = E_2 + Z_2 + E'. \tag{14.5}$$

As the specific head is derived from the energy equation it is frequently, if incorrectly, referred to as the *specific energy*.

Both the total head and the specific head may be represented at a control surface in a flow by measuring a distance $\alpha u^2/2g$ vertically upwards from the free surface of the flow (Fig. 14.2). Joining a series of these points enables the head (or 'energy') line to be drawn for the flow, the gradient of the line being negative if the head of the flow is decreasing due to its doing work either on some turbine or against shear forces. Conversely if energy is added to the flow by a pump the head line will rise.

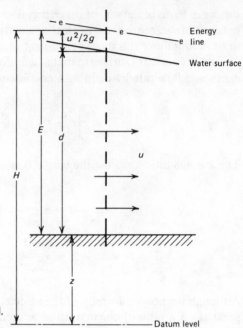

Fig. 14.2 The relationship between mean total head, H, and specific head, E, at a control surface for $\alpha = 1$.

Example

A sluice gate is fitted in a rectangular cross-section channel of width 12 m in a reach where the bed is horizontal. When the depth of water upstream of the gate is 6 m and the depth of water downstream is 1.2 m the volumetric flow rate is 142 m³ s⁻¹. What will be the force on the flow as it passes through the gate, the height of the head line above bed level before and after the gate and the hydraulic power dissipated by the flow?

Solution

The flow under a model sluice is shown in Fig. 14.17 and this may be simplified as shown in Fig. 14.3.

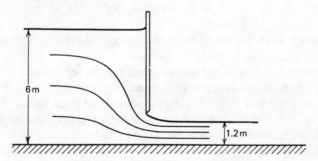

Fig. 14.3 The flow under a sluice gate.

As the channel has constant cross-section and the bed is horizontal then the force momentum equation may be written as

$$M_2 - M_1 = -\text{shear force} + \text{external force}.$$

Now for a rectangular channel the cross-sectional area is $b \times d$ and the depth of the centroid of the cross-sectional area, $\bar{d}$, is $d/2$ so, assuming a uniform distribution of velocity ($\beta = 1$),

$$M = \frac{\rho Q^2}{bd} + \frac{\rho g d^2 b}{2}.$$

Hence $M_2 = 10^3 \left(\frac{142^2}{12 \times 1.2} + 9.81 \times \frac{1.2^2}{2} \times 12 \right) \frac{\text{kg}}{\text{m}^3} \times \frac{\text{m}}{\text{s}^2} \times \text{m}^3$

$$= 1485 \, \text{kN}$$

and $\quad M_1 = 2399 \, \text{kN}$

So, resultant force $= M_2 - M_1 = (1485 - 2399) \, \text{kN} = -914 \, \text{kN}$,
i.e. the force acts in the opposite direction to the flow. Hence the reaction on the gate plus any shear forces on the channel is $914 \, \text{kN}$ in the flow direction.

The level of the head line can be found by calculating the specific head of the flow before and after the sluice. First it is necessary to find the flow velocities before and after the sluice from the continuity equation.

$$u_1 = \frac{Q}{bd_1} = \frac{142}{12 \times 6} = 1.97 \, \text{m s}^{-1}$$

and similarly $\quad u_2 = 9.86 \, \text{m s}^{-1}.$

Hence $\quad E_1 = 6 + \frac{1.97^2}{2 \times 9.81} = 6.20 \, \text{m},$

and $\quad E_2 = 6.16 \, \text{m}.$

Note how the energy line hardly changes in level as the flow passes under the sluice, hence,

$$E' = E_1 - E_2 = 6.20 - 6.16 = 0.04 \, \text{m}.$$

As this is to one significant figure it will give a poor approximation to the actual value, but will give an idea of the orders of magnitude involved. Now the rate at which hydraulic energy is dissipated is $\rho g Q E'$, where

$$\rho g Q E' = 10^3 \times 9.81 \times 142 \times 0.04 \, \frac{\text{kg}}{\text{m}^3} \times \frac{\text{m}}{\text{s}^2} \times \frac{\text{m}^3}{\text{s}} \times \text{m}$$

$$= 60 \, \text{kW}.$$

This is a small proportion of the total energy flow rate (power of the flow) before the sluice as:

$$\text{Power} = 10^3 \times 9.81 \times 142 \times 6.2 = 8600 \, \text{kW}.$$

14.2(d) A further look at the flow force and specific head

The expressions for flow force and specific head for open channels have a remarkable similarity. Both expressions have two terms, one of which is a function of the velocity and increases with velocity, i.e. as the depth decreases, while the second term increases as the depth increases. To be realistic, there can be no negative depth, flow force, or specific energy. Each of the expressions

$$M = \beta\rho\,\frac{Q^2}{a} + \rho g \bar{d} a,$$

and

$$E = \frac{\alpha Q^2}{2ga^2} + d,$$

should have a minimum value at some particular depth. Considering the specific head first, then for a given flow rate, Q, differentiating with respect to the depth, d, gives

$$\frac{\mathrm{d}E}{\mathrm{d}d} = \frac{Q^2\alpha}{2g}\frac{\mathrm{d}}{\mathrm{d}d}\left(\frac{1}{a^2}\right) + \frac{\mathrm{d}d}{\mathrm{d}d} = -\frac{Q^2\alpha}{ga^3}\frac{\mathrm{d}a}{\mathrm{d}d} + 1$$

Considering the cross-section of the flow, see Fig. 14.41, page 336, a change of depth $\mathrm{d}d$ produces a change of area $\mathrm{d}a = b\mathrm{d}d$ if b is the width of the water surface. Hence $\mathrm{d}a/\mathrm{d}d = b$ and then

$$\frac{\mathrm{d}E}{\mathrm{d}d} = 1 - \frac{\alpha Q^2 b}{ga^3},$$

or when the value of E is a minimum, $\mathrm{d}E/\mathrm{d}d = 0$ and

$$\frac{\alpha Q^2 b_c}{ga_c{}^3} = 1, \tag{14.6}$$

where the subscripts c indicate that the area and breadth correspond to the minimum value of E. While the geometry is a little more involved it can also be shown that for a given flow rate the condition for the flow force to be a minimum is similar to equation 14.6 with α replaced by β. See appendix I, page 336.

14.3 Flow in rectangular channels

Equation 14.6 relies only on the original assumptions of section 14.1 and applies to any cross-section of channel so it is not easy to take the analysis any further. If, however, the particular case of a rectangular cross-section channel is considered with the additional assumption that the velocity is uniformly distributed ($\alpha = \beta = 1$), the equations may be simplified as the cross-sectional area of the flow is given by

$$a = bd.$$

If q, the flow per unit width of channel, is defined as

$$q = \frac{Q}{b},$$

the expressions for M, E, and the minimum point become

$$\frac{M}{b} = \frac{\rho q^2}{d} + \frac{\rho g d^2}{2}, \qquad (14.7)$$

$$E = \frac{q^2}{2gd^2} + d, \qquad (14.8)$$

and

$$\frac{q^2}{gd_c^3} = 1. \qquad (14.9)$$

The first two expressions are plotted in Fig. 14.4 and show the general shape of curves with a minimum predicted earlier in this section. The third expression defines d_c, referred to as the *critical depth*, the depth at which both the flow force and specific head are a minimum. Eliminating q^2 in equations 14.7 and 14.8 by the use of equation 14.9, it is possible to produce dimensionless expressions for the flow force and specific head in terms of the ratio of the flow depth to the critical depth, i.e.,

$$\frac{M}{\rho g b d_c^2} = \frac{d_c}{d} + \frac{1}{2}\left(\frac{d}{d_c}\right)^2 \qquad (14.10)$$

$$\frac{E}{d_c} = \frac{1}{2}\left(\frac{d_c}{d}\right)^2 + \frac{d}{d_c}. \qquad (14.11)$$

These expressions are plotted in Fig. 14.5 where it can be seen that both have minimum values of 1.5 at $d/d_c = 1$, i.e. when the depth of flow equals the critical depth. The curves are a contraction of the two families of curves in Fig. 14.4 on to two unique curves.

14.4 The significance of the critical depth

Both Figs. 14.4 and 14.5 show that for a given flow with either a particular value of flow force, or a particular value of specific head, there are two possible depths of flow, one greater than the critical depth and one less than the critical depth. The only exception to this rule is when the flow is actually at the critical depth when the velocity is u_c and so:

$$\frac{q^2}{gd_c^3} = \frac{u_c^2 d_c^2}{gd_c^3} = \frac{u_c^2}{gd_c} = 1$$

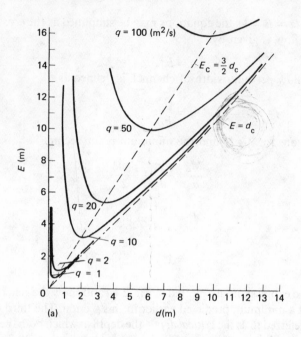

(a)

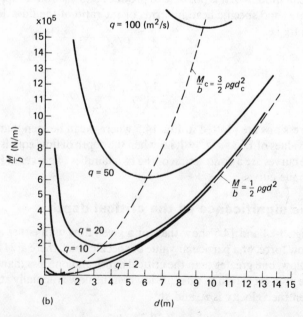

(b)

Fig. 14.4 The flow force, M, equation 14.7, and specific head, E, equation 14.8, for rectangular open channel flows.

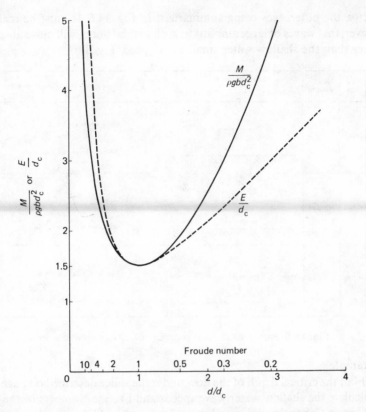

Fig. 14.5 The dimensionless representation of the flow force $M/\rho\, gbd_c^2$, equation 14.10, and specific head E/d_c, equation 14.11, for rectangular open channel flows.

But reference to page 146 will show that this is tantamount to saying that at the critical depth the Froude Number based on the depth of flow is unity. This statement can be extended to flows in channels with non-rectangular cross-sections if $\alpha = 1$ and the Froude number is defined from equation 14.6 as:

$$Fr^2 = \frac{Q^2 b}{a^3 g}.$$

It is shown in Section 14.11 that the velocity of small waves, c, in shallow water in an open channel, is $c = \sqrt{gd}$. Consequently the critical depth occurs when the flow velocity, u_c, is equal to the small wave velocity c. If the flow velocity is less than the wave velocity, i.e. $d > d_c$, then small waves can move upstream as well as downstream, the flow is described as *slow*, *subcritical*, or *tranquil* and the Froude Number is less than unity. The converse is true for *fast*, *supercritical*, or *shooting* flows where the flow velocity is greater than the wave velocity, the differences being summarized in Fig. 14.6. It must be realized, however, that waves of large amplitude, such as tidal bores, will move at speeds greater than the shallow water small wave speed, Fig. 14.7.

Flow Depth	Flow Velocity	Small wave velocity Upstream	Downstream	Froude Number	Type of flow and surface appearance
$d>d_c$	$u<\sqrt{gd}$	$\overleftarrow{\sqrt{gd}-u}$	$\overrightarrow{u+\sqrt{gd}}$ $u\longrightarrow$	$Fr<1$	Slow, subcritical, tranquil. No waves on surface.
$d=d_c$	$u=\sqrt{gd}$	$\sqrt{gd}-u=0$	$\overrightarrow{2\sqrt{gd}}$ $u\longrightarrow$	$Fr=1$	Critical. Stationary waves across flow.
$d<d_c$	$u>\sqrt{gd}$	$\overrightarrow{u-\sqrt{gd}}$	$\overrightarrow{u+\sqrt{gd}}$ $u\longrightarrow$	$Fr>1$	Fast, supercritical, shooting. Stationary waves oblique to flow.

Fig. 14.6 Properties of open channel flows at different depths.

Example

Find the critical depth of the flow under the sluice described on page 274. Calculate the shallow water wave speeds and Froude Numbers of the flow before and after the sluice and describe the type of flow.

Show that the Froude Number of the flow may be expressed in terms of the dimensionless depth d/d_c. What will be the connection between the dimensionless flow force and specific head and the Froude Number?

Solution

As the channel is rectangular then,

$$d_c^{\,3} = \frac{q^2}{g} = \frac{142^2}{12^2 \times 9.81} = 14.3 \text{ m}^3,$$

whence $d_c = 2.43$ m. Note how the flow before the sluice ($d = 6$ m) is at a greater depth than the critical, whereas the flow after the sluice ($d = 1.2$ m) is below the critical depth.

Before the sluice $d = 6$ m whence $\sqrt{gd} = 7.67 \text{ m s}^{-1}$ and u_1 $= 1.97 \text{ m s}^{-1}$. Then the wave speeds are $u_1 \pm \sqrt{gd_1}$, i.e. $+9.64 \text{ m s}^{-1}$ and -5.70 m s^{-1} and,

$$Fr_1 = \frac{1.97}{7.7} = 0.26.$$

Fig. 14.7 Photographs of the Bore in the River Severn. (Upper) The full size bore moving upstream with canoeists surf-riding. (Photograph by E. J. Wynter, Esq.) (Lower) A 1:600 scale model of the river showing a small scale bore. (Photograph by Hydraulics Research Station, Wallingford.)

As the wave velocities have different signs (and $Fr_1 < 1$), the waves will move both upstream and downstream.

After the sluice $d = 1.2$ m, $\sqrt{gd} = 3.43$ m s^{-1}, $u_2 = 9.86$ m s^{-1}, and the wave speeds are $+13.29$ m s^{-1} and $+6.43$ m s^{-1} while:

$$Fr_2 = 2.9.$$

As both the wave velocities are positive (and $Fr_2 > 1$) waves will only move downstream, i.e. in the flow direction.

Hence the flow is slow or subcritical before the sluice gate and fast or supercritical after it.

To express the Froude number in terms of d/d_c, put

$$\frac{q^2}{gd_c^{\,3}} = \frac{u^2 d^2}{gd_c^{\,3}} = \frac{u^2}{gd}\frac{d^3}{d_c^{\,3}} = Fr^2 \left(\frac{d}{d_c}\right)^3 = 1$$

whence $$Fr = \left(\frac{d_c}{d}\right)^{3/2}.$$

Checking the previous results, as $d_c = 2.43$ m,

$$Fr_1 = \left(\frac{2.43}{6}\right)^{3/2} = 0.26,$$

and $$Fr_2 = \left(\frac{2.43}{1.2}\right)^{3/2} = 2.9.$$

As the dimensionless flow force and specific head are both functions solely of the ratio d/d_c then they will both be functions of the Froude Number and will have minima at $Fr = 1$ (See Fig. 14.5).

14.5 Control structures

The fact that upstream of the sluice described in the previous example, the flow is slow, or subcritical, while downstream of the sluice the flow is fast, or supercritical, has an important consequence. Consider the effect of lowering the sluice slightly. The flow area under the sluice will decrease and hence the flow rate passing under the sluice will decrease, while the flow rate before the sluice remains unchanged.

Downstream of the sluice the depth will be reduced immediately and this change of depth will be propagated downstream at a rate equal to the sum of the water velocity and the small wave speed. When the wave has passed the water level will remain fixed by the gate and the flow will, for a short time, remain steady at the reduced rate. It will not be possible for any effects caused by the change of flow downstream of the sluice to be felt at the sluice itself.

Upstream, the initial effect of the fall in the flow rate under the sluice will be a rise in water level just before the sluice to store the extra water and so satisfy t continuity equation. This rise in water level will move against the flow at t small wave speed, slowly increasing the water depths everywhere upstream or the sluice. As the water levels rise the flow rate under the sluice will increase until finally it reaches its original value, at which time the upstream water levels will again become steady. As the upstream water levels rise an increase in the flow downstream of the gate will occur due to an increase of velocity of the flow rather than a change in the depth. If the flow rate remains constant there will be a return to a steady state at new upstream and downstream depths.

Hence under the circumstances where the sluice converts a slow, subcritical flow at a depth above the critical depth to a fast, supercritical flow at a depth below the critical depth the position of the sluice dictates both the upstream and downstream depths and it acts as a *control structure**. While the action of a control structure in altering the flow depths in a steady flow has been described it is also possible to operate adjustable control structures to keep a water level constant, for example in a river, as the flow rate varies. There are also fixed control structures such as weirs where there can be a unique relationship between flow rate and upstream water depth and these fixed structures can also be used for flow measurement.

The common factor in all these devices, to be described in this section, is that a slow, subcritical, flow is accelerated to its critical velocity, or greater and hence it reaches its critical depth. It is possible to predict the performance of such structures theoretically and for simplicity this will be done for control structures in channels with a rectangular cross-section of constant width.

14.5(a) Slow flow over a raised part of the stream bed (Fig. 14.8(a))

A raised part of the bed exerts a force opposing the flow by the pressure on the upstream face of the rise. This force decreases M, the flow force of the stream. If this pressure distribution was known, the flow-force equation could be used to find the relation between q, the upstream speed and depth, and the speed and depth on the top of the rise (i.e. using equation 14.2d). Usually, however, the pressure force is not known and an alternative method must be used.

Use of the energy equation is justified because the streamlines of the approaching flow converge over the raised bed and there is unlikely to be separation of the flow. If only a short length of channel is considered the work done against the shear forces will be small, the energy line may be assumed to be horizontal, and E changes only by the amount z of the rise of the bed.

Thus
$$u_1{}^2/2g + d_1 = u_2{}^2/2g + d_2 + z$$

* Do not confuse a control structure, which is a physical structure controlling the flow with the hypothetical control surface and control volume which are abstractions used as an aid to analyzing flows.

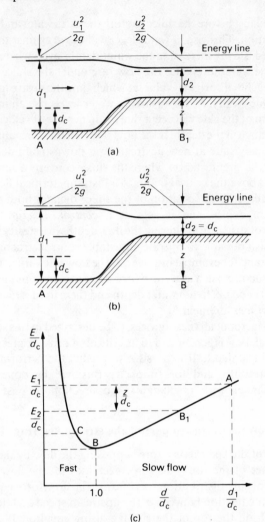

Fig. 14.8 (a) The conditions of slow flow over a raised bed. The rise z of the bed decreases E so that the point A on the specific energy curve representing the upstream conditions must be at a greater E/d_c than the downstream point, B1. (b) If z is large enough, then B lies at the critical point (E_c, d_c), and the rise in bed level acts as a weir controlling the flow.

On the curve of the generalized flow function equation,

$$E/d_c = \tfrac{1}{2}(d_c/d)^2 + d/d_c, \text{ Fig 14.8(c)}$$

so

$$\frac{E_1}{d_c} = \frac{E_2}{d_c} + \frac{z}{d_c}.$$

and the upstream conditions may be represented by the point A on the curve. An increase of z decreases E and therefore E/d_c (for d_c is constant if the width is the same throughout) and the point on the energy curve referring to the cross section of the higher bed level, B_1, will be lower and nearer to the critical depth point, B. If z were made large enough, E_2/d_c would be brought down as far as E_c/d_c, point B, and therefore the depth d_2 would be d_c, Fig. 14.8(b). But E/d_c cannot be reduced any further so that the critical depth is the smallest depth of flow possible over the raised bed under these conditions. If z were increased still further, Fig. 14.9, the depth of flow over the raised bed would remain constant at d_c while the upstream depth (and head E) has to increase as in the case of the closing sluice mentioned earlier in this section. The structure is then acting as a weir controlling the flow.

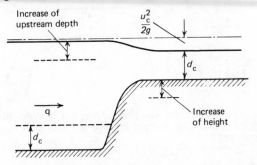

Fig. 14.9 Effect of an increase in height of a weir. Any further increase of height above that shown in Fig. 14.8(b) preserves critical flow over the weir but increases the upstream depth.

Having become critical on a part of a weir, the flow undergoes other changes at the downstream end, where the weir top starts to fall again (Fig. 14.10). The bed height z decreases now so that E increases. The flow accelerates and becomes fast, and the point C on the energy curve (Fig. 14.8(c)) representing the flow now comes on the fast flow side of the critical point. The lower the bed,

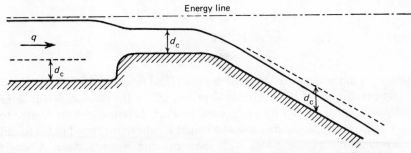

Fig. 14.10 Fast flow produced on a spillway. Slow flow in the approach channel becomes critical at the crest of the weir. As the bed of the downstream slope becomes lower, E increases so that the flow becomes faster and faster.

the greater E and the faster the flow. The depth therefore steadily becomes less. With a downward sloping bed or chute the structure is called a *spillway*. It may be used to spill excess water from a canal or reservoir and, like a sluice, is a way of producing fast flow.

The discharge over a weir controlling the flow in an open channel may be calculated as follows. Consider the longitudinal section of a channel shown in Fig. 14.11. A flow of q per unit width approaches the weir. Downstream of the weir the bed falls away so that the flow has no further obstruction and so the flow reaches its critical depth at some point on the top of the weir.

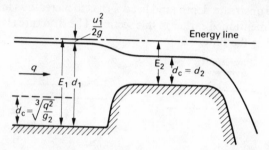

Fig. 14.11 Longitudinal section of a slow flow ($d > d_c$) in an open channel approaching and passing over a weir. The depth over the weir decreases to the critical depth shown by the dotted line.

Since
$$d_c = \sqrt[3]{(q^2/g)}$$
therefore
$$q = u_c d_c = g^{1/2} d_c^{3/2}.$$

A single measurement of d_2 will enable q to be found. This method of measurement is not used much in practice for the surface of the fast-flowing water over the weir is not quite horizontal and so d_2 is difficult to determine accurately.

It is more common to express q in terms of the total specific head E_2 of the fluid, which at the place where the flow is critical, is 1.5 times the depth there. So
$$E_2 = 1.5 d_2 = 1.5 \sqrt[3]{(q^2/g)}.$$
That is
$$q^2 = (\tfrac{2}{3})^3 g E_2^{3}$$
or
$$q = 0.544 g^{1/2} E_2^{3/2}.$$

Now E_2 could be measured by putting a total head tube into the flow, but this is rarely practicable. If the upstream depth d_1 above the channel bed is large compared with $d_2 (= d_c)$, then u_1 is small and $u_1^2/2g$ is negligible compared to d_2. Consequently, the energy line at a height E_2 above the raised bed is for all practical purposes coincident with the upstream water surface. A single measurement of the height of the upstream surface above the weir crest is therefore sufficient to define E_2. This measurement of E_2 is sometimes called the *head over the weir*. Using SI units ($g = 9.81 \text{ m s}^{-1}$), the discharge equation becomes

$$q = 1.70\, E_2{}^{3/2} \text{ m}^{1/2}\text{s}^{-1}. \tag{14.12}$$

or $Q = 1.70\, b\, E_2{}^{3/2} \text{ m}^{1/2}\text{s}^{-1}$, where b is the width of the weir normal to the flow. Such a raised floor is called a *broad-crested weir*, and so long as the flow is critical and parallel somewhere along its length, the above equation holds good. See Fig. 14.12.

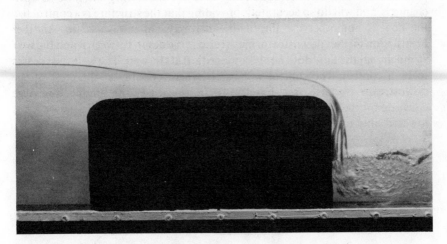

Fig. 14.12 A broad crested weir photographed through the glass side of a laboratory channel. Notice how the flow over the weir crest becomes parallel at a depth of $\frac{2}{3}$ the height of the upstream water surface above the crest.

In practice, when water is the fluid concerned, the combined effect of non-uniformity in the oncoming flow and shear forces of the water on the weir top cause the flow to be a little lower than the theoretical value, and a coefficient $1.68 \text{ m}^{1/2}\text{s}^{-1}$ is commonly used. At large heads ($E > \frac{1}{4}$ the length of the weir) the weir may not be long enough for the flow to become parallel, see Fig. 14.13.

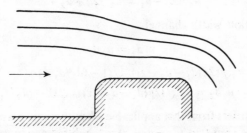

Fig. 14.13 Three successive profiles of the fluid surface over a weir, at increasing discharges. At the high discharges the weir is not long enough to allow the stream to become parallel to the weir surface.

The streamlines become curved, the pressure distribution is no longer hydrostatic, and the velocity distribution is non-uniform. In this case the coefficient is larger. At low heads, the shear forces are relatively more important and the coefficient is slightly less.

14.5(b)   Weirs in series in slow flow

Broad-crested weirs are commonly used for measuring the flow in open channels, and will do so accurately, providing that the structure is a control (i.e. the flow is not otherwise impeded). This may not always be so. Another obstruction farther downstream may increase the depth between it and the weir so much that the flow does not become critical at the weir. The weir then ceases to be a control. The control will be elsewhere possibly at the crest of the downstream obstacle (see Fig. 14.14). In this case, the simple discharge equation for the upstream weir is no longer valid. The water surface over XX is lowered only to the depth d_2, so that

$$E_1 = E_2 + z \text{ as before}$$

where z is the height of XX above the upstream bed.

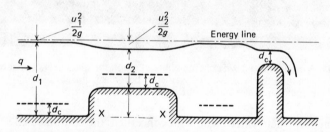

Fig. 14.14 A broad crested weir XX 'drowned' by another obstacle downstream over which there is a critical flow. Note how it is that the structure over which the flow is at critical depth acts as the control surface.

That is
$$u_1^2/2g + d_1 = u_2^2/2g + d_2 + z.$$

and for a constant width channel
$$u_1 d_1 = u_2 d_2.$$

Hence
$$u_2 = \sqrt{(2g(d_1 - d_2 - z))}/\sqrt{(1 - (d_2/d_1)^2)}$$
and
$$q = u_2 d_2 = d_2 \sqrt{(2g(d_1 - d_2 - z))}/\sqrt{(1 - (d_2/d_1)^2)}. \tag{14.13}$$

This equation differs from that applicable when the weir is a control (equation 14.12). It will be seen that $(d_1 - d_2 - z)$ is the difference in water-levels in the approach channel and over the weir. Thus two measurements of water-level are required to find the flow when a weir is not acting as a control and so is described as *drowned*. Since a difference is required, both d_1 and d_2 must be measured more accurately than was the case for the single measurement of d_1 in the critical flow case, if the same accuracy of measurement of the flow rate is required. The practical difficulties of doing so are such that drowned weirs are rarely used for flow measurement.

14.5 (c) The hydraulic jump

The control structures described so far, and most other control structures used in practice, operate by converting a slow flow into a fast flow by accelerating it so that its depth falls to or below the critical depth. Unless there is a method of changing the fast flow back into slow flow it would only be possible to have a single control in one particular open channel regardless of its length. All other structures would affect the depth locally but not control the overall depths in the channel.

Now consider a fast flow, produced by a sluice acting as a control, approaching a rise in the stream bed as shown in Fig. 14.15. The situation bears a similarity to the flow in Fig. 14.14 where a downstream weir is raised until it takes over control from an upstream weir. However in that case the flow before the downstream weir was slow and the effect was similar to the closing of the sluice mentioned in 14.5. With a fast flow before a rise in the bed it is not possible for the small shallow water waves to move upstream and allow the rise to act as a weir and assume control of the flow. The only possibility is for the bed to be raised sufficiently to generate a large wave which will move upstream against the fast flow and produce slow flow upstream. In the unsteady state this strong wave is known as a 'bore', Fig. 14.7. In the photographs it can be seen how in both the model and the prototype the large waves travel upstream faster than the current moves down stream. The less pronounced waves in the model are due to the effects of surface tension. In the steady state when a large wave is

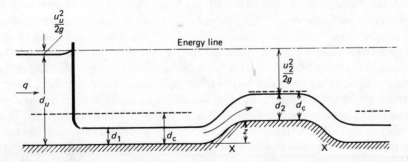

Fig. 14.15 Fast flow (caused by the sluice) passing over a low obstruction XX. The flow becomes deeper over the obstruction and providing that the depth does not reach d_c there, the flow upstream is not affected.

stationary relative to the channel it is called a 'hydraulic jump', Fig. 14.16. Hence if there are two true control structures in an open channel it is inevitable that a hydraulic jump must form between them, the fast flow upstream of the jump being controlled by the upstream control and the slow flow downstream of the jump being controlled by the downstream control structure. Fig. 14.17 shows a hydraulic jump forming after the upstream control of a sluice due to a weir (not visible in the photograph) at the downstream end of the channel.

Fig. 14.16 A hydraulic jump downstream of a weir on a river in flood. The jump is caused by a raised concrete sill across the river downstream of the weir crest. Observe the smooth surface of the water over the weir, where the flow is accelerating, and the rough, air-entraining surface of the jump where there is deceleration.

A diagrammatic representation of the hydraulic jump is shown in Fig. 14.18(a). As with any decelerating flow there is separation and the formation of unsteady eddies in the flow, (see chapters 4 and 12). The unsteadiness of the flow is made obvious in the hydraulic jump by the entrainment of air by the eddies in the flow. Associated with this unsteady flow is a significant fall in the specific head of the fluid due to the work done against internal shear forces. Hence on the specific head curve Fig. 14.18(b) the transition from fast to slow flow is accompanied by a reduction E'/d_c in the specific head of the flow. This ability of the hydraulic jump to reduce the energy of a flow is made use of at the bottom of most spillways where a weir is constructed to form a stilling basin. The jump forms in the stilling basin and after passing through it the water is returned to the river with its energy reduced to a level where it should not cause significant damage by erosion.

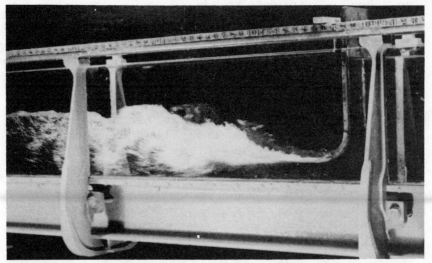

Fig. 14.17 A hydraulic jump in a 10 cm wide glass-sided channel. Fast flow is produced by a sluice (right) and the flow in the jump is observed by a time photograph (1/50 second) of the air bubbles entrained in the water. The tremendous turbulence is clearly seen. Flow from right to left.

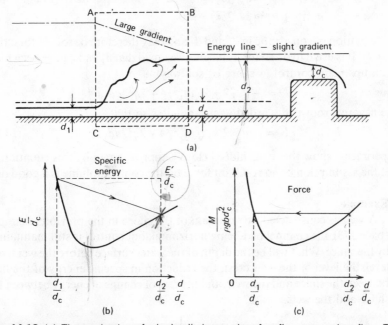

Fig. 14.18 (a) The production of a hydraulic jump when fast flow meets slow flow. The weir has been made so high that slow flow is caused upstream of it. (b) and (c) The representation of the conditions up and downstream of the jump on the specific energy (b) and force (c) diagrams.

If a control volume ABCD is drawn enclosing the jump it is possible to apply the force–momentum equation to the flow through it. As the length of the hydraulic jump is small the shear forces will be relatively small and it can be treated as a rapidly varied depth flow. In the case shown (although not in all cases) there is no structure within the control volume and hence there is no external force applied to the flow within the control volume. If the bed of the channel is horizontal the resultant force in the flow direction is zero, so the change of flow force across the jump must be zero, i.e. $M_2 - M_1 = 0$ or $M_1 = M_2$, and on the flow force diagram Fig. 14.18(c), the jump is represented by a line parallel to the depth axis.

For a rectangular channel of constant width the flow force equation may be written

$$\rho q u_1 + \tfrac{1}{2}\rho g d_1{}^2 = \rho q u_2 + \tfrac{1}{2}\rho g d_2{}^2$$

or

$$\tfrac{1}{2}\rho g(d_1{}^2 - d_2{}^2) = \rho q(u_2 - u_1)$$

$$d_1{}^2 - d_2{}^2 = 2q(u_2 - u_1)/g.$$

Now

$$u_1 = q/d_1 \text{ and } u_2 = q/d_2$$

so that

$$(d_1 + d_2)(d_1 - d_2) = \frac{2q^2}{g}\left(\frac{1}{d_2} - \frac{1}{d_1}\right)$$

or

$$d_1 + d_2 - 2q^2/gd_1 d_2 = 0$$

or

$$d_2{}^2 d_1 + d_1{}^2 d_2 - 2q^2/g = 0$$

This equation is a quadratic in d_1 and d_2 and may therefore be solved for either d_1 or d_2. It is usual to do so for d_2 it being assumed that d_1 and q have been fixed by an upstream control (a sluice or spillway).

Thus

$$d_2 = -d_1/2 \pm \sqrt{(d_1{}^2/4 + 2q^2/gd_1)}$$

The negative root is discarded as meaningless so that

$$d_2 = -d_1/2 + \sqrt{(d_1{}^2/4 + 2q^2/gd_1)} \tag{14.14}$$

Experiments show that d_2 is indeed closely approximated by this equation so that the assumption of no resultant force on the control volume is a good one.

Example

A weir is constructed downstream of the sluice in the previous example, (page 273), the channel bed being horizontal, but control is still maintained by the sluice. What will be the depth of the water surface before the weir crest level, the level of the weir crest, the reduction in specific energy of the flow between the sluice and the weir, and the rate of change of energy between the sluice and the weir?

Solution

This problem can only be solved if it is assumed that the shear forces acting on the flow are negligible compared with the changes in flow force. If this is not so, methods of analysis explained later in this chapter must be used. The flow situation is similar to that in Fig. 14.15, with a hydraulic jump forming between sluice and the weir as in Fig. 14.18(a).

The value for $d_1 = 1.2$ m and $q = 142/12$ m^2 s^{-1}.

Hence from equation 14.13,

$$d_2 = -\frac{d_1}{2} + \sqrt{\frac{d_1^2}{4} + \frac{2q^2}{gd_1}}$$

$$= 4.31 \text{ m.}$$

Note how $d_2 > d_c$, so the flow is slow after the jump and before the weir, at a depth of 4.31 m above the channel bed.

The specific head after the jump $= d_2 + \dfrac{q^2}{2gd_2^2}$

$$= 4.31 + \left(\frac{71}{6 \times 4.3}\right)^2 \frac{1}{2 \times 9.81} = 4.70 \text{ m}$$

But from the previous calculation $d_c = 2.43$ m and the crest of a weir must be $1.5\,d_c$ below the upstream energy line. Hence if Z is the height of the weir crest,

$$Z = 4.70 - 1.5 \times 2.43 = 1.06 \text{ m.}$$

From page 275 the specific head downstream of the sluice was 6.16 m, hence the change of specific head across the jump, E' is given by

$$E' = E_2 - E_1 = 4.70 - 6.16 = -1.46 \text{ m.}$$

The corresponding rate of change of energy, P, is

$$P = \rho g Q E' = 10^3 \times 9.81 \times 71 \times 1.46 = 1017 \text{ kW}$$

Note how this is between thirty and forty times the energy dissipated in the flow under the sluice.

The actual position of a hydraulic jump depends on the equality of the flow forces up and downstream of the jump. If the forces are not equal there must be a resultant force on the control volume ABCD in Fig. 14.18 and the jump will move in the direction of the resultant force. If the level of the weir in the previous example were raised the flow force before the weir would increase and the jump would move towards the sluice which, when the jump reaches it, becomes *drowned*, as in Fig. 14.19. The water level upstream of the sluice then has to rise to keep the flow rate constant and the sluice is no longer a control of the flow. As with the drowned weir, to use the drowned sluice as a flow measuring device it would be necessary, for a given sluice opening, to measure both the upstream and downstream flow depths.

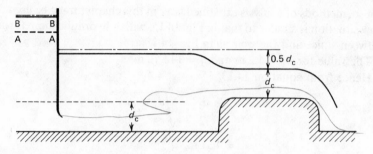

Fig. 14.19 If the weir of Fig. 14.15 is raised more than just necessary to make the depth critical there, then the hydraulic jump may move to the downstream face of the sluice and 'drown' it. The sluice will no longer act as a control as the upstream water level will move from AA to BB and so will be controlled by the water level at the weir.

In the case of long, sloping, channels in which there is gradually varied depth flow the position of a hydraulic jump will depend on the equilibrium of the weight, shear, and external forces. The calculation of the flow force under these conditions is dealt with in Section 14.6. As the formation of a jump in an unlined channel might cause dangerous erosion of the bed and banks, if there is the possibility of a jump forming it is usual to fix its position by means of a step or weir possibly with blocks or boulders to exert extra force on the flow.

14.5(d) Flow passing a drag-force producing object

As well as sluices and weirs, a wide variety of obstacles can be inserted in a stream to produce not only a drag force but also a reduction of specific head of the flow. The piers of a bridge, concrete blocks on the bed, and the bars of a grid to catch refuse (a 'trash-rack') are all examples. Very often it is possible to estimate the drag force of such an arrangement, by measuring the cross-sectional area, finding an appropriate coefficient of drag,* and using the oncoming stream velocity (for method see Chapter 8). When the drag force is known the resultant change of depth (and velocity) of the stream can then be calculated using the flow-force M. If the drag is M' then M is reduced by that amount as the water passes the obstacle. Thus $M_1 = M_2 + M'$ where suffixes 1 and 2 refer to upstream and downstream conditions. Notice, however, that if M' arises because of a single object in a much wider stream (a single pier of a bridge for example), its effect on depths only gradually spreads to include the whole stream, so that the predicted changes apply to a cross-section well downstream where the flow has again become uniform.

Taking $M_1 - M_2 = M'$, dividing throughout by $\rho g d_c^2$,

$$M_1/\rho g d_c^2 = M_2/\rho g d_c^2 + M'/\rho g d_c^2$$

* A useful reference book from which coefficients of drag of many different shapes can be found is S. F. Hoerner, *Fluid Dynamic Drag*.

or $$\tfrac{1}{2}(d_1/d_c)^2 + d_c/d_1 - M'/\rho g d_c^2 = M_2/\rho g d_c^2$$

Knowing d_1, M', and $d_c\,(=\sqrt[3]{(q^2/g)})$, $M_2/\rho g d_c^2$ may be found and the corresponding value of d_c/d_2 obtained from the curve of Fig. 14.5, or by solution of the cubic equation. The former method is demonstrated in Fig. 14.16. The effect of an applied force opposing the flow is to lower the water surface if slow flow approaches but to raise it if fast flow approaches (see Figs. 14.14 and 14.15).

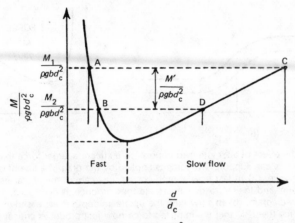

Fig. 14.20 The effect of putting a force of $M'/\rho g d_c^2$ opposing the motion of a stream is to reduce $M/\rho\,g d_c^2$ by AB or CD. Notice that in slow flow d decreases from C to D if there is an opposing force, but in fast flow d increases from A to B.

It is possible that the drag force exerted might be greater than the difference between the force of the flow and the minimum force at the critical depth. The object would then become a control structure, causing a jump in the case of a fast, supercritical flow and a raised upstream depth in the case of a slow, subcritical flow. (Fig. 14.21)

Having found the new depth and velocity, E may be calculated upstream and downstream, and the energy degraded to heat by this drag system $E_1 - E_2$ can be found. The energy line at a place where there is a concentrated drag force always has a sudden downward step in it.

14.5(e) Flow in a narrowing channel

Another way of making changes of water-level in an open channel is by changing the width. As with the case of a weir, the force opposing the stream due to pressures on the forward facing edges of the contractions reduces the flow force M; but since the amount of the reduction is not readily calculable, the flow-force equation cannot be used. For a total discharge Q in the channel, the effect of a narrowing is to increase q, the discharge per unit width. Since

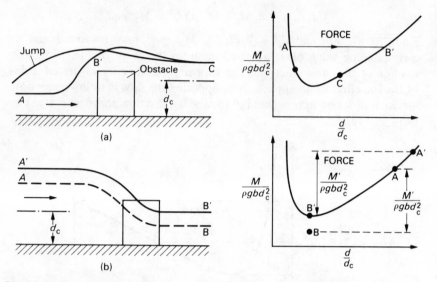

Fig. 14.21 The effect of attempting to impose on a stream a larger force than the upstream conditions can support. The force is applied by the drag of a line of obstacles with gaps between. (a) In fast flow a jump occurs upstream of the obstacles so that the speed past them, and the hydro-dynamic force they produce, is decreased, giving a point C on the force diagram. (b) In slow flow the upstream depth is increased to produce a larger flow-force there, so that the imposed force now just produces critical flow at the obstacles.

$d_c = \sqrt[3]{(q^2/g)}$ the critical depth is increased so that if no energy is added to or taken from the flow, E/d_c is reduced. This effect is precisely the same as that of a rise in the bed, when, as has been described for a weir, the water-level drops for slow flow, or rises for fast flow. If the decrease in width is sufficiently marked, then E/d_c may be decreased as far as $E/d_c = 1.50$; the flow in the narrow part (or *throat*) is then critical, in exactly the same way as occurs over a weir if it is high enough. If the throat is made narrower than the width just necessary for critical flow to take place the throat becomes a control. Then the depth in the throat remains at d_c but the upstream depth increases so that the reduction of E/d_c prescribed by the change in q can take place. This again is similar to the method of controlling a flow by a weir. Fig. 14.22 shows the specific head diagram for a throat in a channel, and the appropriate water-levels, when the oncoming flow is slow.

If it is certain that a throat in a channel has been made narrow enough to become a control with the flow critical, then a single observation of the upstream head E, above the level of the bed, can be used to find Q. By exactly the same reasoning as for a broad-crested weir, the discharge is given by

$$Q = 1.70 \, b_2 \, E^{3/2} \, \text{m}^{1/2} \, \text{s}^{-1} \tag{14.15}$$

where b_2 is the width of the throat.

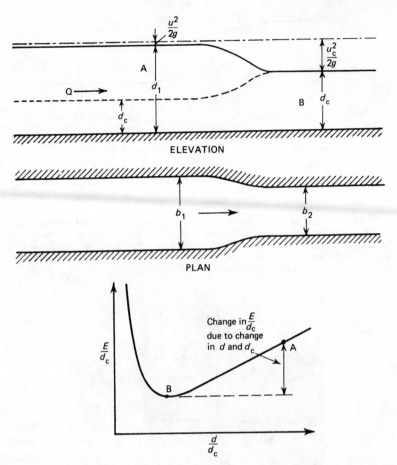

Fig. 14.22 Slow flow approaching a narrowing in the channel so that the discharge per unit width increases from Q/b_1 to Q/b_2. If the narrowing is sufficient the surface is drawn down to the critical depth. On the energy curve the two points A and B represent the upstream and downstream depths.

Downstream of the throat, if the channel is widened again to the original width, the flow will separate from the channel sides with a degradation of energy in the eddies so formed, so that the energy line is lower downstream than it is upstream. An arrangement such as this is used to measure the flow in irrigation and sewage channels and is termed a *Venturi or critical depth flume.* Providing the structure does act as a critical depth flume a single water-level observation gives Q. If there is no assurance that the flow is critical at the throat (because of a high downstream level, or because the narrowing is insufficient), then an additional water-level gauge must be placed at the throat to measure d_2 there. In this case, the more elaborate flow equation of a drowned weir is used

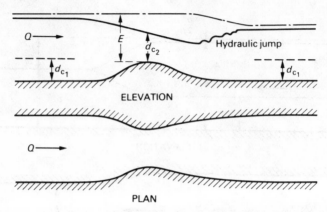

Fig. 14.23 A Venturi flume. The effect of narrowing the channel is the same as the effect of raising the bed. Thus if both occur at the same place the depth is more quickly reduced to the critical depth. Downstream of the weir a hydraulic jump occurs with a degradation of energy. Notice that the critical depth on the hump is greater than that upstream because q has been increased by the narrowing.

Fig. 14.24 A Venturi flume being used to measure the flow in a sewage treatment plant. It is photographed from the downstream side, and the hut contains a water-level recorder. Notice the disturbed flow downstream where a hydraulic jump occurs, and the standing waves at the throat, where critical depth occurs. (Photograph by G. Kent Ltd.)

$$Q = C_d b_2 d_2 \sqrt{(2g(d_1 - d_2))} \left(1 - \left(\frac{b_2 d_2}{b_1 d_1}\right)^2\right)^{-\frac{1}{2}} \qquad (14.16)$$

where both d_1 and d_2 are measured above the channel bed.

One way of ensuring critical depth at the throat is by putting a rise in the channel bed at the throat, thus combining a weir and a Venturi flume. If the flow upstream is slow and downstream is fast, then somewhere the flow must be critical and a simple equation $Q = C_d b E^{3/2}$ can be used. Farther downstream, where the original bed level is regained, the flow may be restrained by downstream obstructions to be slow flow again and there is a hydraulic jump, (see Fig. 14.23). The presence of such a jump is a proof that the flow is fast immediately upstream of it, and so must have passed through d_c. A Venturi flume, Fig. 14.24, *with a jump* is a flume acting under critical conditions.

14.6 The effect of shear stresses on a horizontal flow

The rapidly varied depth flows considered so far in this chapter have occurred in such a short length of channel that it has been reasonable to neglect the shear stresses acting on the flow, and as the bed has been considered horizontal there have been no weight forces. As an introduction to shear forces on open channel flows their effect on a flow in a short length of horizontal channel will be considered.

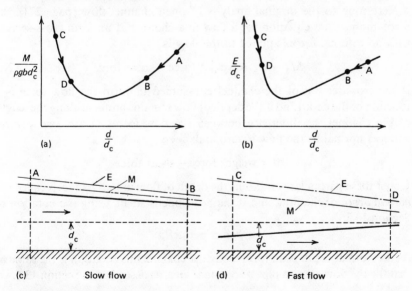

Fig. 14.25 The effect of shear forces on the free surface of a horizontal bed stream. In slow flow (c), the fall in the flow force and energy lines give a decrease of depth as shown in the diagrams (a) and (b), but in fast flow (d), the reverse is true, the depth increasing for the fall in the flow force and energy lines.

The main difference between shear forces and the forces exerted by structures in a horizontal bed flow is that as the shear stresses are relatively small a long length of channel has to be considered before a shear force is exerted on the flow which is comparable to the force exerted by a structure. Although the final effect of the shear force in reducing the flow force of the stream and hence changing its depth will be the same as that of a concentrated force the change of depth will be spread over a far greater length of channel and will produce a *gradually varied depth flow*. The arguments used for predicting the depth using the flow force curve can be used in exactly the same way as Section 14.5(c) and show that in slow, subcritical, flows the depth of flow is decreased downstream, while in the case of fast, supercritical flows the depth increases, as shown in Fig. 14.25.

The effect of shear forces have already been calculated in the preceding chapter on pipe flows. In free surface flows, however, the effects are more complex as the changes of depth caused by the shear stresses will cause changes of velocity which will cause further changes in the shear stresses. These effects will be discussed more fully in Section 14.9, but before this is attempted a free surface flow which is very similar to pipe flow will be considered.

14.7 The normal depth of flow in a channel

Returning to the original analysis of open channel flow (page 271), the force–momentum equation for a flow in a channel of uniform cross-section with no external forces applied to the flow is

$$M_2 - M_1 = \text{weight force} - \text{shear force.} \qquad \text{14.2(e)}$$

If the cross-section of the channel is assumed uniform the flow force is a function of the depth and if the depth of flow remains uniform along the length of the channel the difference between the flow forces at any two control surfaces normal to the flow is zero and:

$$0 = \text{weight force} - \text{shear force}$$

Under these particular conditions the depth of flow in the channel is called the *normal depth* of the flow, and if the bed slope is small, using the notation of Section 14.2:

$$\rho g a x s = \tau_0 x P.$$

As the normal depth is uniform along the channel this equation is the same as that for the flow along a pipe of constant area, as described in Section 13.6, i.e.

$$s = i = \frac{\tau_0}{\rho g m}$$

As a consequence the flow may be considered to have the possibility of laminar, transitional, turbulent and rough turbulent properties (see Section 13.6). The methods of analysis used for pipe flow based on the Chézy, Darcy, or Manning equations can therefore be applied to open channels but with the reminder that in open channels the depth, and any property which depends on the depth, are variables.

Thus in general the friction formulae (of which Manning and Chézy are simple examples of a large family) give the local energy line slope i at one particular cross-section; only under the very special case of normal depth flow do they give the bed slope s. This is of importance if a test is carried out to determine (for example) Chézy's coefficient C; it is essential to ensure that the depth is quite constant all along the selected length of channel. It is fortunate that if s is constant over a great enough length, then however the depth is constrained at the ends by sluices, weirs and the like, the depth of the stream is controlled by the shear stresses to give normal depth in the middle part of the channel. Values of C or n are always quoted for uniform, normal depth.

Example

A trapezoidal channel with a bed slope of 1:1200 has a bed width of 10 m and side slopes of 1:1. If the value of Mannings 'n' is 0.02 and the flow rate 20 m³ s⁻¹, what will be the normal depth of flow and the corresponding Froude Number?

Solution

Mannings formula requires the calculation of the hydraulic radius of the cross section of the flow at normal depth d_0, Fig. 14.26.

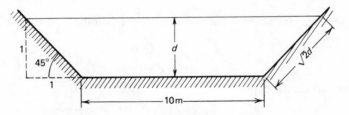

Fig. 14.26 Cross-section of a trapezoidal channel.

Area of cross-section, $a = 10 \times d_0 + 2 \times \frac{1}{2} d_0^2$
$$= 10 d_0 + d_0^2.$$

Perimeter, $P = 10 + 2 \sqrt{2} d_0$.
Then the hydraulic radius, $m = a/P$, and $\bar{u} = m^{2/3} i^{1/2}/n$.

Hence $Q = \bar{u} a = \frac{1}{n} m^{2/3} s^{1/2} a = \frac{1}{n} \frac{a^{5/3}}{P^{2/3}} s^{1/2}$ for flow at normal depth.

Cubing the equation and taking the numerical values to one side

$$\frac{a^5}{P^2} = \frac{(nQ)^3}{s^{3/2}} = (0.02 \times 20)^3 \times (1200)^{3/2} = 2660$$

hence

$$\frac{(10d_0 + d_0{}^2)^5}{(10 + 2\sqrt{2}d_0)^2} = 2660$$

The only way to solve this equation is by successive approximations. As a first estimate take $d_0 = 1$, then the left hand side of the equation becomes

$$\frac{(10 \times 1 + 1)^5}{(10 + 2\sqrt{2})^2} = \frac{11^5}{12.83^2} = 978,$$

which is less than the right hand side. Further estimates must be taken until the equation is satisfied.

$$d_0 = 2 \qquad \text{LHS} = 32482$$
$$d_0 = 1.2 \qquad \text{LHS} = 2444$$
$$d_0 = 1.25 \qquad \text{LHS} = 3002$$
$$d_0 = 1.22 \qquad \text{LHS} = 2656$$

Hence the flow will have a normal depth of 1.22 m, and a normal flow velocity u_0 given by

$$u_0 = \frac{Q}{a} = \frac{20}{10 \times 1.22 + 1.22^2} = \frac{20}{13.69} = 1.46 \text{ m s}^{-1}.$$

For a non-rectangular channel the Froude Number is defined as

$$Fr = \sqrt{\frac{Q^2 b}{ga^3}}$$

Hence at normal depth:

$$Fr_0 = \left(\frac{400 \times 12.44}{9.81 \times 13.69^3}\right)^{1/2} = 0.44.$$

14.8 The relation between normal and critical depth

For a given length of channel with slope s, the resistance in the form of $C, f,$ or n may be specified and will be independent of the flow. At any particular flow rate it is possible to calculate the critical depth, d_c, and the normal depth, d_0. Depending on the ratio of these two depths the normal depth flow in the channel will be fast, supercritical, flow $(d_0 < d_c)$, critical flow $(d_0 = d_c)$ or slow, subcritical, flow $(d_0 > d_c)$, and as Section 14.6 showed this will have a significant effect on the shape of the water surface as the force and energy change. The criterion for flow at critical depth, (subscripts c) in an open channel is:

$$\frac{Q^2 b_c}{a_c^3 g} = 1,$$

while for normal depth flow (subscripts 0) using the Chézy or Darcy equation:

$$Q = C a_0 m_0^{1/2} s^{1/2} = C a_0^{3/2} P_0^{-1/2} s^{1/2} = \left(\frac{2 g a_0^3 s}{f P_0}\right)^{1/2}$$

Eliminating Q between the equations and re-arranging with all the cross-section parameters on one side gives

$$\left(\frac{a_0}{a_c}\right)^3 \frac{b_c}{P_0} = \frac{g}{C^2 s} = \frac{f}{2s}$$

which shows that the ratio between a_0 and a_c is independent of the flow rate. For a rectangular channel

$$\frac{a_0}{a_c} = \frac{d_0}{d_c}$$

In the case of a channel in which the normal depth and critical depth are the same the bed is said to be at the *critical slope* s_c and as $d_0 = d_c$

$$s_c = \frac{P_0}{b_c} \frac{g}{C^2} = \frac{P_0}{b_c} \frac{f}{2}$$

If the channel is broad $P_0 \simeq b_c$ and,

$$s_c = \frac{g}{C^2} = \frac{f}{2}$$

Note that if Manning's equation is used to calculate the channel resistance the expression for s_c is more complex and is dependent on the depth.

The consequence is that the critical slope is specified by the resistance of the channel and is independent of the flow rate. If the bed slope of the channel is less than the critical slope the normal depth flow will be a slow flow $(d_0 > d_c)$ and the channel will be said to have a *mild slope* (M). Conversely a bed slope greater than the critical slope will produce a fast flow at normal depth $(d_0 < d_c)$ and will be referred to as a *steep slope* (S). These categories of slope play as important a part in the understanding of gradually varied depth flow as does the relationship between critical depth and flow force or specific energy described previously, Fig. 14.27.

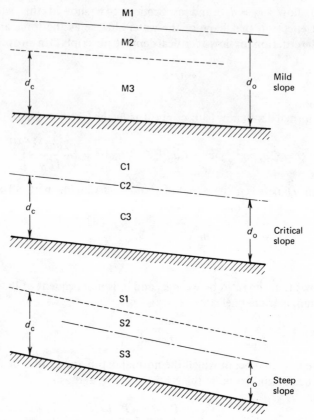

Fig. 14.27 A comparison of mild, critical and steep slopes in an open channel and the corresponding flow categories.

14.9 Gradually varied depth flow

14.9(a) Water surface shapes in gradually varied depth flow

It is now necessary to return to consider a small element of the general open channel shown in Fig. 14.28. If the flow is at a depth, d, $(d \neq d_0)$ at a certain cross-section of the flow it will be necessary to determine the depth at a cross-section a small distance, δx, downstream of the original cross-section, i.e. is δd in Fig. 14.28 positive or negative?

Before becoming immersed in the analysis it is possible to predict the shape of the water surface by quantitative physical argument from the properties of the flow at normal depth and the flow force curve. At normal depth the flow force of the stream remains constant so in Fig. 14.28 the following could be stated

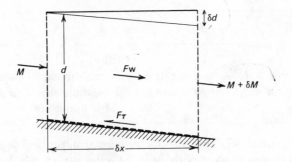

Fig. 14.28 An elementary control volume for flow in an open channel.

a) The weight force balances the shear force, i.e. $F_{\tau 0} = F_{w0}$
b) $\delta M_0 = 0$.

There are now two possibilities, either $d > d_0$ or $d < d_0$, and the consequences of these will be considered together (see table), remembering that if the force–momentum equation is applied to the flow through the control volume

$$M + \delta M - M = F_w - F_\tau,$$

that is,

$$\delta M = F_w - F_\tau,$$

if the channel cross-section does not change.

Flow depth	$d < d_0$	$d > d_0$	Comments
Flow velocity	$u > u_0$	$u < u_0$	by continuity
Shear force	$F_\tau > F_{\tau 0}$	$F_\tau < F_{\tau 0}$	as $F_\tau \propto \frac{1}{2}\rho C_f u^2$
Mass force	$F_w < F_{w0}$	$F_w > F_{w0}$	as mass depends on depth of flow
Hence	$F_\tau > F_w$	$F_\tau < F_w$	as $F_{\tau 0} = F_{w0}$
Then	δM is negative	δM is positive	as $\delta M = F_w - F_\tau$

It is therefore possible to generalise and say that for gradually varied depth flows those with a depth greater than the normal depth have a flow force which increases with distance along the channel, while those with a depth less than the normal depth have a flow force which decreases with distance in the flow direction. The effect of this change of force depends on the relationship of the depth to the critical depth and there are now three possibilities which will be investigated together remembering the shape of the flow-force curve, Fig. 14.5.

Flow depth	$d < d_c$	$d = d_c$	$d > d_c$
Flow type	Fast	Critical	Slow
δM positive $d > d_0$	δd negative	δd indeterminate	δd positive
δM negative $d < d_0$	δd positive	not possible as	δd negative
		M is minimum	

The dependence of the depth on both the normal and critical depths leads to a numbering system to identify the types of water surface which complements the lettering system for the slopes. The first category comprises all depths which are greater than both the critical and the normal depth, the second those depths which lie between the critical and the normal depth (both in the case $d_0 > d_c$ and $d_c > d_0$) and the third in which the flow depth is less than both the critical and normal depths, Fig. 14.27. It is possible to extend the argument to discover whether the rate of change of depth increases or decreases in the downstream direction. The result of this is that it is found that changes of depth near the normal depth are very gradual, while near the critical depth changes of depth are very rapid. The general shapes of the water surface profiles are shown in Fig. 14.29 which include the extra cases of horizontal bed channels, H, and adverse bed slopes, A.

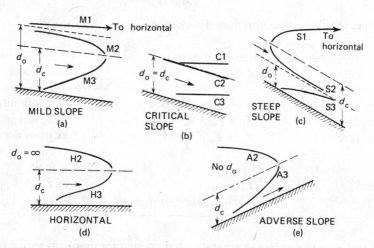

Fig. 14.29 The shapes of gradually varied water surface curves. The diagrams are all drawn with an exaggerated vertical scale compared to the horizontal.

14.9(b) The gradually varied depth flow equation.

To calculate the changes of depth in gradually varied depth flow it is necessary to quantify the argument of the previous Section. The force–momentum equation for the element of channel in Fig. 14.28 can be written in terms of equation 14.2(e) (page 272) as

$$\delta M = \frac{\partial M}{\partial d}\,\delta d = F_\mathrm{w} - F_\tau$$

average P (handwritten annotation)

average area (handwritten annotation)

$$= \rho g\left(a + \frac{\delta a}{2}\right)s\,\delta x - \tau\left(P + \frac{\delta P}{2}\right)\delta x.$$

Neglecting second order terms and taking $\delta \to 0$,

$$\frac{\partial M}{\partial d}\frac{\mathrm{d}d}{\mathrm{d}x} = \rho g a s - \tau P.$$

But it is shown in appendix 1 (page 336) that $\dfrac{\partial M}{\partial d} = \rho g a\left(1 - \beta\dfrac{Q^2 b}{a^3 g}\right).$

Hence

$$\frac{\mathrm{d}d}{\mathrm{d}x} = \frac{\rho g a s - \tau P}{\rho g a\left(1 - \beta\dfrac{Q^2 b}{a^3 g}\right)} = \frac{s - \dfrac{\tau P}{\rho g a}}{1 - \beta\dfrac{Q^2 b}{a^3 g}} = \frac{s - \dfrac{\tau}{\rho g m}}{1 - \beta\dfrac{Q^2 b}{a^3 g}}$$

$$= \frac{s - i}{1 - \beta\dfrac{Q^2 b}{a^3 g}} = \frac{s - i}{1 - \beta Fr^2} \tag{14.17}$$

This equation is perfectly general for any channel of uniform cross-section in which the slope is small and the assumptions of Section 14.1 are valid. Even in a simplified form it is rarely possible to integrate the equation directly and recourse has to be made to numerical methods to produce a solution.

The simplest solution is to calculate the slope of the water surface at the cross section where the depth is known and approximate the curve by its tangent at that point, Fig 14.30 (a). As can be seen from the figure this leads to a constantly increasing error in the estimation of the depth and as a consequence this 'first order' method is rarely used. An improved 'second order' method is shown in Fig. 14.30 (b) where the slope is taken at the start of the element and at the end of the element and averaged to give

$$\frac{1}{2}\left(\left(\frac{\mathrm{d}d}{\mathrm{d}x}\right)_1 + \left(\frac{\mathrm{d}d}{\mathrm{d}x}\right)_2\right) \simeq \frac{d_2 - d_1}{\delta x}.$$

Unfortunately if δx is chosen then d_2 and the slope at d_2 are unknown, and although it is possible to solve such 'implicit' equations the process is time consuming even with the aid of a computer. An easier method is to choose a value of d_2 and make x the unknown thus converting the equation into an 'explicit' form which is easy to solve.

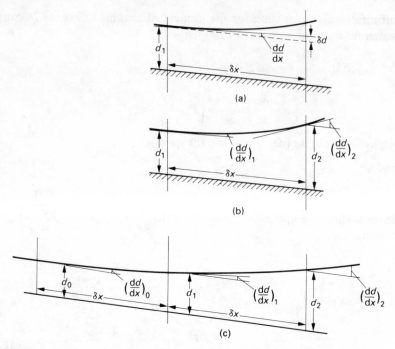

Fig. 14.30 Comparison of (a) first order, (b) second order and (c) parabolic approximations to a water surface profile.

An alternative approximation is to substitute

$$\bar{d} = \tfrac{1}{2}(d_1 + d_2),$$

and then put:

$$\frac{d\bar{d}}{dx} \simeq \frac{d_2 - d_1}{\delta x}.$$

If an implicit method of solution is chosen it is possible to make use of the parabolic approximation Fig. 14.30(c) to the water surface curves which uses the gradients at the two previously calculated water depths d_0, d_1, a distance δx apart, to calculate the unknown depth d_2 at a further distance δx from:

$$\frac{d_2 - d_0}{\delta x} = \frac{1}{6}\left[\left(\frac{dd}{dx}\right)_0 + 4\left(\frac{dd}{dx}\right)_1 + \left(\frac{dd}{dx}\right)_2 \right].$$

14.9(c) Gradually varied depth flow calculations

Before calculations can be performed using equation 14.17 it is necessary to substitute an expression for the hydraulic gradient, *i*. The most common substitutions are derived from the Chézy, Darcy, and Manning equations which result in the expressions

$$\frac{\mathrm{d}d}{\mathrm{d}x} = \frac{s - \dfrac{Q^2 P}{a^3 C^2}}{1 - \beta \dfrac{Q^2 b}{a^3 g}} = \frac{s - \dfrac{f Q^2 P}{2a^3 g}}{1 - \beta \dfrac{Q^2 b}{a^3 g}}$$

and

$$\frac{\mathrm{d}d}{\mathrm{d}x} = \frac{s - \dfrac{Q^2}{a^{10/3}} P^{4/3} n^2}{1 - \beta \dfrac{Q^2 b}{a^3 g}}.$$

(14.18)

In the simplest case of a broad rectangular open channel in which the velocity is assumed to be uniformly distributed ($\beta = 1$) the equations reduce to

$$\frac{\mathrm{d}d}{\mathrm{d}x} = s \frac{1 - \left(\dfrac{d_0}{d}\right)^3}{1 - \left(\dfrac{d_c}{d}\right)^3} \quad \text{for the Chézy and Darcy resistance formulae}$$

and

$$\frac{\mathrm{d}d}{\mathrm{d}x} = s \frac{1 - \left(\dfrac{d_0}{d}\right)^{10/3}}{1 - \left(\dfrac{d_c}{d}\right)^3} \quad \text{for the Manning resistance formula.} \quad (14.19)$$

While this approximation does reduce the effort required in performing a calculation the advent of computers and programmable calculators makes it hardly necessary except for a rapid estimate of flow depths.

Example

The depth of water in a wide irrigation canal is increased to 3 m at one place by a weir. Without this obstruction, the same discharge would produce a stream of uniform depth of only 1.5 m. Bed slope is 1:1000; Manning $n = 0.0214\,\mathrm{s\,m^{-1/3}}$. How far upstream of the weir will the depth be 1.8 m?

Solution

Using Manning's formula for the normal depth, $d_0 = 1.5$ m

$$q = \frac{1.5}{0.0214} \times 1.5^{2/3} \times \left(\frac{1}{1000}\right)^{1/2} = 2.9\,\mathrm{m^2\,s^{-1}}$$

Critical depth $d_c = \sqrt[3]{(q^2/g)} = 0.95$ m

As $d_0 > d_c$ the flow is on a mild slope and because $d_1 > d_0$ then the water surface profile will be in the M1 category with the depth increasing downstream (Fig. 14.31). Divide the range of $d = 3$ m to $d = 1.8$ m into 6 increments each of 0.2 m and carry out a tabular integration working upstream from the weir using the mean depth approximation.

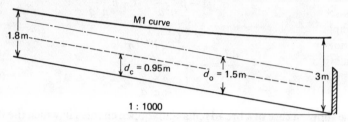

Fig. 14.31 Preliminary information for calculation of the water surface upstream of a weir.

Increment m	Mean d m	$(d_0/d)^{10/3}$	$(d_c/d)^3$	For $s = 0.001$ $\mathrm{d}d/\mathrm{d}l$	Δd m	Δl m
3–2.8	2.9	0.1115	0.0353	0.922×10^{-3}	0.2	217
2.8–2.6	2.7	0.1415	0.0438	0.898	0.2	222
2.6–2.4	2.5	0.182	0.0547	0.866	0.2	230
2.4–2.2	2.3	0.240	0.0705	0.818	0.2	244
2.2–2.0	2.1	0.326	0.0924	0.743	0.2	269
2.0–1.8	1.9	0.453	0.125	0.625	0.2	319

Sum = 1501 m
Say 1500 m

Note   (a) $\mathrm{d}d/\mathrm{d}l$ is positive so depth increases downstream
       (b) $\mathrm{d}d/\mathrm{d}l \to 0$ as $d \to d_0$
       (c) $\mathrm{d}d/\mathrm{d}l$ is the slope of the water surface relative to the stream bed. To calculate the actual slope of the water surface relative to the horizontal the bed slope must be subtracted.
       (d) It is assumed that the variation of n with depth is negligible.

All gradually varied depth flow calculations have to begin at a known depth, usually fixed by a control structure, such as the weir in the previous example. Other such structures are sluices and venturi flumes which will not only act as downstream controls for the slow upstream flow, but will also control the fast downstream flow. It must not be forgotten that a long uniform channel will provide a control in the form of the normal depth. Some simple examples are shown in Fig. 14.32. Care must be taken to ensure that the calculated water surface profile is appropriate to the physical flow conditions imposed by the control structures.

Having found the flow depths in an open channel by means of gradually varied depth flow calculations it is possible to calculate the velocity, flow force, and specific energy at each section of the flow. This is even possible for natural channels although the calculations become more complex as changes in width, cross-section and roughness all have to be taken into account, however the principles remain unchanged. The ability to calculate flow force enables the position of a hydraulic jump to be estimated although results of such calculations should be treated with caution as small changes in flow conditions can make large changes in the position of the jump.

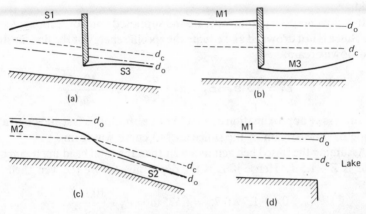

Fig. 14.32 Gradually varied depth flow profiles produced by control structures and long channels. (a) Sluice on a steep slope (b) Sluice on a mild slope (c) Change of slope from mild to steep. (d) Discharge into a lake.

Example

A sluice in a 5 m wide rectangular channel is to pass a flow of 10.15 m³ s⁻¹ when the upstream depth is 9.5 m. The flow continues down the channel which has a slope of 1 : 500 and a value of Manning's n of 0.022. A distance of 500 m downstream of the sluice, a short section of the bed, designed to act as a weir, is raised to a height of 0.4 m. Calculate the water surface profile and the specific energy line of the flow.

Solution

Before starting to consider the control sections it is good practice to calculate the normal and critical depths of the flow.

As the channel is rectangular:

$$d_c = \left(\frac{Q^2}{gb^2} \right)^{1/3} = \left(\frac{10.15^2}{9.81 \times 25} \right)^{1/3} = 0.75 \text{ m}.$$

Using Manning's equation to calculate normal depth,

$$Q = u_0 a_0 = u_0 b d_0 = \frac{b d_0}{n} \, \mathrm{m}^{2/3} \, \mathrm{s}^{1/2}$$

$$= \frac{b d_0}{n} \left(\frac{b d_0}{b + 2 d_0} \right)^{2/3} s^{1/2} = \frac{b^{5/3} d_0^{5/3}}{(b + 2 d_0)^{2/3}} \frac{s^{1/2}}{n}$$

Cubing and separating the variables

$$\frac{d_0^5}{(b + 2 d_0)^2} = \frac{Q^3 n^3}{s^{3/2} b^5} = \frac{10.15^3 \times 0.022^3 \, \mathrm{m}^9}{500^{-3/2} \times 5^5 \, \mathrm{s}^3} \times \frac{\mathrm{s}^3}{\mathrm{m}^6} = 0.0398 \, \mathrm{m}^3.$$

By successive approximation $d_0 = 1.16$ m and hence $d_0 > d_c$; the slope is mild.

It is now necessary to investigate the supposed control sections. Assume the sluice is not drowned and equate the specific energy of the flow upstream and downstream.

$$E_1 \simeq 9.5 \, \mathrm{m} = E_2 = d_2 + \frac{u_2^2}{2g} = d_2 + \frac{10.15^2}{2 g d_2^2}$$

By successive approximation $d_2 = 0.15$ m, i.e. the flow is fast and on to a mild slope so downstream of the sluice an M3 curve will form.

Assuming the raised bed acts as a weir then the total head upstream of the weir is $Z + 1.5 d_c$. Hence if d_3 is the depth just before the weir then

$$0.4 + 1.5 \times 0.75 = 1.525 \, \mathrm{m} = d_3 + \frac{10.15^2}{2 g d_3^2}.$$

Whence $d_3 = 1.42$ m and as $d_3 > d_0$ the water surface upstream of the weir will be an M1 curve. The M3 curve downstream of the sluice and the M1 curve upstream of the weir can only be joined by a hydraulic jump, Fig. 14.33. For the jump to be in equilibrium the flow force of the fast flow must equal the flow force of the slow flow. Hence the jump will form at the point where the flow force curves for the two water surface profiles intersect.

The calculation is best performed by a tabular method either by hand or with the help of a calculator. A suitable increment of depth for the M3 profile which must lie between 0.15 m and 0.75 m will be 0.05 m, while for the M1 curve between depths of 1.16 m and 1.42 m an increment of 0.02 m would be suitable. Using equation 14.18, the M3 curve will be calculated downstream from the sluice (Table 14.1) while the M1 curve will be calculated upstream from the weir (Table 14.2).

At each depth the flow force M and specific head E are tabulated. If the curves of M are plotted downstream from the sluice and upstream from the weir (Fig. 14.34) their point of intersection will be where the hydraulic jump occurs. The complete water surface can then be plotted.

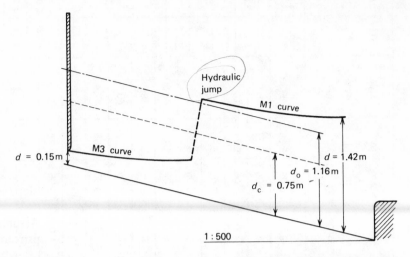

Fig. 14.33 Preliminary information for calculation of the water surface between a sluice and a weir.

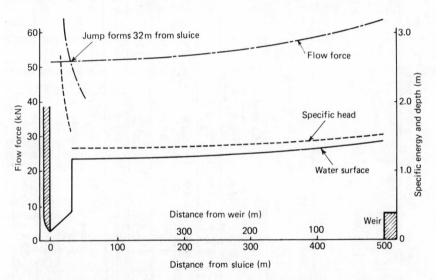

Fig. 14.34 Estimation of the position of a hydraulic jump in an open channel flow.

14.10 Velocity distributions in open channel flow

As mentioned in Section 14.7 the behaviour of the flow in an open channel is similar to the boundary layer flow in a pipe. Because of the size of open channels, such as sewers and canals, it is very rare to have laminar flow conditions in them when there is a significant flow. As with pipes the upper limit

Table 14.1 Calculation of an M3 profile downstream of a sluice gate.

d m	$a = 5d$ m²	$p = 5 + 2d$ m	$\dfrac{Q^2}{a^3}$ s⁻²	$1 - \dfrac{Q^2 b}{a^3 g}$	$\dfrac{Q^2 P^{4/3}\eta^2}{a^{10/3}} = W$	$s - W$	$\dfrac{dd}{dx} \times 10^3$	Mean $\dfrac{dd}{dx} \times 10^3$	Δd m	Δx m	$\Sigma\Delta x$ from sluice m	M kN	E m
0.15	0.75	5.3	244.2	−123	1.202	−1.20	9.75					138	9.48
0.20	1.0	5.4	103	−51.5	0.472	−0.47	9.13	9.44	0.05	5.30	5.30	104	5.45
0.25	1.25	5.5	52.75	−25.9	0.230	−0.228	8.81	8.97	0.05	5.57	10.87	84.0	3.61
0.30	1.50	5.6	30.52	−14.6	0.128	−0.126	8.68	8.75	0.05	5.71	16.58	70.9	2.63
0.35	1.75	5.7	19.22	−8.79	0.079	−0.077	8.71	8.70	0.05	5.75	22.33	61.9	2.06
0.40	2.00	5.8	12.88	−5.56	0.052	−0.050	8.91	8.81	0.05	5.68	28.01	55.4	1.71
0.45	2.25	5.9	9.04	−3.61	0.036	−0.034	9.31	9.11	0.05	5.49	33.50	50.8	1.49
0.50	2.50	6.0	6.59	−2.36	0.026	−0.024	10.0	9.65	0.05	5.18	38.68	47.3	1.34
0.55	2.75	6.1	4.95	−1.52	0.019	−0.017	11.2	10.6	0.05	4.72	43.4	44.9	1.24
0.60	3.00	6.2	3.82	−0.94	0.014	−0.013	13.3	12.3	0.05	4.06	47.46	43.2	1.18
0.65	3.25	6.3	3.00	−0.53	0.011	−0.009	17.8	15.6	0.05	3.21	50.67	42.1	1.15
0.70	3.50	6.4	2.40	−0.22	0.007	−0.005	24.2	21.0	0.05	2.38	53.05	41.4	1.13
0.75	3.75	6.5	1.95	0								41.3	1.12

Table 14.2 Calculation of an M1 profile upstream of a weir.

d	$a = 5d$	$F = 5 + 2d$	$m = a/l$	$\dfrac{Q^2}{a^3}$	$1 - \dfrac{Q^2 b}{a^3 g}$	$\dfrac{Q^2 P^{4/3} n^2}{a^{10/3}}$	$s - W$	$\dfrac{dd}{dx}$	Mean $\dfrac{dd}{dx}$	Δd	Δx	$\Sigma \Delta x$	M	E
m	m²	m	m	s⁻²		$= W \times 10^3$	$\times 10^3$	$\times 10^3$	$\times 10^3$	m	m	from weir m	kN	m
1.42	7.1	7.84	0.91	0.288	0.853	1.13	0.87	1.02	1.0	0.02	20.0	20.0	63.9	1.52
1.40	7.0	7.80	0.90	0.300	0.847	1.18	0.82	0.97	0.95	0.02	21.1	41.1	62.7	1.51
1.38	6.9	7.76	0.89	0.314	0.840	1.22	0.78	0.92	0.90	0.02	22.3	63.4	61.6	1.49
1.36	6.8	7.72	0.88	0.328	0.833	1.28	0.72	0.87	0.84	0.02	23.9	87.3	60.5	1.47
1.34	6.7	7.68	0.87	0.343	0.825	1.33	0.67	0.81	0.78	0.02	25.8	113.1	59.4	1.46
1.32	6.6	7.64	0.86	0.358	0.817	1.39	0.61	0.74	0.71	0.02	28.2	141.3	58.3	1.44
1.30	6.5	7.60	0.86	0.375	0.809	1.45	0.55	0.68	0.64	0.02	31.4	172.7	57.3	1.42
1.28	6.4	7.56	0.85	0.393	0.800	1.52	0.48	0.60	0.56	0.02	35.8	208.5	56.2	1.41
1.26	6.3	7.52	0.84	0.412	0.790	1.59	0.41	0.52	0.47	0.02	42.3	250.8	55.2	1.39
1.24	6.2	7.48	0.83	0.432	0.780	1.67	0.33	0.43	0.38	0.02	52.7	303.5	54.3	1.38
1.22	6.1	7.44	0.82	0.454	0.769	1.75	0.25	0.33	0.28	0.02	72.4	375.9	53.4	1.36
1.20	6.0	7.40	0.81	0.477	0.757	1.83	0.17	0.22	0.16	0.02	123.2	499.1	52.5	1.35
1.18	5.9	7.36	0.80	0.502	0.744	1.92	0.08	0.10					51.6	1.33

for laminar flow is not fixed, but if the Reynolds number is based on the hydraulic radius, $Re = \bar{U}m/v$, then an upper value 2000 is accepted. This means that at a depth of 100 mm a velocity of less than 20 mm s^{-1} would be necessary to ensure laminar flow.

At greater velocities the flow will change to turbulent becoming smooth turbulent, transitional or rough turbulent depending on the Reynolds number of the flow and the relative roughness of the channel. If the flow is fully turbulent the velocity distribution is that given by equation 12.9 on page 208, so, for a hydraulically rough surface,

$$u = 5.75 \left(\frac{\tau_0}{\rho}\right)^{1/2} \log_{10} 33y/k$$

where u is the velocity at a distance y above the bed on which there are roughness elements of effective size k, see Fig. 14.35. It must be remembered that this, and other similar expressions, are only approximations to the real situations. The stress τ_0 of the stream on the bottom is connected to the energy slope i in the same way as for a pipe by the equation

$$\tau_0 = \rho g m i$$
or
$$\tau_0/\rho = g m i$$

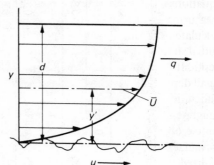

Fig. 14.35 Logarithmic velocity distribution in a rough bedded stream of depth d. The mean velocity $\bar{U}$ occurs at a height y' above the bed where $y' = 0.4 d$.

Substituting into the velocity distribution above,

$$u = 5.75 \, (g m i)^{1/2} \log_{10} 33y/k.$$

A further simplification is possible if the stream is wide compared with its depth, for then the hydraulic radius m is nearly the actual depth d.

Then
$$u = 5.75 \, (g d i)^{1/2} \log_{10} 33y/k.$$

The requirement for any practical friction formula is to correlate $\bar{U}$, the mean velocity, with d and i. Now experimentally, it has been found that a single observation of velocity u, if taken at a height $y' = 0.4d$ gives a value which is the same as $\bar{U} = q/d.$*

---

* See Appendix II to this chapter for the demonstration that $y' = 0.4d$.

It is therefore permissible to put $\bar{U}$ in the velocity distribution equation with $y = y' = 0.4\,d$.

Thus
$$\bar{U} = 5.75\,(dgi)^{1/2}\,\log_{10} 33 \times 0.4d/k$$

$$= 5.75\,g^{1/2}\left(\log_{10}\frac{13.2\,d}{k}\right)(di)^{1/2}.$$

Compare this result with Chézy's empirical equation for the same wide stream, namely,
$$\bar{U} = C\,(di)^{1/2}$$

and it will be seen that

$$C = 5.75\,g^{1/2}\,\log_{10} 13.2d/k \qquad (14.20)$$

or
$$f = 2gC^{-2} = (4.1\,\log_{10} 13.2\,d/k)^{-2} \qquad (14.21)$$

Though the difficulty of assessing k without experiment remains, this link between the empirical equation and the more rational boundary layer equations is of value to the engineer. If friction experiments have been made by measuring $i = s$ on a length of a stream at normal depth, then f, or C, can be calculated and the roughness k found. This roughness of the bed will not alter with d so that the value of k may now be used to determine f or C for any other depth. It may, for example, be practicable to carry out friction experiments at small depths when it is quite impracticable to do so at larger depths because of difficulties of measuring large discharges. It will be seen that for the same roughness, f decreases and C increases somewhat with d: this increase was, of course, observed experimentally in the very early days of empirical energy slope formulae, but the simple form of the logarithmic formula above was not realized until boundary layer theory was applied to open channels.

While the analogy (*above*) of the two-dimensional turbulent boundary layer throws some light on experimentally determined coefficients such as C, there are other motions in open channels which also affect the friction. These motions arise because the shear stresses are not the same at every point on the perimeter of a channel, even though the roughness may be the same. The differences of τ are most noticeable near the corners of a rectangular or trapezoidal channel, although they also exist near the junction of water surface and the side walls. As shown in Fig. 14.36 the presence of two friction-producing surfaces near the corner A causes the contour lines of equal velocity to curve, so that the velocity gradient is less on the diagonal than elsewhere. With smaller shear stresses near A than near (for example) B, less energy is taken from water which travels downstream near the corner than from water which travels near the centreline of the channel. Thus the longitudinal gradient of energy is rather smaller at A than at B. Two cross sections can now be

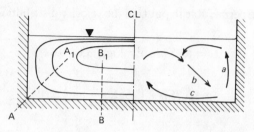

Fig. 14.36 Slow, stirring actions in a straight channel, due to less friction near the corners. (Left) Isovels show smaller velocity gradients at AA_1 than BB_1. (Right) Resultant pressures give flows into A down the diagonal, but away from A to B and from A to the surface.

compared, one upstream of the other, but both having the same arrangement of velocity contours. The upstream section has only flow at right angles to it. The downstream section will have water near A which has not lost as much energy as that near B; consequently the pressure will not have fallen quite as much at A as at B: the pressure at A will therefore be a little higher than at B.

This pressure difference starts motions across the bed away from the corners, and to preserve continuity water must flow into A along a diagonal (Fig. 14.36). Another motion exists up the walls. In this way sets of spiral motions are set up which have the effect of giving a slow, 'stirring' action to the whole flow, assisting the mixing action of turbulent eddies. The additional mixing is in effect an increase in the momentum transfer so that the average shear stresses are increased. The values of f and C found from open channel experiments have a tendency to be higher than those found from pipe experiments of the same relative roughness. Much depends on the precise cross-sectional shape, and on differences of roughness across the bed. Any discontinuity of the bed starts these rotational flows, which are in addition to those caused by bends (see Section 11.8). All rotational flows cause increases of friction, and they are the principal reason why the energy gradient in a meandering river is greater than that in a straight river. Experiments to determine the rotational flows are difficult to arrange, because these relatively slow, steady motions are masked by the turbulence also present.

These complications in river flows give uncertainties in the values of coefficients to be used in calculations, and it is probably not realistic to rely upon an accuracy better than $\pm 15\%$ in a coefficient unless some experimental evidence is available from the river itself. Estimates of coefficients from the appearance of the bed are often made, but quite small variations in roughness and straightness may give surprising results. A useful collection of photographs of river beds, with the accompanying hydraulic data, is shown in *U.S. Geological Service, Water Supply Paper*, No. 1849.

14.11 Water surface waves

In Section 14.4 it has already been shown that in any form of open channel flow the shallow water wave speed has an important bearing on the physical properties of the flow. This is not the only form of wave which is of consequence to the civil engineer. Structures such as oil production platforms are built far out to sea and have to withstand the forces due to ocean waves. Coastlines are in need of protection in many places which are already inhabited, and reclamation of land from the sea is becoming increasingly common. Larger harbours are being built, often in areas which would have been considered too exposed in the past. Hence the deep-water ocean waves and their properties as they approach the land are not just of interest to the oceanographer and geographer. There are also plans to extract power from the waves and the structures proposed will be large pieces of Civil Engineering even when compared with other forms of power generating plant. This section will start by deriving an expression for the speed of a shallow water wave and will then touch briefly on the properties of deep water waves.

Consider a channel of uniform depth d with still water in it. If a disturbance is made in the water, perhaps by a paddle moving slightly as in Fig. 14.37(a), a single smooth-sided wave will immediately move off, travelling at a nearly constant speed but gradually decreasing in height. The movements of the water can be seen experimentally by observing small solid particles or oil bubbles suspended in the water, (see Fig. 5.2, p. 61). It is found that the water under the wave moves in the same direction as the wave, though the water before and after the wave is stationary. The wave can be divided into two parts as in Fig. 14.37(b), the leading part where acceleration takes place, and the trailing part where there is deceleration. Because only one wave need appear at any one time, due to a single movement of the paddle, it is called a *solitary* wave.

Now the forces causing the accelerations and decelerations are entirely the hydrostatic forces caused by the depth at the crest, d_2, being greater than the still water depth d_1. Fig. 14.37(b) shows the force diagram for both these depths, and the resultant force accelerating (or decelerating) the water is $\rho g(d_2^2/2 - d_1^2/2)$ in the direction of motion. It is further observed that the whole of the water in the depth d_2 is put into motion and not just the upper portion above the still water depth. The velocity of the water u_2 so induced produces a discharge of $u_2 d_2$ which is required to provide the volume of water necessary to raise the water surface from d_1 to d_2 over a length ct in a time t.

Thus
$$u_2 d_2 t = c(d_2 - d_1)t$$

where c is the velocity of the *wave*. But the total mass of water put into motion in the time t to a speed u_2 is $\rho\, c\, t\, d_2$ so that the *rate* of change of momentum is

$$\frac{1}{t}\rho\, c\, t\, d_2\, u_2$$

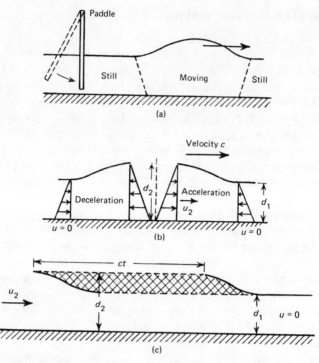

Fig. 14.37 Solitary waves in an open channel. (a) Wave is produced by the movement of a paddle but then travels without other forces. (b) Wave may be divided into two symmetrical halves, with force systems given only by the hydrostatic forces. (c) Travel of the leading half of the wave involves a supply of water to fill the shaded space, which must be caused by the induced velocity of the water under the crest.

or substituting for u_2 above, $\rho c^2 (d_2 - d_1)$.

This rate of change of momentum must be equated to the force producing it so that

$$\rho g (d_2{}^2 - d_1{}^2)/2 = c^2 (d_2 - d_1)\rho$$

or $$c^2 = g(d_2 + d_1)/2.$$

If the height of the wave is small compared to the depth of water $d_2 \to d_1 \to d$

or $$c = \sqrt{(gd)}.$$

The speed of propagation of a small solitary wave is therefore proportional to the square root of the depth. Solitary waves can occur in rivers as a result of sudden floods, or of accidents to sluices or weirs, or indeed of any disturbance of the steady flow. If the height of the wave is large compared with the depth it will move with a greater velocity than a small wave.

In solitary waves, which travel at a speed depending on the depth, the water is, firstly, accelerated and moved in the direction of progress of the wave, then decelerated to a standstill after the wave has passed overhead. The surface water must also rise and fall in this period, so that the paths of particles near the surface are lines which have a hump in them, while particles near the bottom, where there is less rise and fall, follow nearly straight line paths (Fig. 14.38(c)). All particles are displaced during the passage of a solitary wave so that fluid is moved forward by the wave. When the wave reaches the end of the channel it may either break on a sloping beach, or be reflected by a wall to return in the opposite direction. In the latter case the particles of water will be replaced to their original position as the wave passes over on its return passage. But in the former case the water moved forward by the wave is piled up on the beach as the wave breaks so that the water-level at the beach is rather higher than elsewhere. This condition is unstable and a slow return motion commences to return this water uniformly all along the channel. The particles already considered have therefore a slow backward movement following the fast forward movement as the wave passes over. The waves of the sea as they approach the land become a series of solitary waves, each one of which is completely independent of its neighbours. There is a steady outward current (undertow) to replace the water brought forward by the waves.

When waves are travelling in deep water their properties are quite different from those of solitary waves. In general, the waves of the deep sea can be regarded as a succession or *train* of waves, one wave following another at more or less regular intervals. The interval of distance from one wave crest to the next is the wave-length L: the interval of time between succeeding wave crests passing a fixed point is the wave-period T. The motions of the water under each wave interact with the motions of the neighbouring wave so that the particle movements become less and less like those of the solitary wave as the depth of water becomes greater. A full analysis of the motions involved is too advanced for this volume, and a much less rigorous and complete explanation is given here.

In general there is always on the sea surface a number of wave trains each of different wave-length, period and height, each superimposed on the others. The combination forms a *spectrum* of wave-lengths. On a windless day the smooth-sided 'swell' waves are composed of a spectrum which is restricted to a narrow range of wave-lengths. When a strong wind blows the spectrum is widened, and there are waves present which vary in wave-length from under a metre to hundreds of metres. Each wave travels at its own speed c, which may depend upon its wave-length L, height h, the depth of water d, and since a rise or fall of the surface involves a gravitational force, the acceleration g. Thus c may depend on L, h, d, and g. So by the method of dimensions (Chapter 10) one possible way in which the variables may be grouped is

$$c = (gL)^{1/2} \phi(d/L)(h/L).$$

Some advanced theoretical work (see Lamb's *Hydrodynamics*) shows that if the waves are long compared with the height, that is h/L is small, then the value of h/L does not affect the velocity c. Furthermore, it can be shown that

$$c^2 = (gL/2\pi) \tanh (2\pi d/L).$$

If d/L is large, as in the deep ocean, then

$$c^2 = gL/2\pi,$$

and if d/L is small, that is, in shallow water, then $c^2 = gd$, as has already been demonstrated.

It will be seen that if the spectrum of the waves on the sea consists of one wave-length only then all waves travel at the same speed. If there are a selection of wave-lengths present, then the longer, faster ones are continually overtaking the smaller ones. As two wave crests coincide, an instantaneous wave is produced the height of which is the sum of their amplitudes. It is this ever-changing appearance of the sea due to the overtaking that is so evident when the wind is creating waves. Despite some approximations made in the theoretical analysis, experiments show that the speed of waves is in excellent agreement with the above equations. If T is the period of the waves, defined as the time taken for a crest or trough to travel a distance equal to L at a speed c,

then $$T = L/c = L \sqrt{(2\pi/gL)} = \sqrt{(2\pi L/g)}.$$

Thus reliable estimates of wave-lengths can be made by timing the wave crests past a fixed object. The wave-lengths so found can be used for model studies so that the model waves are truly to scale.

The motions in the water if d is great compared with L, prove to be such that all particles describe circular paths or orbits in deep water, making one complete circle in each wave period T (Fig. 14.38(a)). The diameters of the orbits decrease downward by an exponential law, being equal to the trough-to-crest height h for surface particles, and being only $h/535$ at a depth equal to the wave-length. If the water is not deep, but there is a solid bottom at a finite depth, then the orbits are distorted into ellipses as they approach the bottom. Adjacent to the bottom the particles can only move linearly, of course, as was the case of the solitary wave. It is the rapid decrease of the water motions downwards that allows small particles of sand and silt to remain unmoved on the sea bed, despite the raging of storms overhead (see Fig. 14.38 (a) and (b), and Fig. 5.2, p. 61).

As with the solitary wave the speed of a deep water wave does not depend on the wave height if the height is small compared to the wave-length. Deep water waves of large height do move faster, but there is a limit above which the height of a wave cannot increase or the wave will break. The wave height is even more important when the energy of waves is considered. A wave possesses both potential and kinetic energy, the former because the water has been raised or lowered from its equilibrium (still water) level, the latter because of the velocity

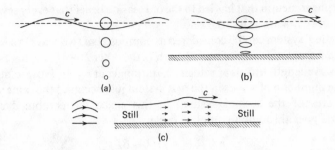

Fig. 14.38 A comparison of water movements below different sorts of waves. (a) Deep-water waves have motions which are circles whose radii decrease exponentially downwards. (b) Deep-water waves approaching the shore have motions which are ellipses more distorted near the bottom than at the top, where they are nearly circles. (c) Solitary waves create fluid motions only under themselves and there is stationary fluid before and after them. Thus the paths of particles are lines in one direction and are 'humped' as shown on the left. There is a slower return motion only if the wave breaks subsequently on a beach.

of the orbital motions under the wave. In all, the total energy of a travelling wave train is

$$E = \tfrac{1}{8}\rho g h^2 L \text{ per wave,}$$

for a unit width of wave across the direction of motion.

Though the waves themselves travel at a speed c relative to the water in which they are produced, it is by no means certain that the energy which they possess is carried forward at the same speed. The question can be resolved by considering the combination of two wave systems of slightly different wave-lengths. If there are just two wave-lengths in a particular spectrum, the two trains of waves combine together to 'beat' in the same way as electromagnetic or acoustic wave systems will do. Over a length of time of several wave periods, the combined wave height (amplitude) observed at one stationary point builds up to a maximum, then decreases to zero and then increases again (Fig. 14.39).

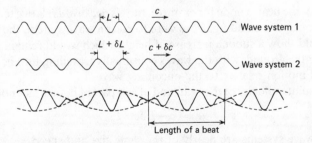

Fig. 14.39 Two wave systems of unequal wave-lengths and speeds 'beat' together to form a more elaborate combined system.

It is this phenomenon that has led to the old seaside belief that every seventh (or fifth, or third?) wave is a large one.

A 'beating' system can be considered as composed of two wave systems each of constant amplitude, one of wave-length L, the other $L + \delta L$. Each wave of the shorter wave-length travels at velocity c, the longer at $c + \delta c$. Now to make one 'beat' the number n of waves in the first system just occupies the same space as $n - 1$ waves of the second system. In this space the combined system's amplitude goes through a complete cycle.

That is
$$nL = (n-1)(L + \delta L)$$

or ignoring δL compared with L, then $L = n\delta L$.
Thus the distance from one end of the beat to the other is

$$nL = L^2/\delta L.$$

A beat of waves will possess a certain amount of energy, being the sum of the energies of all the waves in the beat. No energy can pass the nodes (the points where there is no wave motion) so that if the waves are to transfer any energy from one place to another, the beat itself must move, though not necessarily at the speed of either train of waves. The beat is formed in the time that one longer wave completely overtakes one smaller wave, taking the whole length of the beat to do so. Since the relative velocity between the two sizes of wave is δc, and the distance which the faster wave has to travel relative to the slower one is L in order to overtake, then the time of overtaking is $t = L/\delta c$. But the effect of the overtaking on the combined pattern at a point moving with the speed of the slower waves is to cause the amplitude to undergo a complete cycle of change from zero to maximum (sum of the two amplitudes) to zero again. Thus in this time of overtaking, a complete beat, from node to node, has passed the moving point. Since the beat is $L^2/\delta L$ long, and it passes the point in time $L/\delta c$, its speed, *relative to the moving point*, is

$$\frac{L^2}{\delta L} \bigg/ \frac{L}{\delta c} = L\delta c/\delta L.$$

Any part of the beat, a node for example, travels *backward* relative to the waves. (If a stationary, or standing, wave were to meet a wave system, the combined wave would show a sudden increase of height which would remain stationary relative to the ground, with the waves continually travelling into it, giving it a backward motion relative to the oncoming waves.)

The absolute velocity of the beat (and of any part of it, such as a node) is then

$$c' = c - L\delta c/\delta L,$$

or if the wave systems are nearly of the same size and speed

$$c' = c - L\, dc/dL.$$

So far no mention has been made of the law connecting c and L. In fact the preceding analysis for beat speed c' is perfectly general for any sort of wave motion, fluid, electromagnetic or acoustic. If the waves are deep water ones

$$c = \sqrt{(gL/2\pi)} = KL^{1/2}$$

so
$$dc/dL = \tfrac{1}{2}KL^{-1/2}$$

and thus
$$\begin{aligned}c' &= KL^{1/2} - L\tfrac{1}{2}KL^{-1/2}\\ &= \tfrac{1}{2}KL^{1/2}\\ &= \tfrac{1}{2}c.\end{aligned}$$

So that in deep water, a beat travels at half the speed of the individual waves. Now if the two wave systems above are only infinitesimally different in wavelength and so in speed, then a small portion of the beat can be regarded as a *group* of waves all of the same length and height, such as might be caused by a laboratory wave-making machine which has worked for a few strokes and then stopped. This group will travel at the beat speed c' even though the waves themselves travel at $2c'$: they have a relative velocity forward through the group, so that if attention is given to one particular wave it will appear initially from the back of the group, run forward and then disappear at the front of the group as another appears at the back.

Since a group represents a constant amount of energy that has been given to the fluid by some means (for example, by the wind on the sea, or by an oscillating plate in a laboratory tank), the speed at which this energy is transferred is the speed c' of the group. Therefore the rate at which energy is passing a given point is the product of the wave energy per unit length, and the speed at which this energy is approaching the point.

So Energy per unit length of water surface
and per unit width across wave direction $= \tfrac{1}{8}\rho g h^2 L/L$

and Rate of arrival of energy or Power per
unit width across the waves
$$\begin{aligned}&= \tfrac{1}{8}g\rho h^2 c'\\ &= \tfrac{1}{8}g\rho h^2\,\frac{c}{2}\\ &= \tfrac{1}{8}\rho g h^2\left(\frac{g}{2\pi}\right)^{1/2}\frac{L^{1/2}}{2}\end{aligned}$$

This power in a deep water wave system may be considerable. For example, waves of $h = 3$ m and $L = 64$ m would carry forward a power of 55 kW per metre width of wave across its direction of travel. Thus no less than 55 kW could be obtained per metre if the whole of the oncoming power could be extracted from the waves, leaving a plane surface. This power has come from the wind which has acted upon the sea surface to form the waves.

Recently there has been effort put into the design and model testing of a number of devices to extract energy from sea waves. It has been demonstrated that it is possible to produce power this way but so far it has not been sufficiently attractive economically for prototype structures to be built.

No mention has yet been made about the way in which waves are modified as they approach the shore. As the water depths decrease, a train originally of deep water waves becomes more nearly a succession of solitary waves whose speed is governed by the depth and not by the wave-length. The orbital movements within the waves remain circular for particles near the water surface but are distorted in lower layers to ellipses until at the solid surface of the sea bed the motions are merely straight line oscillations. With decreasing depths of water both the wave speed and the speed at which the energy travels are reduced so that energy is travelling out of the water at the shallower end of a fixed length at a lower rate than it is arriving at the deeper end. The wave energy, $\frac{1}{8}g\rho h^2 L$ per wave, in such a length therefore increases, while because the waves are decelerating, the distance between waves, originally L, also decreases. The height h therefore increases and so does the *steepness ratio* h/L. When this ratio approaches a limiting value of about 1:10, when the depth is about 1.2 h, the wave is too high for its length and crashes over as a breaker. The resulting eddies and turbulence degrades most of the wave energy into heat, leaving the kinetic energy of the broken water as it runs up the beach.

14.12 Free surface flow models

The examples of different kinds of open channel flow given in this chapter are mainly for idealized conditions. Real rivers, estuaries, coastlines, and the modifications to them caused by engineering works, are rarely so simple and because of the possible expense of errors in design it is always desirable to predict, in advance, the effect of such modifications. Experiment on small-scale models is the only practical way of making these predictions and it is therefore necessary to know scaling laws for converting the observed discharges, velocities, water-levels and wave heights in the model to those in the prototype. Dimensional analysis (Chapter 10) can be used to determine the necessary grouping of variables for model experiments, see Fig. 14.40.

The magnitude of the forces F exerted on the water in the river may depend in some way on the mean velocity U, on the size of the river characterized by a length or depth measurement l, on the fluid viscosity μ and density ρ and on the gravitational acceleration g. The last must be included because any changes in water-level involve a gravity force tending to cause an acceleration or deceleration. The length measurement must be taken in the same place in the model as in the prototype so that the ratio l_m/l_p gives the scale of the model (suffix m for model, p for prototype). These variables U, l, μ, ρ and g are precisely the same as those applicable to the case of the drag force of a ship, as

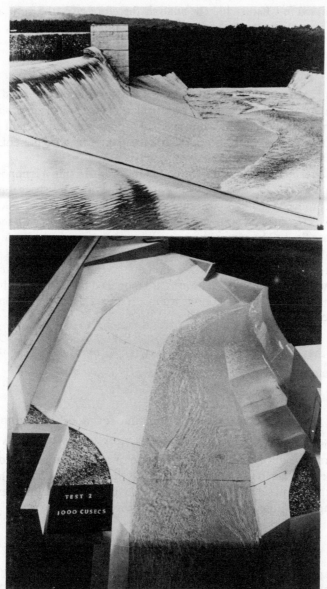

Fig. 14.40 (a) Flow over Sasuma Dam, Kenya. This is an old structure for a water-supply reservoir, recently raised to provide more storage. The layout is complicated, giving two fast streams combining at the foot of the spillway. A completely new structure would probably be laid out quite differently. (b) A 1:40 scale model in the laboratory, showing the complicated wave pattern downstream of the spillway. A model of this sort is the only way of determining this pattern and to find the heights of the water surface at all points with the complex geometry of the weir crests. (Photograph by Wimpey Laboratories Ltd.)

Flow in open channels

was explained in Chapter 10. The analysis therefore gives the same answers, that one way of grouping the variables is

$$F = U^2 l^2 \rho \phi (Ul\rho/\mu), \ (U^2/lg)$$
$$F = U^2 l^2 \rho \phi_1 (Re, \, Fr).$$

The term $F/U^2 l^2 \rho$ is therefore a function of the Reynolds number and also the Froude number. Experiments with a particular-shaped model might show that one or the other of these non-dimensional groups are unimportant for that shape, but in general both shear and gravitational forces are present, so that both Re and F must be used.

As has already been shown for the case of the drag of a ship it is impossible to ensure that $Re_m = Re_p$ and $Fr_m = Fr_p$ at the same time; the shear forces cannot be reproduced to the same scale as the gravitational forces unless quite impossible viscosity scales are used. The same difficulty arises with models of free surface flows. The compromise usually made is to scale the model so that the gravitational forces are correct but to ignore the shear forces. By this method the local changes of surface level due to changes in bed level, width (rapidly varied depth flow), and waves are modelled correctly, but the gentle slopes due to shear forces are modelled incorrectly. For this compromise it is only necessary to have

$$U_m/(l_m g)^{1/2} = U_p/(l_p g)^{1/2}$$

in order that

$$\phi_1 (U_m/(l_m g)^{1/2}) = \phi_1 (U_p/(l_p g)^{1/2})$$

and therefore

$$F_m/U^2{}_m l^2{}_m \rho_m = F_p/U^2{}_p l^2{}_p \rho_p$$

The scaling law for velocities is therefore

$$U_m/U_p = (l_m/l_p)^{1/2} \text{ as } g \text{ is usually the same for model and prototype.}$$

All velocities in the model are less than those in the prototype by the ratio of the square root of the model scale. The ratio of the discharges in model and prototype may then be found: a discharge Q is the product of U and the cross-sectional area, and since l characterizes every length, horizontal and vertical, in the arrangement, then l^2 characterizes an area, so that $Q \propto Ul^2$

or

$$Q_m/Q_p = \frac{l_m^2}{l_p^2} \cdot \left(\frac{l_m}{l_p} \right)^{1/2} = (l_m/l_p)^{5/2}.$$

It is the large discharge ratio obtainable with the usual model scales that makes possible the modelling of large rivers and estuaries in the laboratory. Thus if a river whose discharge is $10^5 \ \mathrm{m^3 \, s^{-1}}$ is to be reproduced by a 1:100-scale model, a discharge of $1 \ \mathrm{m^3 \, s^{-1}}$ only will be required in the model.

Time in the model is scaled by the ratio of length divided by velocity so

$$t_m/t_p = \frac{l_m}{l_p} \left(\frac{l_p}{l_m} \right)^{1/2} = (l_m/l_p)^{1/2}$$

and hence time periods are much shorter allowing models of unsteady flow to be tested in much shorter times. For example the 1 : 100 scale model would have a time scale of only 1/10th that of the prototype, e.g. a 12 hour tidal period would be completed in 1.2 hours.

Having ignored the part played by shear forces (by not attempting to make the Reynolds number the same in the model as in the prototype) it is not surprising that there is good agreement between model and prototype if the portion of the flow chosen for study is short, and the changes of water-level great (as in rapidly varied depth flow). Models of flow over weirs and spillways, through contractions and expansions, and past obstructions of various shapes give good representations of the prototype conditions, and changes in the flow caused by modifications to the boundaries may be confidently extrapolated to the prototype size. It is otherwise, in gradually varied depth flow, the stream is long with only minor changes of level: now the shear forces on the boundaries of the flow are important. If the scaling law $Q_m/Q_p = (l_m/l_p)^{5/2}$ is used, there is poor agreement between corresponding measurements of velocity and discharge on model and prototype because the important scaling of shear forces is ignored.

Fortunately for the engineer there is a way around this difficulty, using the hydrodynamic roughness k of the solid boundaries. If k is regarded as a variable in the dimensional analysis, it is found that

$$F/U^2 l^2 \rho = \phi(Re)(Fr)(l/k).$$

The group l/k, called the *relative roughness*, is introduced in the same way as it occurred in the analysis for flow in pipes. It is well known that making a surface more rough alters the shear forces in rough turbulent flow. The shear forces in the model can therefore be increased by making k/l larger than in the prototype, and vice versa.

One way of calibrating a Froude scaled river model is to pass a given discharge through it and to measure the water levels at a number of cross sections. These levels are then compared with those of the prototype river at the discharge given by the Froude law, that is

$$Q_p = Q_m (l_p/l_m)^{5/2}$$

If these levels agree tolerably well, using the scale (l_p/l_m), the shear forces are scaled satisfactorily, and the modifications desired by the engineer may be placed in the model. The new water-levels may now be measured and scaled up to the prototype by the same law. If the levels in the model do not agree well with those in the prototype, then the roughness of the bed and sides of the model are changed until by a process of trial and error the correct water surface profile is obtained. The modifications can now be inserted as before. Models where roughness adjustment is required can never be relied upon to such an extent as those where only the Froude law is required for scaling, because it cannot be definitely assured that the modifications themselves (walls, jetties, bridge piers, etc.) do not change the shear forces in a different manner with the roughness present from the way in which the full-size modifications affect the prototype shear forces (which do not rely on such roughnesses).

It is good practice to make two models of a river, if time and money permit, each being at a different scale. One should be as large as the laboratory space permits, the other smaller. If the water-levels and discharges of the smaller model scale up satisfactorily to those in the larger model, according to the Froude law, then it is fairly certain that shear forces are modelled correctly, and that the same Froude law scaling will be satisfactory for the extrapolation to the prototype. The smaller model must be large enough so that surface tension forces do not become important.

A difficulty sometimes found with free surface models is that at the scale prescribed by restrictions in laboratory space the depths and changes of water-level are so small that they are difficult to measure accurately. Further, the combination of small depth and low speed may give laminar flow in the model so that the turbulent flow in the prototype is unrealistically modelled. Both difficulties can be overcome by exaggerating all vertical measurements compared with the horizontal, so that all depths, gradients and water surface heights are made to a larger scale than the horizontal measurements. The model is thus not geometrically similar to the prototype. However, if U and d are made so that the ratio $U/(gd)^{1/2}$ is the same in the distorted model as in the prototype it will be certain that the same type of flow (fast or slow) exists in both at the corresponding places. In this expression d is a typical *depth* measurement and is not a length measurement in the horizontal plane. If this is done then

$$U_m/(gd_m)^{1/2} = U_p/(gd_p)^{1/2}$$

or

$$U_m/U_p = (d_m/d_p)^{1/2}.$$

The scaling law for velocity in a distorted model depends therefore on the square root of the vertical scale, and is independent of the horizontal scale.

The discharge ratio must involve both vertical and horizontal scales because a typical cross-section area of a river depends on the product of a vertical depth and a horizontal length. Thus

$$Q \propto Uld$$

and so

$$Q_m/Q_p = (d_m/d_p)^{1/2}(l_m/l_p)(d_m/d_p)$$

$$= l_m/l_p(d_m/d_p)^{3/2}$$

where l_m/l_p is the horizontal scale and d_m/d_p is the vertical scale. If the discharge as given by the above scaling law flows through a distorted scale model, then the changes of depth due to gravitational forces will be reproduced truly to the vertical scale: the changes due to frictional forces can be approximated if necessary by adjustment of the roughness as before.

Distorted scale models are often used successfully to predict changes in water levels due to projected structures. They are less successful if at the same time they are used to predict changes in the sandy bed. Because of the distortion, all the slopes of the sides and of sandbanks are far greater than in the prototype. Thus sand may fall down a slope in such a model where it would not fall in the prototype. Also the pattern of secondary currents (14.10) may be different from that in the river: since the direction of the flow of the water near the bed controls the direction of the sand flow, the model may give quite misleading information about sand movements. In general, a distorted scale model is never regarded as quite so conclusive in its predictions as a natural (undistorted) scale model, if the latter can be made large enough.

14.13 Sand movements in a model

Though a time scale for water movements may be derived from the Froude relation, it should not be thought that the same scale can be used for sand movements in a model with a non-rigid bed. So little is in fact known about the action of water on silt and sand that no certain scaling law for time is yet possible, and the rate of growth of sandbanks or scour holes cannot be inferred from models.

However, certain broad· principles of sand bed models seem to be established. First, it is not necessary to scale down the size of the sand particles in the same proportion as the gross features of the river have been scaled down; over a wide range of sand sizes and densities the erosion or deposition patterns are nearly the same, so long as the experiments are allowed to continue until a steady state is reached. Second, the effect of the sand on the water movements is largely controlled by the large-scale features of the sandbanks, and not by the properties of the individual grains; thus it is possible to conduct experiments with a fixed bed moulded to the shape of the present sandbanks, and to observe the motion of a small number of grains thereon, and so to infer the stability of the bank in this form. The difficulty arises here that in some parts of the model the water speeds may be too small to give sand movement, whereas there may be movement in those places in the prototype.

Third, the final erosion and deposition patterns found locally around solid structures are well and truly scaled according to the linear scale of the structure; but large-scale sandbanks in a long length of alluvial river or estuary are far less certainly scaled. An example of this uncertainty is the bed movement of a small stream over an extensive uniform sand bed, producing a model meandering river; it would be a matter of complete chance if the model resembled at any stage any real alluvial river (the Mississippi, for instance) for the meanders may well be started by adventitious collapses of the river bank or by unexpected inhomogeneities of the river walls.

Thus models to predict sand movements prove to be difficult and uncertain in extrapolation. Replicate experiments are necessary, and several different sands may be tested until one proves to give movements on the unmodified river in accordance with experience. Sands of reduced specific gravity, e.g. perspex or pumice, instead of quartz, may also be used so as to increase bed movements and to hasten experiments.

Example

The site for a peak-load hydro-electric power station is just upstream of the junction of a river with a tidal estuary, and models are to be made to determine if the irregular flow from the turbines affects the river bed near the power station and the sandbanks further down the estuary. Select suitable scales for the models to fit a laboratory space, 20 m × 10 m, the largest available pump giving a flow of $0.06 \, \text{m}^3 \, \text{s}^{-1}$. The river is 70 m wide and 3 m deep with a maximum flow of $580 \, \text{m}^3 \, \text{s}^{-1}$. The estuary is 5 km long to the open sea, averages 1 km wide and has a maximum tidal current of $2.06 \, \text{m} \, \text{s}^{-1}$ (4 knots). The lowest bed level in the estuary is 12 m below mean sea level and the tidal range is ± 2.5 m.

Solution

At least two models will be necessary, both made to the Froude scaling law. One model should be of the whole estuary from the sea to above the power station to investigate the effect on the sandbanks, while the other model should be of the immediate neighbourhood of the power station to study the motion of the bed near the turbine exits.

For the estuary model a distance of 5 km must be modelled in a length of 20 m so the maximum horizontal scale will be $\dfrac{5000}{20} = 250 : 1$. If this were adopted as the vertical scale the water depth near the turbines would only be $3/250$ m, i.e. 12 mm, which would be far too small for accurate modelling. Hence a larger vertical scale will have to be used, say $50 : 1$. The maximum water depth would then be $\dfrac{12 + 2.5}{50} = 0.29$ m, the river depth 60 mm and the tidal range $\pm 2.5/50 = \pm 50$ mm, which is probably about as small as can be accepted. To give equal Froude numbers in the model and prototype the velocity scale must be the square root of the vertical scale, i.e. $\sqrt{50} = 7.07 : 1$. The maximum tidal velocity will be $\dfrac{2.06}{7.07} = 0.29 \, \text{m} \, \text{s}^{-1}$. Taking the mean depth of the model estuary to be 0.15 m and a mean velocity of $0.15 \, \text{m} \, \text{s}^{-1}$, if the kinematic viscosity of water is 10^{-6} m^2 s^{-1} a typical Reynolds number in the model will be

$$R = \frac{Ud}{v} = \frac{0.15 \times 0.15}{10^{-6}} = 22\,500$$

which is well above the value which corresponds to laminar flow in open channels.

$$\text{The time scale of the flow} = \frac{\text{horizontal scale}}{\text{velocity scale}}$$

$$= \frac{250}{7.07} = 35.4:1$$

Hence a 12 hour tidal cycle in the prototype will correspond to $12 \times 60/35.4$ = 20.4 minutes in the model.

The river model should not be distorted as the bed may well be affected by the turbine exit jets. The choice of scale would have to be a compromise between including as much of the river as will affect the flow near the jets, but to keep the scale as large as possible to make the motion of the bed realistic. The flow of water available for the model might limit the scale of the model and should be checked first.

As discharge = velocity × area,

discharge scale = (linear scale)$^{5/2}$,

i.e. the maximum linear scale $= \left(\dfrac{580}{0.06}\right)^{2/5} = 39.3:1$

If a linear scale of 40:1 were chosen the width of the model river would be 1.75 m and a length of 800 m of the river could be represented. The depth of the model would be 75 mm which is reasonable for modelling bed movements.

$$\text{Mean river velocity} = \frac{580}{70 \times 3} = 2.76 \text{ m s}^{-1}, \text{ and hence}$$

$$\text{model velocity} = \frac{2.76}{\sqrt{40}} = 0.44 \text{ m s}^{-1}$$

$$\text{The model Reynolds Number would be } \frac{75}{10^3} \times \frac{0.44}{10^{-6}} = 33\,000$$

which is well above the value required for turbulent flow.

While the calculations show that it would be possible to construct models of the prototype, great care would be necessary to ensure that the tidal generator was not placed too close to the model to give distorted tidal motions. Because of the exaggerated scale in the estuary model any information gained on the motion of sandbanks would have to be interpreted with great care. The final decision on the river model scale would have to take into account the length of river which should be included both up and downstream of the turbine exits to represent the flow correctly.

14.14 Density currents

A stream of water, with its upper boundary always at atmospheric pressure, is a particular case of a heavier fluid (water) flowing beneath a lighter one (air). There are many important phenomena which are more general cases of the same sort, the density difference being much less than that between water and air. For example, in estuaries the heavier salt water often moves (due to tidal action) in the opposite direction to the lighter fresh water above, there being a sharp interface between the two fluids. A lock of a dock system filled with fresh water gives a strong surface outflow when the sea gates are opened, as salt water flows in to displace fresh water above; and in low-lying country (like the Netherlands) the locks bringing ships from the sea to the inland waterway discharge an undercurrent of salt water which may cause biological damage or interfere with water supplies. In these cases the underflow may well carry silt which can be deposited in inconvenient and unexpected places. Air streams in ventilating systems, for example, are also prone to this *density stratification*, hot air moving differently from the colder air beneath. In general, the phenomena of open channels are reproduced in these 'density' flows, but at much lower speeds. If the depths of the layers concerned are large enough, then the analogy is close, the effective gravitational acceleration now being $g\dfrac{\Delta\rho}{\rho}$, where $\Delta\rho$ is the finite difference of density between the layers and ρ the mean density; thus the critical velocity and the speed of small solitary (long) waves is now $\left(gd\dfrac{\Delta\rho}{\rho}\right)^{1/2}$

If the layers are thinner, then the speed of these waves is affected by two horizontal boundaries, the one at the interface, the other at the free surface with the atmosphere. More complicated expressions are needed for analysis of these situations.

In most cases, the two different-density fluids have a tendency to mix (e.g. brine with fresh water). This is caused only to a trivial extent by molecular diffusion and mixing, and is predominantly caused by turbulent eddies at the interface. Density layers therefore become less pronounced, if there is relative motion, as the difference of gravitational pull on the two layers gradually decreases. Thus non-uniform flows are usual. The tendency for mixing to occur is expressed by the value of another non-dimensional quantity, the Richardson number, which is the ratio of the momentum flow rate that tends to cause eddies (see Section 12.12 for a discussion of this effect in the Reynolds number) and the gravitational forces that tend to keep heavier fluid in lower positions and to stop it from being carried away in eddies into the lighter fluid (where it mixes). While the eddy-producing forces on a parcel of fluid in a particular situation can be expressed as $\rho u^2 l^2$ (where l is a representative distance measurement), the gravitational force which resists the parcel of heavy fluid from moving out of its layer into the lighter one (and vice versa), and then mixing, is $g\Delta\rho l^3$. The ratio of these forces, $\dfrac{\rho u^2 l^2}{g\Delta\rho l^3} = \dfrac{u^2}{g\dfrac{\Delta\rho l}{\rho}}$, is called the Richardson number, Ri.

If the Richardson numbers are the same for two flows, each within boundaries which are geometrically similar to each other but of different size (one being the model of the other), then the mixing will be the same in the two cases, and similar interfaces and density and velocity profiles will be observed. Thus equality of Ri is another criterion for satisfactory modelling of density-layered flows.

Since $\left(gl\dfrac{\Delta\rho}{\rho} \right)^{1/2}$ is the proportional to the speed of small, long waves in a two-layer system, it can be seen that the square root of Ri is equivalent to a Froude number, with the addition of the fraction $\Delta\rho/\rho$. It therefore should have a significance with regard to waves. This can be explained by analogy to the production of waves on the sea by wind blowing over it. The lighter upper fluid, moving relatively to the lower fluid, causes waves to form: with high enough relative speed, the waves 'break', detaching parcels of heavy fluid which are thrown up into the lighter one. As these parcels move up they gradually acquire the speed of the lighter fluid (and so act as a mixing agent for momentum), but their outer surfaces mix more slowly, heavy with light fluid, to form a mixture of intermediate density. High values of Ri ($Ri^{1/2}$ is sometimes called the densimetric Froude Number) give large waves, which break furiously and mix momentum and density rather well if the fluids are miscible (i.e. mix); low values of Ri give better mixing of momentum than density, with neither very powerful. If the fluids are imiscible, a further variable, the surface tension, may have to be modelled.

A more detailed review of these phenomena is given in Section 26 of *Handbook of Fluid Dynamics* (McGraw-Hill 1961), entitled 'Stratified Flow', by D. R. F. Harleman.

Appendix I: The minimum value of the flow force

To find the minimum value of the flow force for a particular flow rate, equation 14.2, must be differentiated with respect to the depth, d, and equated to zero.

As
$$M = \beta\rho Q^2 a^{-1} + \rho g \bar{d} a$$

$$\frac{dM}{dd} = \beta\rho Q^2 \frac{d}{dd} a^{-1} + \rho g \frac{d}{dd}(\bar{d}a)$$

$$= -\beta\rho Q^2 \frac{1}{a^2}\frac{da}{dd} + \rho g\left(a\frac{d\bar{d}}{dd} + \bar{d}\frac{da}{dd}\right) \qquad (14.21)$$

Consider a cross-section of the flow in which the depth increases by an increment δd (Fig. 14.41).

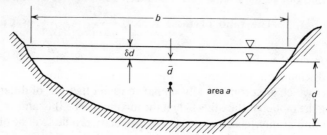

Fig. 14.41 The effect of an increase of depth on the geometric properties of a cross-section of an open channel flow.

As $\delta a = b\delta d$ then in the limit $\dfrac{da}{dd} = b$.

To find $\dfrac{d\bar{d}}{dd}$ take the first moment of area about the new water surface

$$(a + b\delta d)(\bar{d} + \delta\bar{d}) = a(\bar{d} + \delta d) + b\delta d\frac{\delta d}{2}$$

which, neglecting second order terms becomes

$$a\delta\bar{d} + b\bar{d}\delta d = a\delta d.$$

Re-arranging and taking the limit as $\delta \to 0$

$$\frac{\delta\bar{d}}{\delta d} \to \frac{d\bar{d}}{dd} = \left(1 - \frac{b\bar{d}}{a}\right).$$

Substituting in equation 14.21,

$$\frac{dM}{dd} = -\beta\rho Q^2 \frac{b}{a^2} + \rho g(a - b\bar{d} + b\bar{d})$$

$$= -\beta\rho Q^2 \frac{b}{a^2} + \rho g a$$

$$= \rho g a\left(1 - \beta\frac{Q^2 b}{a^3 g}\right)$$

$$= \rho g a(1 - \beta Fr^2)$$

Hence the force of the flow will have a minimum value at the same depth as the specific energy, i.e. at $Fr = 1$, $d = d_c$, if $\alpha = \beta = 1$.

Appendix II: The mean velocity in a wide stream

In Section 14.10 the experimental fact was given that the mean velocity $U (= q/d)$ is observed in a wide stream at a height above the bed $y' = 0.4\,d$, where d is the total depth of the stream. The result may also be inferred by integration, assuming a logarithmic velocity profile.

Thus, putting $q = \bar{U}d = \int_0^d u\,dy$ and inserting

$$u = 5.75\,(\tau_0/\rho)^{1/2}\log_{10} 33y/k$$

and integrating, it is found that

$$\bar{U} = 5.75\,(\tau_0/\rho)^{1/2}\{\log_{10} 33/k + \log_{10} d - 1/2.303\}.$$

Now suppose $\bar{U}$ occurs at a height y' above the bed. Then

$$\bar{U} = 5.75\,(\tau_0/\rho)^{1/2}\log_{10} 33y'/k,$$

and comparison with the former equation for $\bar{U}$ shows that

$$\log_{10} 33/k + \log_{10} d - 1/2.303 = \log_{10} 33y'/k.$$

But since $$1/2.303 = \log_{10} e.$$

Then $$\log_{10} 33d/ke = \log_{10} 33y'/k$$

or $$y' = d/e = 0.37\,d.$$

Problems

1. Find the downstream height of a hydraulic jump occurring on a level bed when the upstream depth is 1 m and the velocity is $10\ \mathrm{m\ s^{-1}}$. What is the depth of water upstream of the sluice producing the 1 m depth of flow?
   Ans. 4.04 m: 5.97 m.

2. A jet, 1.3 m deep, with total head 10 m, issues from a sluice gate 6 m wide and then flows along a horizontal frictionless channel to a narrow culvert. Determine a safe width for the culvert so that the sluice is not 'drowned'.
   Ans. 1.9 m.

3. A horizontal channel is 3 m wide and 2 m deep. A little farther downstream the width is 2.4 m, and the bed is 30 cm *higher*, but the water surface is 12 cm *lower*. What is the flow? What must be the difference of bed levels at the two sections if the surface is to remain level? What are the Froude numbers of the flow?   *Ans.* $7.5\ \mathrm{m^3\ s^{-1}}$: 0.5 m: 0.28: 0.50: 0.54.

4. A rectangular channel 3.2 m wide has a horizontal bed. It is desired to measure the flow through it by a contraction in width which is to produce critical flow at the narrowest section (a Venturi flume). If the upstream depth is not to exceed 1.8 m when the flow is $8\ \mathrm{m^3\ s^{-1}}$, what should be the width at the throat? What effect would be caused if the throat were narrower? If it was desired to ensure critical flow by a low, streamlined weir 40 cm high at the throat, what now must be the width?
   Ans. 1.8 m: 2.56 m.

5. A frictionless channel of constant width sloping downward at 1 in 20 is fed by a sluice as broad as the channel and 1.6 m open (discharge coefficient 0.6). Find the depth 15 and 50 m from the sluice when the upstream stagnation level is 6.5 m. Find also the greatest length of channel sloping 1 in 10 upwards which will just not drown the sluice. How will friction affect the results?   *Ans.* 0.90 m: 0.79 m: 64 m.

6. (*a*) A structure is required in an open channel to control the depth at a certain point. With sketches explain where the structure must be placed if the flow at that point is (i) fast, and (ii) slow.

   (*b*) For the conditions of the figure, find the discharge if $z = 1$ m.

   (*c*) If the sill in the figure is raised until $z = 2$ m, find the new value of y_1 if the discharge of (*b*) remains constant.   *Ans.* 1.195 m^2 s^{-1}; 4.51 m.

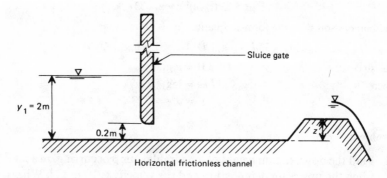

Horizontal frictionless channel

7. Assuming that shearing forces in a turbulent stream are caused solely by the eddies, show that Chézy's C in a wide river which has rough walls and bottom is

$$C = 18 \log_{10} \frac{13 \times \text{depth}}{\text{sand size}} \text{ m}^{1/2} \text{ s}^{-1}$$

   Plot this curve in the form $\log C$ against $\log d/k$ and find the equations to tangents of this curve at $C = 33, 44$, and 67. Compare the equations of the tangents with the empirical formulae of Lacey ($\bar{u} = Lm^{3/4} i^{1/2}$), Manning ($\bar{u} = n^{-1} m^{2/3} i^{1/2}$) and Blasius ($h = A (Re)^{0.25} l\bar{u}^2/2gd$) and draw conclusions about the range of validity of each. Compare Strickler's formula ($\bar{u} = Ak^{-1/6} m^{2/3} i^{1/2}$) with the general equation and show that it is an approximation valid in a small range only.

8. A channel 60 m wide sloping 1:1000 conveys 280 m^3 s^{-1} and Chézy's coefficient is $C = 47$ m$^{1/2}$ s^{-1}. What is the normal depth?   *Ans.* 2.14 m.

9. Define and contrast the normal and critical depths in a river. A long wide channel of bed slope 1:2000 has a Chézy C of 45 m$^{1/2}$ s^{-1}. A flow of 5 m^3 s^{-1} per metre width is to be maintained at normal depth by a sluice-gate at the downstream end of the channel. Find the normal and critical

depths of the flow and the category of the slope. Calculate the opening of
the gate.   *Ans.* 2.91 m: 1.37 m: Mild: 0.73 m.

10. Field data from a concrete lined irrigation ditch flowing at normal depth
are:

Bottom width 1 m: side slopes 45° width of water surface 3 m: discharge
$3 \text{ m}^3 \text{ s}^{-1}$.

Level of bed at an upstream point A	256.42 m O.D.
Level of bed at a downstream point B	261.17 m O.D.

A and B are 4750 m apart.

What are the values of the coefficients in Manning's and Chézy's
formulae? What are the practical difficulties in carrying out these
measurements accurately?   *Ans.* $n = 0.014 \text{ s m}^{-1/3}$: $C = 65.5 \text{ m}^{1/2} \text{ s}^{-1}$.

11. A flow of $12 \text{ m}^3 \text{ s}^{-1}$ per metre width enters a long open channel with a
poor concrete surface equivalent to gravel of 0.5 cm diameter. Determine
the eventual depth of flow in the channel if the bed-slope is (i) 1 in 1600 and
(ii) 1 in 49.

If the channel were narrowed at one place, sketch the water surface
profile for both cases (i) and (ii).

(*Hint.* In a turbulent boundary layer the velocity, u, varies according to

$$u = 5.75 \, (\tau_0/\rho)^{1/2} \log_{10} 33y/k$$

where τ_0 is the boundary shear stress, y is the distance from the bed and k is
the equivalent roughness.)

Ans. $d = 3.6$ m, 1.22 m; $d_c = 2.45$ m, so one is fast, other slow.

12. The figure shows two reaches of a long broad open channel carrying a
steady flow with $d_c = 0.7$ m. Calculate the normal depths of flow and the
depths at the point of change of slope. Plot the water surface profiles and
estimate the position of any hydraulic jump.

Ans. $d_0 = 1.26$, 0.56 m a) $d = 0.7$ m b) $d = 1.26$ m
Hydraulic jump in b) approximately 80 m up steeper slope.

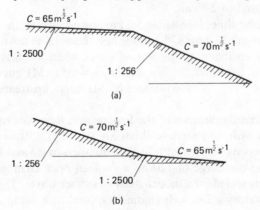

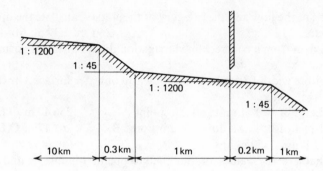

13. A channel 10 km long, with a slope of 1:1200, ends with the arrangement of bed slopes shown in the figure. The channel is trapezoidal in cross-section with a bottom width of 10 m and side slopes of 1:1, except close to the sluice where the section is 10 m wide and rectangular.

Calculate the critical and normal depths of flow in the trapezoidal channel if Manning's n is 0.02 and the flow rate 19 m^3 s^{-1}.

Assuming that the sluice is fully open sketch a likely water surface profile in each of the reaches of the channel, carefully marking those sections which could control the flow.

Sketch the steady water surface profiles which could result from the sluice being set at smaller and smaller openings. In each case indicate the likely control sections of the flow.   *Ans. d_c = 0.7 m, d_0 = 0.45, 1.18 m.*

14. A long, wide open channel with a Manning's n of 0.02 and a slope of 6×10^{-4} carries a flow of 12.4 m^3 s^{-1} for each metre of its width. A sluice with a coefficient of contraction of 0.6 is installed in the channel.

a) Estimate the water surface profiles upstream and downstream of the sluice when its opening is 1.6 m.

b) What will be the water surface profiles when the sluice opening is increased to 2.75 m?

c) A second sluice is installed 5.2 km downstream of the first sluice. If the flow is still 12.4 m^3 s^{-1} and the first sluice 1.6 m open, what will be the depth upstream of the second sluice when the first sluice ceases to control the flow?   *Ans. d_0 = 4 m. a) M1 curve upstream, M3 downstream b) M1 curve upstream, sluice downed.*
    c) d = 7.7 m.

15. Discuss the significance of the Froude and Reynolds numbers as force-ratios in hydraulic models. What are the chief differences between two flows in open channels with Froude numbers of 0.5 and 1.5? The width of 10 m and discharge of 120 m^3 s^{-1} of an open channel are both to be accurately scaled for a model-study of surface waves. There is available in the laboratory a 1 m wide channel supplied by a pump with a maximum

discharge of $0.25 \text{ m}^3 \text{ s}^{-1}$. What is the linear scale ratio of the largest model which can be built, and what modifications of the apparatus will be necessary? *Ans.* Scale. 1:11.8.

16. The discharge of a river model passes into a channel 15 m long and 1.4 m wide, and is measured by a broad crested weir 0.7 m wide at the end of the channel. The maximum weir head is 20 cm. How long will the system take to settle down when starting from zero flow? Between readings the flow is changed by 20 per cent; how many readings can be expected in an hour? The error of the flow is not to exceed $\frac{1}{2}$ per cent.

    Hint. Search for an exact integral solution of the reservoir equation.
    Ans. 131 s: about 35.

17. A long, wide canal has a water flow 1.8 m deep at 1.2 m s^{-1} in it. Accidentally, a control gate at the downstream end is shut suddenly. Calculate the height of the consequent surge, and the speed at which it travels upstream.

    Hint. Use the technique of the pressure wave computation in 14.11, reducing the problem to a steady flow hydraulic jump by imposing the wave velocity backwards on the whole system. *Ans.* 0.55 m: 3.94 m s^{-1}.

15
Work and power in fluid flows—pumps, fans, and turbines

15.1 Transfer of mechanical energy in fluid flows

The common factor in the last three chapters on boundary layers, pipe flows, and open channel flows is that there have been resultant forces opposing the motion and in each case the energy of the fluid has decreased as it has moved in the flow direction. While in some cases the fluid may have possessed the necessary energy to achieve the desired flow, (as with the water in the reservoir, Section 13.10, page 243), there is not always a balance of energy available and flow required. In some cases work has to be done on a fluid by means of a pump or fan to provide the required energy, alternatively a fluid flow with excess energy may do work on a turbine or windmill to generate power, often in the form of electrical power. The aerodynamic and hydraulic machines which have been devised to perform these tasks range from the bucket and windlass in a well to the turbines and pumps of the present day, and are a tribute to the constructive ingenuity of man. Essentially the machines can be divided into two groups, the positive displacement devices such as the piston pump, gear pump etc., and the impeller devices such as the propeller, and axial or centrifugal pump, fan, or turbine. The performance of this latter group of impeller machines, sometimes referred to as 'rotodynamic' machines, will be the concern of this chapter. There will be no attempt to explain how to design impeller machines: firstly because the details of the flow in these machines are so complicated that the understanding of them is not complete and the design is empirical as well as theoretical. Secondly, very few engineers are involved as designers although many have to choose and use pumps, fans, and, to a lesser extent, turbines.

15.2 Types of impeller machines

For convenience impeller machines can be divided arbitrarily into two groups, the axial flow type, and the radial (or centrifugal) flow type. As will be seen later there is in fact a continuous range of types of machines which joins the two basic groups, the performance of all the machines depending on the properties of vortex motions described in Chapter 11

342

The most familiar axial flow machines are the windmill, propeller, and single stage fan, which can be either uncontained or in a duct. These have developed into the axial flow pump and multi-stage axial flow compressor (see Fig. 15.1). The rotors may be considered to consist of a number of aerofoils each of which generates a circulation and hence a lift force which does work on the flow, or has work done on it in the case of a turbine or windmill. The calculation of the flow through the rotor is more complicated than with a simple aerofoil as the velocity of the aerofoil is different at different radii, requiring the blade to be twisted to keep the aerofoil section at a suitable angle of incidence. The blades are often short relative to their chord and the flow about each blade is affected by the flow about the other blades, which complicates the analyses still further due to the importance of secondary flows. In spite of these analytical difficulties modern axial flow machines are able to do work on or receive energy from a fluid with an efficiency which may be as high as 90 % if they are used in situations where relatively small quantities of energy are supplied or extracted from relatively large flows. Typical uses are as cooling fans, fans in ventilation systems, drainage or irrigation pumps, windmills, and Kaplan turbines in hydro-power schemes where the differences in water levels are only a few metres but the flows are large.

In contrast to the axial flow machines the centrifugal machines are most appropriate for the addition or removal of relatively large quantities of energy to or from relatively small flows. Francis turbines in hydro-power schemes, borehole pumps, central heating pumps, vacuum cleaner fans etc. are examples of the use of centrifugal machines, (see Fig. 15.2). It is even possible to operate centrifugal pumps in series if more energy is to be added than can be accomplished by a single pump impeller.

When running, a centrifugal machine may be considered to produce a forced vortex, Section 11.5, with a radial flow of fluid whose pressure and kinetic energy are greater at the periphery of the rotor than at the centre. The change of energy of the fluid can be shown to be equal to the change of circulation as it passes through the impeller. The change in kinetic energy of the fluid requires the use of a complex outer casing in which the fluid can accelerate, in the case of a turbine, or decelerate and increase its pressure in the case of a pump. Similar casings are sometimes used to accelerate a flow before it enters an axial flow turbine and hence such installations, in common with centrifugal machines, can be large compared with the impeller diameter, (see Figs. 15.3 and 15.4).

Between the true axial flow and radial flow machines there are mixed flow pumps and turbines while for generating power from a small flow of water with a high head there is the Pelton wheel in which individual jets at atmospheric pressure hit vanes or buckets mounted on the perimeter of a large wheel, (see Fig. 15.5).

(a)

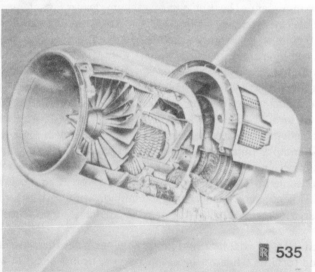

(b)

Fig. 15.1 (a) A 0.6 m diameter axial flow water pump impeller with a dimensionless label $N_s = 0.57$. (Photograph by Gwynnes Pumps Ltd.) (b) A section of a modern axial flow gas turbine showing the large number of aerofoil shaped blades in the compressor at the inlet and the turbines at the outlet. (Courtesy Rolls-Royce Ltd.)

Fig. 15.2 A radial flow pump rotor with an inlet diameter of 1.14 m from a power station circulating water system. Dimensionless label $N_s = 0.24$.

15.3 Performance of impeller machines

The objective of an impeller machine is to do flow work on, or take energy from a flowing fluid as economically as possible. Hence the performance is usually expressed in terms of the flow rate, the energy change in the flow, the power produced or required, and the efficiency of the process. A set of performance curves for a typical impeller machine, an axial flow fan, is shown in Fig. 15.6. The independent variables which determine the performance will be fixed by the design of the machine, i.e. its shape, size, and rotational speed, and the properties of the flowing fluid. With so many variables involved, the method of dimensional analysis described in Chapter 10 is particularly appropriate for handling the information. A convenient way of arranging the analysis is to tabulate the variables (Table 15.1) with their dimensions, showing the independent variables from which the dimensionless groups are to be constructed.

The machine is represented by the shape and the diameter of the impeller, D, the motion by the rotational speed of the impeller, N, (expressed in revolutions), and the fluid by its density, ρ. These three independent variables contain the dimensions of length, time, and mass, respectively and are used

Fig. 15.3 A radial (centrifugal) flow pump. The upper part of the casing has been lifted to show the rotor, which is narrow compared with its diameter and so has a low value for the dimensionless label N_s.

to construct the eight dimensionless groups shown in the fifth column of Table 15.1. As $N \times D$ represents a linear velocity the final three dimensionless groups in the table can be recognised as forms of the Reynolds number, Mach number squared, and a cavitation number.

In most impeller machines the Reynolds Numbers are high and their values change little over the working range of the machine. While the Mach Number is of great importance in the design of aircraft propellers and gas turbines, high Mach Numbers are avoided in other applications as they are invariably accompanied by unsociable noise levels. In the case of liquids the compressibility effects on the performance of pumps and turbines are negligible although as has been shown in Section 13.12 they are of considerable significance in the design of the piping system. Finally the onset of cavitation predicted by the cavitation number tends to produce a discontinuity in the performance which efforts are made to avoid (Section 15.8). Hence the effects of the fluid properties tend to be of secondary importance in the operation of a particular machine, but they are of great importance when conducting model tests to predict the performance of prototype machines.

Fig. 15.4 A scale model of the axial flow Kaplan turbines for the Owen Falls power station, Uganda, showing rotor, guide vanes and a section of the scroll case and draft tube. Notice how large these last two items are compared with the rotor; a small reduction of the rotor size may make a large reduction in the civil engineering work in building the scroll case and draft tube. The turbine has a dimensionless label value $N_s = 0.39$. (Photographs by Boving & Co. Ltd.)

The general expression derived from the dimensional analysis is

$$\phi_1\left[\frac{P}{\rho N^3 D^5}, \frac{\gamma}{\rho N^2 D^2}, \frac{\psi}{\rho N D^3}, \eta, \text{shape}, \frac{\mu}{\rho N D^2}, \frac{K}{\rho N^2 D^2}, \frac{\sigma}{\rho N^2 D^2}\right] = 0$$

and can be reduced for a particular shape of machine to

$$\phi_2\left[\frac{P}{\rho N^3 D^5}, \frac{\gamma}{\rho N^2 D^2}, \frac{\psi}{\rho N D^3}, \eta\right] = 0,$$

over most of its operating range. In the case of incompressible flows it is convenient to work in terms of head, H, rather than pressure, and volumetric flow rate, Q, rather than mass flow rate. Hence by substituting $\gamma = \rho g H$ and

Fig. 15.5 A Pelton wheel rotor. The buckets are cast integrally with the wheel, and are made as large as possible so that a jet of large cross-sectional area may be used, developing the maximum power from a given size wheel.

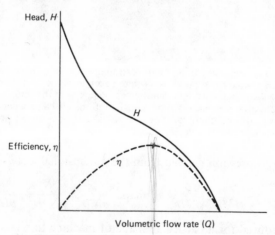

Fig. 15.6 Performance curves for an axial flow fan.

$\psi = \rho Q$ and simplifying, the resulting equations for the performance of a particular shape of machine are

$$\phi_3 \left[\frac{P}{\rho N^3 D^5}, \frac{gH}{N^2 D^2}, \frac{Q}{ND^3}, \eta \right] = 0,$$

Table 15.1 Variables affecting the performance of a pump, fan, or turbine.

Variables	Property	Symbol	Dimension	Dimensionless group
	Power	P	$M\,L^2T^{-3}$	$\dfrac{P}{\rho N^3 D^5}$
All	Pressure change	γ	$M\,L^{-1}T^{-2}$	$\dfrac{\gamma}{\rho N^2 D^2}$
Dependent	Mass flow rate	ψ	$M\,T^{-1}$	$\dfrac{\psi}{\rho N D^3}$
	Efficiency	η	1	η
Machine Properties				
All	Size (diameter)	D	L	–
Independent	Shape	–	1	Shape
	Rotational speed	N	T^{-1}	–
Fluid Properties				
Independent,	Density	ρ	$M\,L^{-3}$	
all	Viscosity	μ	$M\,L^{-1}T^{-1}$	$\dfrac{\mu}{\rho N D^2}$
remainder	Compressibility	K	$M\,L^{-1}T^{-2}$	$\dfrac{K}{\rho N^2 D^2}$
dependent	Vapour pressure	σ	$M\,L^{-1}T^{-2}$	$\dfrac{\sigma}{\rho N^2 D^2}$

and the variables are also related by

$$\eta P = \rho g Q H \quad \text{for a pump,}$$

or
$$P = \eta \rho g Q H \quad \text{for a turbine.}$$

As the power and efficiency depend on the product of the flow rate and head, the latter are the only remaining independent variables and can be related by

$$\frac{gH}{N^2 D^2} = \phi_4\left(\frac{Q}{ND^3}\right).$$

It is conventional to express the power and efficiency in terms of the flow rate as

$$\frac{P}{\rho N^3 D^5} = \phi_5\left(\frac{Q}{ND^3}\right),$$

and
$$\eta = \phi_6\left(\frac{Q}{ND^3}\right).$$

When plotted, these functions are referred to as 'unit' performance curves, although it is more appropriate to consider them as dimensionless coefficients

C_Q, C_H, and C_P, which are analogous to the drag and lift coefficients of Chapter 8. Then, if the effects of the variation of Reynolds, Mach, and cavitation numbers are neglected, the performance of a particular *shape* of machine may be expressed in the form of three dimensionless curves

$$C_H = \phi_4(C_Q),$$
$$C_P = \phi_5(C_Q),$$

and

$$\eta = \phi_6(C_Q).$$

Fig. 15.7 shows a set of dimensionless performance curves for a centrifugal pump installed in a pipe system, derived from measurements taken at three different speeds. If such information is to be used care must be taken to ensure that the curves are dimensionless and the units used are consistent, e.g. if gH is expressed in $m^2 s^{-2}$ then the value for $N^2 D^2$ must also be $m^2 s^{-2}$, not, for example, in mm^2 minute^{-2}. Even over the limited range of speeds at which the particular machine was tested the curves for each coefficient, although similar, are not identical, probably due to the variation of Reynolds Number which was expected to have little effect. Fortunately the head coefficient curve, which is the most important in describing the hydraulic performance of the machine, shows the smallest variation with speed and hence can be used with considerable certainty to predict pump performance at other speeds and scales. Note how the efficiency of the pump is less than would be expected from manufacturers figures obtained from tests on a rig carefully designed to allow a pump to give of its best.

Example

A pump of 0.2 m impeller diameter and of the same shape as that whose performance is shown in Fig. 15.7 is to be run at a speed of 50 rev s^{-1} in a similar installation. Estimate the head, power, and efficiency curves for the pump.

Solution

The unit performance curves show some variation with speed, probably due to Reynolds Number effects. In the case of the 0.3 m pump,

$$Re = \left(\frac{\mu}{\rho N (0.3\,\text{m})^2}\right)^{-1},$$

while for the 0.2 m pump,

$$Re = \left(\frac{\mu}{\rho \times 50\,\frac{\text{s}}{\text{rev}}\,(0.2\,\text{m})^2}\right)^{-1}.$$

Hence the speed of the 0.3 m pump for equal Reynolds Numbers is given by

$$N = 50\frac{\text{rev}}{\text{s}} \times \frac{(0.2)^2}{(0.3)^2} = 22.2\frac{\text{rev}}{\text{s}}.$$

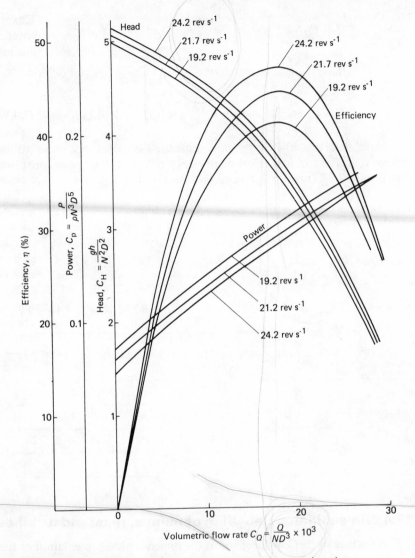

Fig. 15.7 The performance of a 0.3 m diameter centrifugal pump plotted as dimensionless coefficients.

The values of the coefficients from the 21.7 rev s^{-1} test would be most appropriate and these are shown in Table 15.2, lines 1–4.

To obtain the performance figures for the particular pump size and speed, (0.2 m, 50 rev s^{-1}), the coefficients must be multiplied by the appropriate powers of N and D in consistent units.

$$Q = C_Q ND^3 = C_Q \times 50\frac{\text{rev}}{\text{s}} \times (0.2\text{ m})^3 = 0.4\,C_Q\,\text{m}^3\,\text{s}^{-1}$$

$$H = \frac{C_H N^2 D^2}{g} = C_H \times 50^2\frac{\text{rev}^2}{\text{s}^2} \times \frac{0.2^2\,\text{m}^2}{9.81\,\text{m s}^{-2}} = 10.2\,C_H\,\text{m}$$

$$P = C_P \rho N^3 D^5 = C_P \times 10^3\frac{\text{kg}}{\text{m}^3} \times \left(50\frac{\text{rev}}{\text{s}}\right)^3 (0.2\text{ m})^5 = 40\,C_P\,\text{kW}$$

Note that the efficiency being dimensionless does not have to be converted. The required values of efficiency, flow, head, and power are given in lines 4–7 of Table 15.2, and plotted on Fig. 15.8.

Table 15.2 Conversion of performance coefficients to performance figures for a 0.2 m pump running at 50 rev s^{-1}

$\frac{Q}{ND^3} = C_Q \times 10^{-3}$	0	4.2	8.3	13.1	1.7.3	21.7	26.1	28.3
$\frac{gH}{N^2D^2} = C_H$	5.17	4.97	4.71	4.27	3.79	3.23	2.43	1.97
$\frac{P}{\rho N^3 D^5} = C_P$	0.084	0.098	0.114	0.133	0.147	0.164	0.171	0.177
$\eta(\%)$	0	21	34	42	44.5	43	37	32
$Q \times 10^{-3}/(\text{m}^3\text{s}^{-1})$	0	1.7	3.3	5.2	6.9	8.7	10.4	11.3
$H/(\text{m})$	52.7	50.7	48.0	43.5	38.6	32.9	24.8	20.1
$P/(\text{kW})$	3.4	3.9	4.6	5.3	5.9	6.6	6.8	7.1

15.4 Dimensionless labelling of pumps, fans, and turbines

There is one further consequence of the dimensionless presentation of the performance of pumps, fans and turbines which is of great convenience when selecting a machine to perform a particular duty. Although there are a number of performance criteria by which the suitability of a machine can be judged, such as the ability to produce a wide range of flow rates, or to cope with rapidly varying head, it is usually possible to define the flow and head under which a machine will be required to run for most of its life. It is then reasonable to choose a design which is working near its maximum efficiency at this duty.

Consider the dimensionless performance curves for a particular shape of machine, as in Fig. 15.9. Drawing a vertical line through the point of maximum

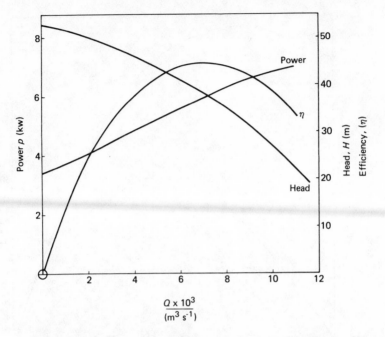

Fig. 15.8 The performance of a 0.2 m centrifugal pump at 50 rev s⁻¹.

efficiency defines particular numerical values of $\dfrac{Q}{ND^3} = \hat{C}_Q$, $\dfrac{gH}{N^2D^2} = \hat{C}_H$ and

$\dfrac{P}{\rho N^3 D^5} = \hat{C}_P$, called the *design values*. For a pump or fan $\hat{C}_Q$ and $\hat{C}_H$ define the *design point*, while for a turbine the design point is usually defined by $\hat{C}_H$ and $\hat{C}_P$. With the numerical values at the design point known for a machine it is possible to use the two equations to eliminate the size of the machine, D. So, in the case of a pump, as

$$\hat{C}_Q = \left(\frac{Q}{ND^3}\right)_{\max\eta} \quad \text{and} \quad \hat{C}_H = \left(\frac{gH}{N^2D^2}\right)_{\max\eta},$$

then
$$\frac{\hat{C}_Q{}^2}{\hat{C}_H{}^3} = \left(\frac{Q^2}{N^2D^6} \times \frac{N^6D^6}{g^3H^3}\right)_{\max\eta} = \left(\frac{N^4Q^2}{g^3H^3}\right)_{\max\eta}$$

If a number N_s is defined such that

$$N_s = \left(\frac{N^4Q^2}{g^3H^3}\right)^{1/4}_{\max\eta} = \left(N\frac{Q^{1/2}}{(gH)^{3/4}}\right)_{\max\eta} = \frac{\hat{C}_Q{}^{1/2}}{\hat{C}_H{}^{3/4}}$$

then N_s has a unique value for each design of pump regardless of its size. The value of N_s is therefore a dimensionless label for the pump design which will

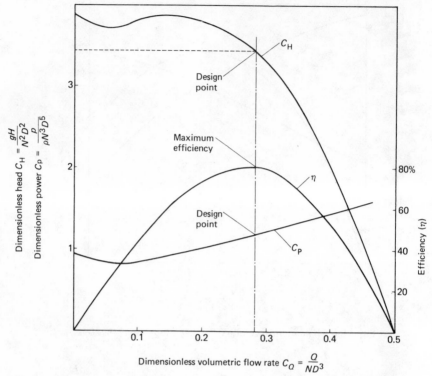

Fig. 15.9 Dimensionless performance curves for a centrifugal pump.

enable it to be compared with other designs of pump to determine the most suitable for a particular duty.

The quantitites required to evaluate N_s for a particular duty at maximum efficiency are the flow, Q, the head, H, and the speed, N, at which the pump is to be driven. As many pumps are driven by alternating current electric motors, whose speed is usually just below the frequency of the supply (50 hertz or 60 hertz) divided by an integer, there is a limited choice of speeds available to the pump operator. In the case of large pumps the speed may be constrained further by structural considerations. As most water turbines are used to generate alternating current the same restriction on speed also applies as to pumps. The dimensionless label for turbines is usually in the form

$$N_s = N \left(\frac{P}{\rho}\right)^{1/2} (gH)^{-5/4}$$ where P and H are the power and head at maximum efficiency.

The effectiveness of the dimensionless labelling of machines can be seen when the physical shapes of the rotors of machines with different values of N_s are compared and it is found that they fall naturally into a progression from centrifugal to axial flow pumps and turbines, (see Fig. 15.10).

Pelton wheel.
$N_s = 0.008$

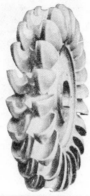

Pelton wheel.
$N_s = 0.021$

Francis turbine impeller (radial flow)
$N_s = 0.058$

Mixed flow turbine impeller
$N_s = 0.27$

Mixed flow turbine impeller
$N_s = 0.33$

Kaplan turbine impeller (axial flow)
$N_s = 0.38$

Kaplan turbine impeller (axial flow)
$N_s = 0.83$

Fig. 15.10 (a) Dimensionless specific speed and shape for turbines. Specific speed based on revolutions. (Courtesy of Escher Wyss.)

Radial flow pump impeller
$N_s = 0.08$

Radial flow pump impeller
$N_s = 0.17$

Mixed flow pump impeller
$N_s = 0.33$

Axial flow pump impeller
$N_s = 0.50$

Axial flow pump impeller
$N_s = 0.83$

Fig. 15.10 (b) Dimensionless specific speed and shape for pumps. Specific speed based on revolutions. (Courtesy of Escher Wyss.)

A more comprehensive guide to the choice of pumps is given in the 'Manual of British Water Engineering Practice' (see References page 381). There is, however, one warning which must be given. The dimensionless label is usually referred to as the *specific speed* and quoted values of N_s have a bewildering variety of units used in their calculation and are rarely dimensionless or unitless. Do not use information on specific speeds unless you are sure of its units.

Example

Calculate the dimensionless label which specifies the pump performance shown in Fig. 15.9. What diameter of impeller and driving speed should be used to deliver $0.1 \ \mathrm{m^3 \ s^{-1}}$ against a head of 8.0 m? If pumps were available with diameters in 50 mm steps, and direct drive a.c. motors with speeds of $2950 \ \mathrm{rev/n} \ \mathrm{min}$ (where n is an integer), what would be an appropriate combination of pump and motor and what would be the efficiency of the pump?

Solution

A line drawn through the point of maximum efficiency parallel to the η axis cuts the performance curve at the design point where $\hat{C}_\mathrm{H} = 3.4$ and $\hat{C}_\mathrm{Q} = 0.28$. The non-dimensional label, N_s is given by

$$N_\mathrm{s} = \frac{\hat{C}_\mathrm{Q}^{1/2}}{\hat{C}_\mathrm{H}^{3/4}} = \frac{0.28^{1/2}}{3.4^{3/4}} = 0.21 = N \frac{Q^{1/2}}{(gH)^{3/4}}$$

which corresponds with a mixed flow type of machine as in Fig. 15.10.

Hence for the pump to work at maximum efficiency under the condition specified

$$N = 0.21 \frac{(9.81 \times 8.0 \ \mathrm{m^2 \ s^{-2}})^{3/4}}{(0.1 \ \mathrm{m^3 \ s^{-1}})^{1/2}} = 17.6 \frac{\mathrm{m^{3/2} \ s^{-3/2}}}{\mathrm{m^{3/2} \ s^{-1/2}}} = 17.6 \ \mathrm{rev \ s^{-1}}$$
$$= 1056 \ \mathrm{rev \ min^{-1}}$$

At maximum efficiency $\dfrac{Q}{ND^3} = 0.28$,

so
$$D = \left(\frac{Q}{0.28 \ N}\right)^{1/3} = \left(\frac{0.1 \ \mathrm{m^3 \ s^{-1}}}{0.28 \times 17.6 \ \mathrm{s^{-1}}}\right)^{1/3} = 0.273 \ \mathrm{m}.$$

The optimum pump would have a diameter of 0.273 m and run at $1056 \ \mathrm{rev \ min^{-1}}$. The choice would therefore be between pumps of diameter 0.25 m and 0.3 m diameter with motor speeds of 2950, 1475, 983 or $738 \ \mathrm{rev \ min^{-1}}$, but neither pump would run at their maximum efficiency or give the exact head required at the flow rate specified. Hence the outlet of the pump may have to be throttled, thus wasting energy and giving a still lower installed efficiency.

Considering first the smaller, 0.25 m, pump driven at a speed of $2950 \ \mathrm{rev \ min^{-1}}$ which, at the required flow rate, would operate at a value of

$$\frac{Q}{ND^3} = \frac{0.1 \ \mathrm{m^3 \ s^{-1}}}{\dfrac{2950 \ \mathrm{rev}}{60 \ \mathrm{s}} \cdot 0.25^3 \ \mathrm{m^3}} = 0.13$$

Reference to the performance curves shows that this value of Q/ND^3 corresponds to values

$$\eta = 55\% \text{ and } \frac{gH}{N^2D^2} = 3.85,$$

and hence $H = 3.85\dfrac{N^2D^2}{g} = \dfrac{3.85 \text{ s}^2}{9.81 \text{ m}}\left(\dfrac{2950}{60}\dfrac{\text{rev}}{\text{s}}0.25 \text{ m}\right)^2 = 59.3 \text{ m}.$

Power required to drive the pump $= \dfrac{1}{\eta}\rho g Q H = \dfrac{10^3}{0.55} \times 9.81 \times 0.1 \times 59.3 \text{ W}$

$$= 106 \text{ kW}.$$

Power delivered to the water at the specified flow and head

$$= \rho g Q h$$
$$= 10^3 \times 9.81 \times 0.1 \times 8 \text{ W}$$
$$= 7.8 \text{ kW}$$

Hence overall efficiency of installation $= \dfrac{7.8}{106} \times 100 = 7.4\%$

The repetitive process of calculating the performance of each combination of pump and motor is most easily done by tabulation with the aid of a programmable calculator and the results are presented in Table 15.3. The table shows that decreasing the speed too much (below 1475 rev min^{-1} for the 0.25 m pump and below 983 rev min^{-1} for the 0.3 m pump) does not enable sufficient head to be generated. Increasing the speed too much produces an unnecessarily high head and a low installed efficiency. For the particular duty the 0.3 m diameter pump driven at 983 rev min^{-1} and requiring a power of 11.5 kW to give an efficiency of 68% would be the best choice with a 0.25 m

Table 15.3 Sizes and speeds of pumps to perform a particular duty.

Pump diameter D m	Speed N rev min^{-1}	$C_Q = \dfrac{Q}{ND^3}$ for $Q = 0.1$ m^{3}s^{-1}	Values from curves η	$\dfrac{gH}{N^2D^2}$	Pump Head H m	Pump power input kW	Overall efficiency %
	2950	0.13	55	3.85	59.3	106	7.4
0.25	1475	0.26	80	3.55	13.7	16.8	46.8
	983	0.39	56	2.1	3.6	–	–
	2950	0.08	33	3.75	83.2	247	3.2
0.3	1475	0.15	62	3.88	21.5	34	23.0
	983	0.23	78	3.71	9.1	11.5	68
	738	0.30	79	3.30	4.6	–	–

pump at 1475 rev min^{-1} consuming 16.8 kW a possible, if less efficient, second choice.

Notice how much more power is required if the pump is not closely matched to the duty.

15.5 The matching of impeller machines and pipe systems

The example in the previous section showed how the most suitable pump of a particular design can be chosen to perform a set duty, and it was assumed that any mis-match in the head would be corrected by the use of a valve as an energy dissipator. In most practical situations this approach could be unnecessarily wasteful as the duty specified is a minimum requirement. If the performance of the pump exceeded the duty the increase in flow rate might result in the pump only having to run for part of the time to deliver the water required, the pump being switched on and off by an automatic control system. To find the actual flow rate in the installation it is necessary to consider the performance of both the pump and the pipe or open channel system into which it discharges, i.e. the variation of head with flow rate in each case.

The most convenient way of finding the 'operating point' of a pump and pipe installation is to plot the head requirement of the flow on the performance curve of the pump. The head required will be the sum of the change in total head of the flow plus the energy losses in the pipe installation. The head can therefore be expressed, for a limited range of Reynolds Numbers, as

$$h = Z + KQ^2.$$

This curve cuts the pump performance curve at the operating point, as in Fig. 15.11, thus determining the actual flow rate in the installation.

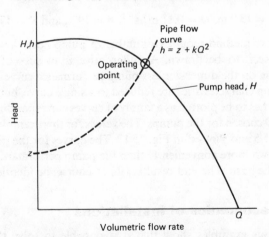

Fig. 15.11 Determining the flow in a pipe and pump installation.

Example

The pumps selected in the previous example to perform the duty of raising the head of a flow of $0.1 \text{ m}^3 \text{ s}^{-1}$ by 8 m are to be considered as an alternative to the doubling pipe in the example in Section 13.15. What would be the flow rate in the pipe and the power required by a pump:
a) of 0.3 m diameter running at 983 rev min^{-1},
b) of 0.25 m diameter running at 1475 rev min^{-1}?

Solution

It is first necessary to find the equation which describes the characteristic of the pipe installation in terms of the head which is to be provided by the pump. In the arrangement proposed it is only necessary for the pump to provide the extra head, h, required by the increase in flow from $0.08 \text{ m}^3 \text{ s}^{-1}$, which was provided by the existing head of 14.2 m, to $0.1 \text{ m}^3 \text{ s}^{-1}$.

So $$h = klQ^2 - Z = 2{,}221 \text{ s}^2 \text{ m}^{-5} \, Q^2 - 14.2 \text{ m}, \quad \text{(see page 259)}.$$

The dimensionless pump head-flow curve can be re-plotted from the figures in Table 15.4 as two curves, one for each pump, (Fig. 15.12). From Table 15.4 the pipe head-flow curve is then plotted on the same axes and the intersections with the two pump characteristics give the two operating points.

The 0.3 m diameter pump will operate at a head of 9.1 m, a flow of $0.102 \text{ m}^3 \text{ s}^{-1}$ and the corresponding efficiency of 78%.

$$\text{Power required} = \frac{1}{\eta} g Q H = \frac{1}{0.78} \times 10^3 \times 9.81 \times 0.102 \times 9.1 \text{ W} = 11.7 \text{ kW}.$$

Similarly for the 0.25 m diameter pump the values will be

$$H = 13.0 \text{ m}, \, Q = 0.11 \text{ m}^3 \text{ s}^{-1}, \, \eta = 79\% \text{ and } P = 17.8 \text{ kW}.$$

The above method requires a number of pump head-flow curves, which may intersect, to be drawn. An alternative is to present the pipe-flow information on the dimensionless pump performance curves. By doing this all the pump characteristics are reduced to a single curve, but the pipe head-flow curve has to be plotted as a family of curves corresponding to the values of N and D chosen for the pumps. The values for this example are calculated in Table 15.5 and plotted in Fig. 15.13. The curves for the pipe installation can be drawn more conveniently than the pump performance curves in the previous diagram. The end results will, of course, be identical.

15.6 The economics of installations

The previous examples show that it is possible to solve the problem of increasing a flow rate in a water main by at least two ways. The first, requiring

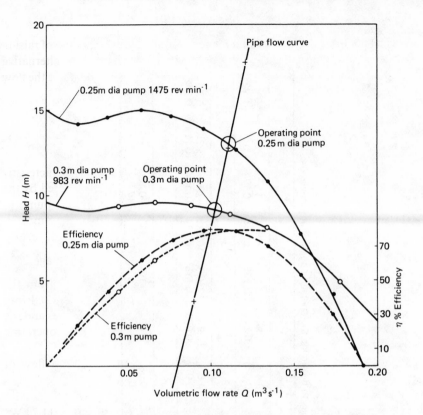

Fig. 15.12 Pipe flow resistance and the operating points of two geometrically similar pumps.

an extra pipe, would involve the laying of 514 m of 0.24 m diameter pipe with the costs of the excavation and reinstatement to be added to the cost of the materials. The second alternative is to install a pump whose costs will include not only the building and pipework but also the electrical power consumed when the flow is required. If the flow were not required continuously a third alternative might be to build a storage reservoir at the outlet end of the pipe if space were available.

The decision on which solution is appropriate will depend on the first cost of the installation, the expected life, maintenance costs, running costs and interest rates. The full implication can only be realized after an economic analysis, but some simple calculations can often indicate if one alternative is obviously superior.

Table 15.4 Pump performance and pipe flow values to determine the operating point of an installation.

$D = 0.3$ m $N = 983$ rev min^{-1}

H/m	9.6	9.1	9.4	9.6	9.4	8.9	8.1	6.9	4.9	2.7	0
Q/m^{3}s^{-1}	0	0.022	0.044	0.066	0.088	0.111	0.133	0.155	0.177	0.20	0.22

$D = 0.25$ m $N = 1475$ rev min^{-1}

H/m	15.0	14.2	14.6	15.0	14.6	13.9	12.7	10.8	7.7	4.2	0
Q/m^{3}s^{-1}	0	0.019	0.038	0.058	0.077	0.096	0.115	0.134	0.154	0.173	0.192

Pipe flow $h = 2.221Q^2 - 14.21$ m

Q/m^{3}s^{-1}	0	0.02	0.04	0.06	0.08	0.09	0.10	0.11	0.12
h/m	−14	−13.3	−10.6	−6.2	0	3.8	8.0	12.7	17.8

Table 15.5 Dimensionless pump performance and pipe flow for two pumps.

Dimensionless pump performance

$C_H = \dfrac{gH}{N^2D^2}$	3.9	3.7	3.8	3.9	3.8	3.6	3.3	2.8	2.0	1.1	0
$C_Q = \dfrac{Q}{ND^3}$	0	0.05	0.1	0.15	0.2	0.25	0.3	0.35	0.4	0.45	0.5
$\eta\%$	0	24	44	62	74	79	79	70	53	30	0

Pipe flow expressed in dimensionless form
$D = 0.3$ m $N = 983$ rev min^{-1}

Q/ND^3	0	0.05	0.09	0.14	0.18	0.20	0.23	0.25
gh/N^2D^2	-5.8	-5.4	-4.3	-2.5	0	1.54	3.25	5.14

$D = 0.25$ m $N = 1475$ rev min^{-1}

Q/ND^3	0	0.05	0.10	0.16	0.21	0.23	0.26	0.29	0.31
gh/N^2D^2	-3.7	-3.5	-2.8	-1.6	0.0	0.98	2.07	3.29	4.62

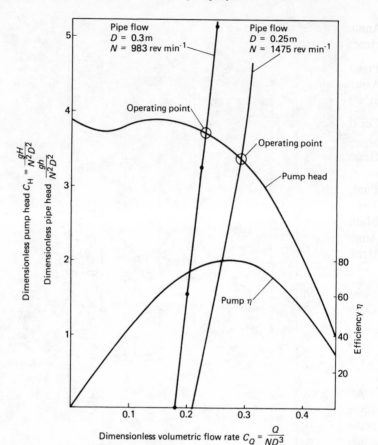

Fig. 15.13 Determination of pump operating points using dimensionless curves.

Example

Compare the costs of the doubling pipe and the installation of pumps if the cost of pipe is £20 per metre laid and its life is expected to be 50 years. The cost of pumps may be taken to be £200 for each kilowatt installed, with a 20 year life, a maintenance cost of 10% per annum and power purchased at 4 p per kW hour. If the supply is required for 12 hours a day, 5 days a week and 48 weeks a year, which will be the cheaper installation if the interest rate is
  a) 10%,
  b) 20%?
Neglect the effects of inflation.

Solution

Pipe costs

The capital cost of the pipe = £20 × 514 = £10 280
The annual replacement charge will be 2% of this   = £205.6

Annual interest charges at n% $\qquad$ = £102.8n
Hence approximate total cost of pipe each year $\qquad$ = £(205.6 + 102.8n)

Pump costs
Assume that the 0.3 m diameter pump is installed and throttled to deliver 0.1 $m^3 s^{-1}$.

For the pump the running time each year will be $12 \times 5 \times 48 = 2880$ hours

Hence power costs $= 2880 \times 11.7 \times \dfrac{4}{100} h \times kW \times \dfrac{£}{kWh} = £1348$

Pump capital cost $= £200 \times 11.7 = £2340$
The annual replacement charge will be 5% of cost $\qquad$ = £117
Maintenance cost at 10% of installed cost $\qquad$ = £234
Annual interest charges at n% $\qquad$ = £ 23.4n
Hence approximate total cost of pump each year
$\quad = £(1348 + 117 + 234 + 23.4n)$ $\qquad$ = £(1699 + 23.4n)

Comparative costs
Costs for the two rates of interest may now be tabulated

Rate of interest	10%	20%
Cost of pipe	£1234	£2262
Cost of pump	£1933	£2167

At the lower interest rate the pipe would appear to have a distinct advantage, while at the higher interest rate the pump would appear to be a better investment. However a small rise in power prices could soon offset this advantage, but if capital were difficult to raise, regardless of the interest rate, the pump may be a more readily obtainable solution as its initial cost is less than one quarter of that of the pipe.

15.7 Installation of impeller machines

The designer of a fan, pump, or turbine will put great effort into producing a machine with a high efficiency, although as the previous sections have shown it is still the responsibility of the engineer to choose a machine appropriate to the duty it is to perform, or the cost of the power will be more than it need have been. The task of the engineer does not end with the selection of the most economical machine as the detailed design of the hydraulics of the installation is also of importance. The machine will have been designed on the assumption that the flow at entry is both steady and uniformly distributed, and the development of the machine will have been done on test rigs which produce this type of flow. In practice the machine may have to be installed in a pipe or duct arrangement whose shape is dictated by considerations other than the flow requirements. It will almost certainly be necessary to provide valves so that the machine can be isolated from the installation for repair or replacement, other valves may be required to connect machines in parallel or to alternative supply or delivery ducts, and filters or strainers may have to be provided to protect the machine. The size and shape of the supply and delivery pipes or ducts may differ from the inlet and outlet size and shape of the machine and it will then be necessary to design transition sections which give uniform entry conditions and do not waste energy.

Centrifugal flow and mixed flow machines can tolerate considerable non-uniformity in entry flows with only a small drop in efficiency. Flow non-uniformity at the entrance to an axial flow machine causes changes in blade incidence which can reduce the efficiency of the machine seriously and also can tend to cause vibration of the blades which may then generate noise or even structural failure. In general it pays to take considerable pains over the design of inlet arrangements for impeller machines even to the extent of performing model tests for complicated arrangements such as the inlet manifold for a number of water turbines, (see Fig. 15.14).

Fig. 15.14 A model of an inlet manifold for a hydro-power station containing five turbines.

As an example, the simple pump installation shown in Fig. 15.15 would provide the optimum flow conditions. The pump inlet is submerged so the pump will always be full of liquid and not require *priming*, i.e. filling with liquid

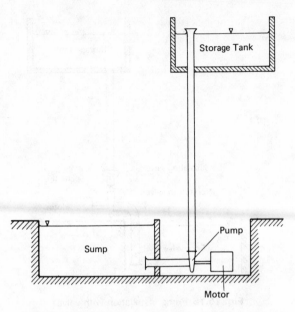

Fig. 15.15 A simple pump installation without valves.

so that flow can be sucked into the pump when it is started. With an inlet well below the free surface of the liquid in a large sump the flow velocities in the sump will be small and there will be little chance of vortices forming which cause air to be sucked into the pump. There are no valves to cause non uniformity of flow before the pump or head loss after the pump. The disadvantages of the arrangement are that the pump cannot be emptied, or removed without draining the sump. If the sump is below ground level extra excavation is required to accommodate the pump in a dry well which itself may require a small pump to remove leakage. The outlet requires a longer length of pipe than is necessary to reach the tank and also makes the pump deliver liquid to a higher level than required when the tank is not full.

A more usual arrangement is shown in Fig. 15.16, but this requires the addition of a non-return valve in the delivery pipe, and possibly a foot valve and some means of priming the pump. Alternative solutions are to have the pump and electric motor submerged in the sump which increases the cost of the motor, or to use a vertical axis pump with the pump submerged and the motor above sump water level.

15.8 Cavitation in impeller machines

The flow through a pump only has energy added to it in the impeller. Any increase in kinetic energy as the flow enters the inlet pipe or any increase of

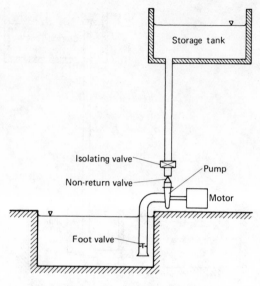

Fig. 15.16 Pump installation with valves.

potential energy as the flow rises and any energy dissipated by shear stresses before the flow reaches the impeller can thus only be provided by flow work expended by the fluid itself, i.e. by a drop in pressure as the liquid flows towards the impeller. If, at a particular flow rate, the pressure anywhere in the liquid falls to the vapour pressure of the liquid it will form vapour cavities in that locality. As the flow moves through the rotor the pressure of the liquid will rise above the vapour pressure and the cavities will collapse, the process being known as cavitation (page 7). The formation of cavities will reduce the volume of liquid flowing through the machine, change the streamline pattern from the design condition, and thus reduce the efficiency. The collapse of the cavities will be accompanied by noise and may even result in mechanical damage to the machine.

To avoid cavitation it is important to design machines to have small entry velocities, to have short entry pipes with the minimum of energy losses, and to install the machine with the impeller well below the free liquid surface if possible. It is not always possible to prevent a machine cavitating, but it is possible to reduce any noise and mechanical damage by introducing a small quantity (1% or 2% by volume) of air into the flow to absorb some of the energy of the collapsing cavities.

Example

A pump situated 4 m above sump water level begins to cavitate when delivering $0.56 \, \text{m}^3 \, \text{s}^{-1}$ against a total head of 5 m with a speed of

1200 rev min^{-1} and power input of 30 kW. What will be the maximum speed at which the pump may be run before it cavitates when it is installed 2 m below the water level if it is to work at the same efficiency? What will then be the flow rate and head produced by the pump?

Assume that the atmospheric pressure is equivalent to a head of 10 m water and the absolute water vapour pressure at the temperature of the sump water is equivalent to 1.3 m head of water.

Solution

The absolute pressure head, h_1, at inlet to the pump can be found by applying the Energy equation between the pump inlet and water surface,

$$h_a = h_s + h_1 + \frac{u^2}{2g} + k_1 \frac{u^2}{2g}$$

where h_a = atmospheric pressure head of water and $k_1 u^2/2g$ represents the head lost in the suction pipe, (Fig. 15.17).

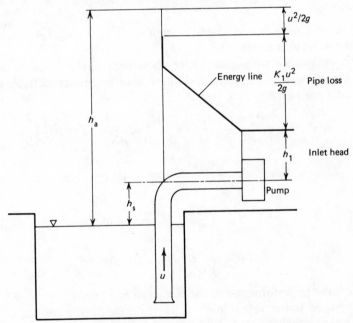

Fig. 15.17 Energy line diagram for the suction pipe of a pump.

Within the pump there will be a further fall in the pressure head, $k_2 Q_1^2$, before the flow has work done on it in the impeller. Hence, if h_2 is the minimum pressure head in the flow,

$$h_2 = h_1 - k_2 Q_1^2 = h_a - h_s - (k_1 + 1)\frac{u^2}{2g} - k_2 Q_1^2,$$

370 Work and power in fluid flows—pumps, fans, and turbines

which may be simplified to,

$$h_2 = h_a - h_s - k_3 Q_1{}^2,$$

where k_3 represents all the pressure changes which depend on the kinetic energy of the flow.

As the vapour pressure of water is 1.3 m absolute, and atmospheric pressure head is 10 m then, if the pump is just starting to cavitate,

$$h_2 = 1.3\,\text{m} = 10\,\text{m} - 4\,\text{m} - k_3 \times 0.56^2\,\text{m}^6\,\text{s}^{-2}$$

so

$$k_3 = \frac{4.7}{0.56^2} = 15\,\text{m}^{-5}\,\text{s}^2.$$

In the new installation, if the piping arrangements are assumed to be similar, the flow rate is Q_2, and h_s is now $+2$ m. When the pump starts to cavitate

$$h_2 = 1.3\,\text{m} = 10\,\text{m} + 2\,\text{m} - 15\,Q^2{}_2\,\text{m}^{-5}\,\text{s}^2$$

whence

$$Q_2 = 0.845\,\text{m}^3\,\text{s}^{-1}.$$

As the new pipes will probably be shorter as h_s is less, this should be a conservative estimate for Q.

To work at the same efficiency the operating point of the pump must be the same in both installations, i.e. the flow, head, and power coefficients must be the same in both cases. Hence

$$\frac{Q_1}{N_1 D^3} = \frac{Q_2}{N_2 D^3}, \frac{gH_1}{N_1{}^2 D^2} = \frac{gH_2}{N^2{}_2 D^2}, \frac{P_1}{\rho N_1{}^3 D^5} = \frac{P_2}{\rho N_2{}^3 D^5},$$

so

$$N_2 = \frac{Q_2}{Q_1} N_1 = \frac{0.845}{0.56} \times 1200 = 1810\,\text{rev min}^{-1},$$

smilarly

$$H_2 = \left(\frac{N_2}{N_1}\right)^2 H_1 = 11.4\,\text{m},$$

$$P_2 = \left(\frac{N_2}{N_1}\right)^3 P_1 = 103\,\text{kW}.$$

Note how the performance of the pump can be increased by siting it lower relative to the water surface level or, for the same duty, a smaller, cheaper, pump run at a higher speed could be used.

The avoidance of cavitation in water turbines follows similar principles to that in pumps although it is at the impeller outlet of the turbine where cavitation is most likely to occur. The flow leaving a turbine has considerable kinetic energy and if this flow is decelerated in a *draft tube* the pressure at the impeller outlet can be decreased and hence the power output of the turbine increased. The pressure at the outlet of the impeller must not be allowed to fall below the water vapour pressure but it may be possible to place the turbine at a lower level (as in the case of the pump in the previous example) and thus prevent cavitation. However this may increase the difficulties of maintenance and the cost of the excavation and structure. Hence the final level of the turbine is fixed not only by its design but also by the relative cost of capital and the income which could be produced from the sale of the extra power generated by lowering the turbine and so increasing its output.

Example
A one twentieth scale model propeller turbine develops 4.87 kW when running at 1466 rev min^{-1} under a head of 5.25 m. The model starts to cavitate when the pressure head at the turbine exit falls to 8.6 m below the atmospheric pressure head of 10.4 m.

The full size prototype is to deliver 16.4 MW at a head of 22 m at an expected efficiency of 90%. What is the specific speed of the turbines, the speed, and the flow through the prototype?

The prototype turbine has an exit diameter of 3 m and the draft tube is expected to recover 75% of the kinetic energy rejected by the turbine. What is the maximum height the prototype turbine exit can be mounted above the tailrace water level if it is not to cavitate at full load when the atmospheric pressure is 900 mb?

Assume that for the model the water vapour pressure at 20°C is equivalent to 1.3 m of water, while the temperature of the water in the prototype will be 5°C.

Solution
At the point of maximum efficiency the specific speed of model and prototype will be very nearly the same, hence

$$N_s = \frac{N}{(gH)^{5/4}} \sqrt{\frac{p}{\rho}} = \frac{1466}{60} \frac{\text{rev}}{\text{s}} (9.81 \times 5.25 \text{ m}^2 \text{ s}^{-2})^{-5/4}$$

$$\times \left(\frac{4.78 \times 10^3 \text{ kgm}^2.\text{s}^{-3}}{10^3 \text{ kgm}^{-3}}\right)^{1/2}$$

$$= 0.387.$$

For the prototype, knowing the specific speed allows the speed of rotation to be calculated from

$$N = 0.387 (gH)^{5/4} \sqrt{\frac{\rho}{p}} = 0.387 (9.81 \times 22)^{5/4} \sqrt{\frac{10^3}{16.4 \times 10^6}} \frac{\text{rev}}{\text{s}}$$

$$= 2.5 \text{ rev s}^{-1} = 150 \text{ rev min}^{-1}.$$

If the efficiency of the turbine is expected to be 90% and Q is the flow rate then

$$p = \eta \rho g Q H$$

$$Q = \frac{p}{\eta \rho g H} = \frac{16.4 \times 10^6 \, \text{N} \, \text{m} \, \text{s}^{-1}}{0.9 \times 9.81 \times 10^3 \times 22 \, \text{N} \, \text{m}^{-3} \, \text{m}} = 84.4 \, \text{m}^3 \, \text{s}^{-1}.$$

Cavitation in the model will occur when the pressure head at some point in it drops to an absolute value of 1.3 m and this occurs when the absolute pressure head at the turbine exit is h_e, where

$$h_e = 10.4 - 8.6 = 1.8 \, \text{m}.$$

The minimum pressure head at some point in the model turbine under the test conditions is therefore $1.8 - 1.3 = 0.5$ m below that at the turbine exit.

The pressure head in the prototype turbine will be a quantity h_p corresponding to the head of 0.5 m in the model below the pressure head at its exit. If the prototype is operating under similar flow conditions to the model then the heads in prototype and model will scale linearly, i.e.

$$\frac{h_p}{22 \, \text{m}} = \frac{0.5 \, \text{m}}{5.25 \, \text{m}},$$

so

$$h_p = 2.1 \, \text{m}.$$

When the flow in the prototype is just cavitating, the absolute pressure head at turbine exit must be $2.1 \, \text{m} + h_v$, where h_v is the absolute vapour pressure.

Now as the vapour pressure of a liquid is proportional to its absolute temperature, for the prototype flow at 5°C,

$$h_v = 1.3 \times \frac{278}{293} = 1.23 \, \text{m}.$$

Hence the absolute pressure head at the turbine exit must not fall below $2.1 + 1.23 = 3.33$ m. But as the atmospheric pressure is 900 mb,

$$\frac{P_a}{\rho g} = 900 \times \frac{10^5 \, \text{N}}{1000 \, \text{m}^2} \times \frac{1}{9.81 \times 10^3 \, \text{Nm}^{-3}} = 9.17 \, \text{m}.$$

As the turbine has an exit diameter of 3 m then the exit velocity of the flow, u, is given by

$$u_1 = \frac{84.4 \times 4}{9\pi} = 11.94 \, \text{m} \, \text{s}^{-1},$$

and

$$\frac{u_1^{\,2}}{2g} = 7.27 \, \text{m}.$$

If z_1 is the elevation of the turbine exit the total head, H, of the flow there relative to the water level in the tailrace when the prototype is just about cavitate is,

$$H_1 = 3.33\,\text{m} - \frac{p_a}{\rho g} + \frac{{u_1}^2}{2g} + Z_1 = (3.33 - 9.17 + 7.17)\text{m} + Z_1 = 1.43\,\text{m} + Z_1.$$

But this must equal the head loss in the draft tube,

$$1.43\,\text{m} + Z_1 = 0.25\frac{{u_1}^2}{2g} = 0.25 \times 7.27\,\text{m} = 1.82\,\text{m}$$

Whence $Z_1 = 1.82 - 1.43 = 0.39\,\text{m}$.

The exit of the turbine should not be placed more than 0.39 m above the lowest tailrace water level if the flow is not to cavitate under the conditions specified.

15.9 Non-steady flow in machine installations

In hydro-electric installations the turbines are often supplied with water through a long pipeline or tunnel. High and perhaps destructive pressures can be developed if there is a change in the discharge, which can occur quite suddenly if the electrical load is taken off the generators: the turbines start to race, and the governors react, closing the guide vane openings, reducing the flow as quickly as possible without causing water-hammer (Section 13.12). The consequent deceleration of the water in a tunnel or pipeline several km long might prove disastrous if there was not a *surge tank*, a vertical shaft of steel or tunnelled in rock, placed near the turbines (Fig. 15.18). The water rises in the shaft to provide the hydrostatic pressure which decelerates the fluid in the tunnel. The short length of pipe or penstock from the surge tank to the turbine is not protected from high pressures, and must therefore be strongly built. When the water comes finally to rest the water level in the tank is, of course, the same as that in the reservoir.

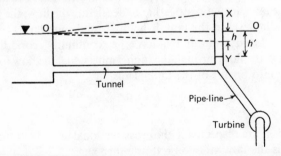

Fig. 15.18 A *surge tank* on a hydro-electric scheme. If the flow to the turbine is suddenly stopped the water can rise in the surge tank to X, thus limiting the pressure on the tunnel. When the turbine starts the level falls to Y, to give an accelerating force on the water in the tunnel due to the pressure head *h'*.

The surge tank also holds a supply of water to operate the turbine when starting, when an accelerating pressure must be applied to the water in the tunnel. The water surface in the surge tank falls below the no-flow level by an amount h' in order to provide this accelerating force on the upstream water. As the speed in the tunnel approaches the steady flow value, the surge tank water-level rises until it is below reservoir level by only the friction head h in the tunnel. The diameter of the surge tank, and the level of the intersection with the tunnel must be chosen so that there is sufficient water in the tank to drive the turbine during the time when the water in the tunnel is accelerating. Surge tanks are often large structures, sometimes 15 m or more in diameter, but it is economical to build them rather than to make the tunnel stronger.

Although after a change in turbine load the water level in the surge tank eventually comes to a new equilibrium level, it does so only after a number of oscillations about this mean. The reservoir, tunnel and surge tank act like a U-tube in which the water level has been disturbed. If the electrical load also has an oscillation which has the same period as that of the hydraulic system, resonance will occur and the amplitude of the oscillations grow until damage is caused. For this reason, and also to determine if a surge tank is large enough, it is necessary to estimate the oscillations by calculation at the design stage, and to determine the damping of the system.

At one instant of time when the surge tank level is z below the reservoir, and the water speed in the tunnel is u, the head h due to the shear stresses may be estimated by the Darcy law, and written $h = ku^2$. At equilibrium, then $z = h = ku^2$, and there would be no acceleration of the water in the tunnel. But if the system is not in equilibrium, the difference between z and ku^2 is a head tending to accelerate or decelerate the flow, according to the slow closure rule of Section 13.13 so that

$$ku^2 \pm z = -\frac{l}{g}\frac{du}{dt}$$

In this equation the positive sign applies when the friction head and the rise of the level in the tank are acting together to decelerate the flow; the negative sign applies when the level in the tank is below equilibrium level.

As well as this dynamical relation, the 'continuity' condition must be satisfied; that is, the flow through the tunnel must equal the flow to the turbines plus the flow into or out of the tank. Thus,

$$uA_t = A_s\frac{dz}{dt} + Q$$

where A_t and A_s are respectively the cross-sectional areas of tunnel and surge tank, and Q is the flow still going to the turbine.

The two differential equations must be solved simultaneously to find how z varies with t. An approximate 'step by step' method is usually necessary and one of the many possible computations is shown in the following example.

Example

In a small hydro-electric scheme the proposed tunnel is 1.2 m dia and has $f = 0.01$. At 150 m along the tunnel from the reservoir there is a simple, open surge tank 3.6 m dia. The steady full load flow to the turbines is $2.3\ m^3\ s^{-1}$.

Show how to estimate the maximum rise of water level in the surge tank after the full flow has been suddenly rejected by the turbines.

Solution

Full load tunnel speed $= v_0 = 2.3/\frac{1}{4}\pi \times 1.2^2 = 2.04\ m\ s^{-1}$

Full load head loss in tunnel $= kv_0{}^2 = \dfrac{4 \times 0.01 \times 150 \times 2.04^2}{1.2 \times 2 \times 9.81}$

$$= 1.06 \text{ m below still water level.}$$

After full load rejection $Q = 0$, so for a small but finite change Δz, the continuity condition gives

$$z = \frac{A_t}{A_s}\, v\, \Delta t = \frac{v}{9}\, \Delta t$$

where Δt is the small but finite time during which Δz takes place, and v is a mean tunnel velocity during this time.

The dynamic conditions are rearranged to give

$$\Delta v = -\frac{g}{l}\,(z \pm kv^2)\Delta t$$

where Δv is the small but finite change of tunnel speed during Δt, and z is the mean height of water in the surge tank during this interval.

These equations must be solved for successive intervals each of Δt. The choice of a numerical value for Δt is a matter of experience and of trial; a value of $\Delta t = 5$ seconds is suitable in this case, though for larger installations a longer time is appropriate.

Thus $\qquad \Delta z = \dfrac{5}{9}v \quad$ and $\quad \Delta v = -0.328(z \pm 0.254v^2)$

(a) Period 0 to 5 seconds after load rejection: *first approximation*: put v as the velocity at the beginning of the interval but z as the water level at the end of the interval.

Then $\qquad \Delta z_1 = 5 \times 2.04/9 = 1.13\ m$

so $\qquad z = -1.06 + 1.13 = +0.07\ m$ at end of interval

and $\qquad \Delta v_1 = -0.328(0.07 + 0.254 \times 2.04^2) = -0.37\ m\ s^{-1}$

or $\qquad v = 2.04 - 0.37 = 1.67\ m\ s^{-1}$ at end of period.

second approximation: mean $v = \frac{1}{2}(1.67 + 2.04) = 1.85\ m\ s^{-1}$

then $\qquad \Delta z_2 = \frac{5}{9} \times 1.85 = 1.03\ m$

so $\qquad z = -1.06 + 1.03 = -0.03\,\text{m}$

and mean $\qquad z = -1.06 + \dfrac{1.03}{2} = -0.55\,\text{m}$

and $\qquad \Delta v_2 = -0.328(-0.55 + 0.254 \times 1.85^2) = -0.11\,\text{m s}^{-1}$

or $\qquad v = 2.04 - 0.11 = 1.93\,\text{m s}^{-1}$

third approximation: same method as second approximation
using mean $\qquad v = \frac{1}{2}(2.04 + 1.93) = 1.99\,\text{m s}^{-1}$
gives $\qquad z = +0.04\,\text{m}$ and $v = 1.88\,\text{m s}^{-1}$.

fourth approximation: same method as second and third,
gives $\qquad z = +0.03\,\text{m}$ and $v = 1.89\,\text{m s}^{-1}$.

(*b*) Period 5 to 10 seconds after load rejection:*first approximation*: same as first approximation for the first time interval
gives $\qquad z = +1.08\,\text{m}$ and $v = 1.24\,\text{m s}^{-1}$ at end of period.

Second and third approximations as before, using mean values of z and v finally give

$$z = +0.98\,\text{m} \quad \text{and} \quad v = 1.48\,\text{m s}^{-1}$$

(*c*) Successive 5 second periods are similarly computed and a curve drawn of z against time. The maximum height is about 1.85 m above still water level at about 22 seconds after load rejection.

If only the first approximation method is applied to the same Δt periods, without the refinements of the later approximations, a maximum is found of 1.7 m at about 21 seconds after rejection. The close agreement of this result with that of the more exact approximations has been observed in many cases of this sort. The later history of water level in the surge tank is sketched (not to scale) in the figure below. With $f = 0.01$, the system is unusually heavily damped.

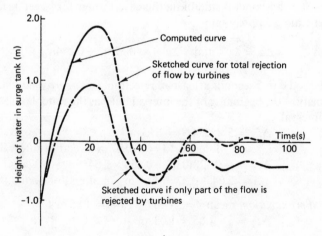

While few pump installations handle flows and heads of the magnitude of those associated with hydro-power schemes the problems of unsteady flow are similar if smaller in scale. The closing of non-return valves can often generate water hammer (see Section 13.12) while the starting and stopping of pumps causes pressure surges in the delivery pipe line. It is rarely convenient to build an open top surge tank in a delivery main and this duty is usually performed by an air reservoir connected to the outlet of the pump. Water can flow into the reservoir and compress the air when the pump is started and out of the reservoir reducing the air pressure as the pump stops. Water mains are not usually designed to withstand the high pressure associated with hydro-power penstocks and therefore have greater risk of collapse as the pressure falls when the pump is shut off. To safeguard against this some mains are fitted with relief valves which allow air to enter the pipe if the pressure in it falls much below the atmospheric pressure.

15.10 Conclusion

The design of impeller machines is a well developed area of engineering, the high efficiency of present designs being due to experience based on complex three dimensional flow theory. However, for many practical purposes, the user of an impeller machine may consider it as a 'black box' whose performance can be represented by dimensionless performance curves. This enables the suitability of the many designs of machine available to be assessed for a particular duty. At this stage it is often important to consider economic factors, particularly the relationship between capital and running costs, not only at the time the project will be completed, but also in the future.

When a choice of machine has been made it is necessary to design the installation into which it is to be built with great care so that the hydraulic performance achieves the values which were used in the economic analysis. Care must also be taken in the design to ensure that the machine is safeguarded against the effects of both slow and rapid changes of flow, power failure, cavitation and even the ingress of foreign matter from fine silt to tree trunks. Only when the machine has been shown to provide the hydraulic and economic performance required will the designers efforts have been justified.

Problems

1. In a centrifugal pump test the discharge pressure gauge reads $700 \, \text{kN} \, \text{m}^{-2}$ and the suction gauge $35 \, \text{kN} \, \text{m}^{-2}$. Both gauges read above atmospheric pressure, and their centres are at the same level. The diameters of the suction and delivery pipes are 80 mm and 50 mm respectively. If oil, sp. gr. 0.85, is being pumped at $0.5 \, \text{m}^3 \, \text{min}^{-1}$, what is the power supplied to the pump, assuming an efficiency of 75 per cent?   *Ans.* 8.15 kW.

2. Sea water, sp. gr. 1.03, is to be circulated through condensers by a propeller pump 1.2 m diameter. It is found that a scale model of the pump 0.25 m diameter gives its best efficiency when pumping $0.097 \, \text{m}^3 \, \text{s}^{-1}$ of fresh water against a head of 4.20 metres when running at $2060 \, \text{rev min}^{-1}$. What should be the speed of the full size pump to deliver $1.55 \, \text{m}^3 \, \text{s}^{-1}$ and what pressure difference will it generate?   *Ans.* $295 \, \text{rev min}^{-1}$: $20 \, \text{kN m}^{-2}$.

3. It is desired to pump $2.7 \, \text{m}^3$ of water per minute against a head, including friction, of 15 m. The only available pump, when tested at $600 \, \text{rev min}^{-1}$, gave the following results—

Pressure head produced	15	16.2	16.2	14.4	12.6	10.2	7.5	4.5	m
Water flow	0	0.45	0.9	1.35	1.58	1.8	2.03	2.25	$\text{m}^3 \text{min}^{-1}$
Efficiency	—	30	61	81	85	80	67	47	%

At what speed should it be driven to do the job, and what input power will be required?   *Ans.* $823 \, \text{rev min}^{-1}$: 70 per cent efficiency: 9.48 kW.

4. On a contractor's remote site there are several similar pumps available, all diesel driven and with the following characteristics at $1200 \, \text{rev min}^{-1}$.

Head	(m)	30	33	34	34	32	28	19	7
Discharge	$(\text{m}^3 \, \text{s}^{-1})$	0	0.1	0.15	0.2	0.25	0.3	0.35	0.4
Efficiency	(%)	0	26	38	48	57	60	59	32

The diesels cannot be effectively run at speeds more than 20 % different from $1200 \, \text{rev min}^{-1}$. How could these machines be used to deliver $0.5 \, \text{m}^3 \, \text{s}^{-1}$ against a head, including pipe friction, of 50 m? If diesels use $0.25 \, \text{kg/kW h}$, what is the likely fuel consumption?

Comment on the advantages and disadvantages of separate delivery pipes against one large, common delivery pipe.

5. A hydraulic turbine develops 9000 kW under a head of 10 m at a speed of $90 \, \text{rev min}^{-1}$ and gives an efficiency of 92.7 per cent. Calculate the water consumption and the specific speed. If a model 1/10 full size is constructed to operate under a head of 8 m, what must be its speed, power and water consumption to run under the similar conditions to the prototype? How would the model efficiency compare with the prototype?   *Ans.* $99.1 \, \text{m}^3 \, \text{s}^{-1}$: $N_s = 0.46$. $805 \, \text{rev min}^{-1}$: 64·5 kW: $0.885 \, \text{m}^3 \, \text{s}^{-1}$.

6. A fan has the characteristic $H-Q$ curve of the figure. It supplies air to a long length of ducting for which the friction law is

$$H \, (\text{cm of water gauge}) = 0.01 \, Q^2 ((\text{m}^3 \, \text{s}^{-1})^2).$$

What will be the air discharge at a fan speed, N, of $900 \, \text{rev min}^{-1}$? If the air discharge must be increased by 50 % summarize all the advantages and disadvantages of

(*a*) introducing a second fan in parallel with the first;
(*b*) driving the existing fan with a different motor.   *Ans.* 38·3 m³ s⁻¹.

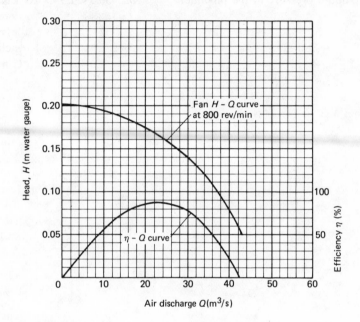

Air discharge Q(m³/s)

7. A centrifugal pump with a 20 cm suction pipe was tested at sea-level with its axis 1.2 m *below* water-level in the suction sump. It just commenced to cavitate at 2440 rev min⁻¹ when working at peak efficiency, and it then was giving 0.12 m³ s⁻¹ with a *total* head of 10.2 m.

The pump is eventually to be used on a plateau at 4000 m above sea-level, pumping from a reservoir with the water-level 3·5 m below its axis. At what speed would cavitation set in, and what will be the total head and discharge if it is to be used at maximum efficiency? Ignore suction pipe head loss.

Density of air at 2000 m is 0.93 kg m⁻³
Water-vapour pressure is 14 kN m⁻²
At sea-level atmospheric pressure is 101 kN m⁻²

Ans. 985 rev min⁻¹: 1·67 m: 0.049 m³ s⁻¹.

8. Explain the significance of Q7, Chapter 5 in the design of pumps or turbines.
9. In a Pelton wheel installation, a pressure tunnel 9.5 km long and cross-sectional area 7.5 m² conducts water from a reservoir to a surge tank, from which a steel pipeline 700 m long and 2 m diameter leads to the turbines. The nozzles are 300 m below, and the surge tank top 15 m above the reservoir

level. If the machines are giving 10 000 kW at 80 per cent overall efficiency (includes pipe head losses), what is the minimum time in which the valves may be closed so that water does not spill from the surge tank? What is the maximum pressure in the pipeline?   *Ans.* 36.5 s: $1.94 \times 10^3 \, \text{kN} \, \text{m}^{-2}$.

Bibliography

Introductory Books

Water in the Service of Man. H. R. Vallentine. Penguin Books 1967.
　　An easy to read and informative introduction to the subject.
Man and Water. N. A. F. Smith. Peter Davis, 1976.
　　An historical introduction to the efforts made by man to control water resources.
History of Hydraulics. Hunter Rouse and S. Ince. Dover 1957.
　　A scholarly history of the development of the theory of hydraulics.
Shape and Flow. A. H. Shapiro. Heinemann. 1964.
　　Many excellent photographs of flow patterns and a simple introduction to aspects of hydraulics.

Fluid mechanics

Advanced Fluid Mechanics. A. J. Raudkivi and R. A. Callander. Arnold 1975.
　　An extension of many of the topics covered in the present text on a more mathematical basis.
Fluid Dynamics. G. K. Batchelor. Cambridge University Press 1967.
　　A wide survey of many aspects of fluid mechanics.
Hydrodynamics. H. Lamb. Cambridge University Press 1962.
　　The mathematical treatment, mainly of ideal fluid flows. The classical reference book on this subject.
Boundary Layer Theory. H. Schlichting. McGraw-Hill 1979.
　　A comprehensive introduction to advanced boundary layer theory.
Turbulence. P. Bradshaw. Springer Verlag 1976.
　　One of the few books available on this topic which is thorough but not difficult to follow
Fluid Mechanics—A Laboratory course. M. A. Plint and L. Böswirth. Griffin 1978.
　　Descriptions of and readings from many basic experiments in fluid mechanics especially useful to assist students with laboratory work.

Civil Engineering Hydraulics

Manual of British Water Engineering Practice. Institution of Water Engineering and Scientists 1969.
　　An engineers handbook with much practical information and summaries of theory. Frequently re-published to keep information up to date.

381

Open Channel Flow. F. M. Henderson. Macmillan. 1966
   An extension of the subject of open-channel flow well explained and arranged for learning.
Open Channel Hydraulics. V. T. Chow. McGraw Hill 1959.
   A handbook containing a great quantity of information, but not an elementary book.
Computational Hydraulics. M. B. Abbott. Pitman. 1979.
   A review of a whole new field of applied fluid mechanics.
Coastal Engineering. R. Silvester. Elsevier. 1974.
   Two volumes of information on one of the most important parts of hydraulics rarely covered in books.
Loose Boundary Hydraulics. A. J. Raudkivi. Pergamon. 1976.
   An introduction to the study of one of the more demanding topics in civil engineering hydraulics.
Centrifugal and Axial Flow Pumps. A. S. Stepanoff. Wiley. 1957.
   An introduction to the design and detailed description of the use and performance of pumps.

Flow measurement

Weirs and Flumes for Flow Measurement. P. Ackers et al. Wiley, 1978.
Continuous Measurement of Unsteady Flows. G. P. Katys. Pergamon 1964.
Flow Measurements and Meters. A. Linford. Pergamon, 1961.
The Measurement of Airflow. E. Ower and R. C. Pankhurst. Pergamon, 1977.

Journals

Water Power and Dam Construction. I. P. C. Press.
   A technical magazine giving up to date information on the engineering aspects of hydraulics.
Journal of the International Association of Hydraulic Research.
   Mainly research papers but also reviews of research in specific areas.
Proceedings of the American Society of Civil Engineers. Hydraulics Division.
   Research papers covering a very wide range of topics associated with hydraulics.
Proceedings of the Institution of Civil Engineers.
   Papers cover both construction and design (Part I) and research (Part II) in all branches of civil engineering.

Index